PROTOPLASMATOLOGIA
HANDBUCH DER PROTOPLASMAFORSCHUNG

BEGRÜNDET VON

L. V. HEILBRUNN · **F. WEBER**
PHILADELPHIA   GRAZ

HERAUSGEGEBEN VON

M. ALFERT · **H. BAUER** · **C. V. HARDING** · **P. SITTE**
BERKELEY   TÜBINGEN   ROCHESTER   HEIDELBERG

MITHERAUSGEBER

W. H. ARISZ-GRONINGEN · J. BRACHET-BRUXELLES · H. G. CALLAN-ST. ANDREWS
R. COLLANDER-HELSINKI · K. DAN-TOKYO · E. FAURÉ-FREMIET-PARIS
A. FREY-WYSSLING-ZÜRICH · L. GEITLER-WIEN · K. HÖFLER-WIEN
M. H. JACOBS-PHILADELPHIA · N. KAMIYA-OSAKA · D. MAZIA-BERKELEY
W. MENKE-KÖLN · A. MONROY-PALERMO · A. PISCHINGER-WIEN
J. RUNNSTRÖM-STOCKHOLM · W. J. SCHMIDT-GIESSEN

BAND VI

F 1

W0042894

THE MEIOTIC SYSTEM

1965

SPRINGER-VERLAG

WIEN · NEW YORK

THE MEIOTIC SYSTEM

BY

B. JOHN and **K. R. LEWIS**
BIRMINGHAM   OXFORD

WITH 195 FIGURES

1965

SPRINGER-VERLAG

WIEN · NEW YORK

ISBN-13: 978-3-211-80733-0   e-ISBN-13: 978-3-7091-5748-0
DOI:10.1007/978-3-7091-5748-0

TITEL-NR. 8745

Protoplasmatologia
VI. Kern- und Zellteilung
F. Die Chromosomen in der Meiose
1. The Meiotic System

The Meiotic System

By

Dr. BERNARD JOHN

Department of Genetics, The University, Birmingham, England

and

Dr. KENNETH R. LEWIS

Botany School, The University, Oxford, England

With 195 Figures

Contents

Part Two: Analytical

Introduction

When the study of heredity and variation first came to be treated as a scientific subject—and this, one must remember, was only just over a hundred years ago—there was an unfortunate separation between the disciplines of cytology and experimental breeding. This separation was based partly on a lack of understanding and partly on a lack of the desire to understand. Even WILLIAM BATESON, the first apostle of mendelism in England, had a blind spot for cytology and for many years dogmatically refused to believe that MENDEL's determinants were transmitted and distributed by the chromosomes.

This separation between cytology and experimental breeding is one which persists, in a measure, even today, simply because there are two quite different, though complementary, techniques available for the study of heredity and variation. On the one hand, one can study directly the structure and behaviour of the actual vehicles which transmit the genetic determinants from one generation to the next. This is the method employed by those who study genetics through a microscope. The alternative method is that used by the experimental breeder who, in default of being able to watch the hereditary factors segregate from each other directly, is obliged to examine the constitution of the germ cells indirectly by sampling, and usually at random, the products of a controlled mating.

The importance of the breeding method needs no emphasis, though there still remains a large number of organisms to which this method has not been applied and indeed will not be easy to apply. But in this day and age when cytology has, to say the least, become unfashionable it is as well to remind ourselves of the importance of the cytological method. The distinctive importance of this method lies in its relatively easy application not to a small number of species and individuals but to many. And it is equally applicable to homozygotes that show no segregation, hybrids and chromosome mutants that yield no offspring and organisms, like man, where experimental breeding is either difficult or impossible. And one learns most by applying the cytological method to the study of meiosis.

The meiotic cycle varies considerably in the details of its constituent stages from species to species and, in some cases, the differences involve more than matters of detail. We have chosen to start our account of the meiotic system by considering the process in the male of the grasshopper *Chorthippus*. As was first recognized by JANSSENS (1924) and later by BĚLAŘ (1929) and DARLINGTON (1936) meiosis in this genus can be studied more completely and more thoroughly than in any other organism. For this reason we are using it to describe the basic processes involved in this critical nuclear cycle.

Our choice of starting subject serves to highlight a principle which we have adopted liberally throughout this monograph. The hallmark of a good review is often held to be that all available facts should be considered and considered in an "impartial" manner. And the measure of impartiality demanded is that one should treat each and every observation as though it commanded equal respect. But it is often obvious that different observations, no less than those who make them, cannot be treated equally. Some observations are unquestionably inferior to others, at least for particular purposes. This is especially true in cytology because of the enormous variation in chromosome size and behaviour.

Part One

Descriptive

I. Meiosis in Male *Chorthippus*

In the species of the genus *Chorthippus*, as in a majority of the genera in the sub-family Truxalinae, the mitotic chromosome complement of the male includes seventeen members. Sixteen of these are autosomes and can be matched into eight homologous pairs; the seventeenth chromosome is unpaired and represents a sex or X-chromosome. All these elements are individually distinguishable (Fig. 1). Six of the autosomes are metacentric and these are the longest chromosomes in the complement (pairs L 1–L 3); the remainder are acrocentric and include one pair of small chromosomes (S 8) and a range of medium pairs (M 4–M 7). The unpaired X-chromosome is the longest of the acrocentrics.

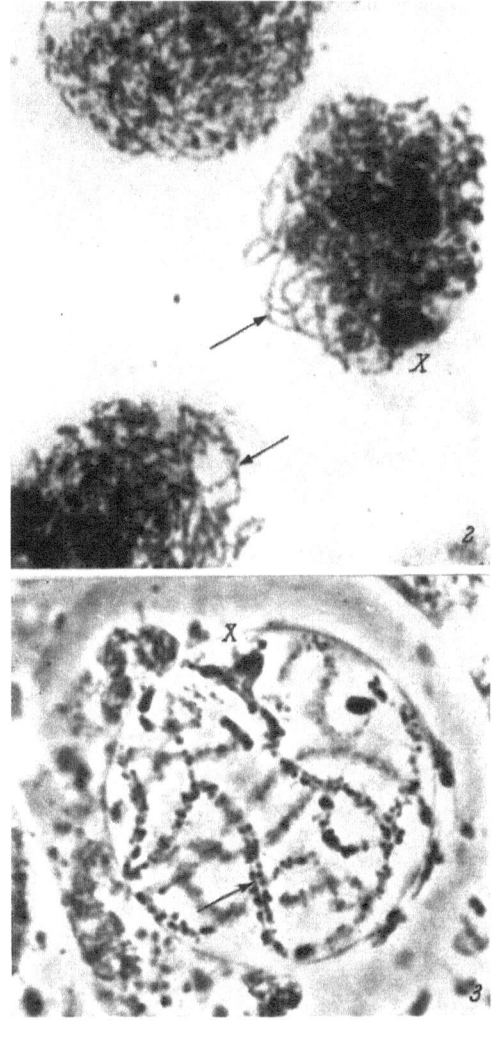

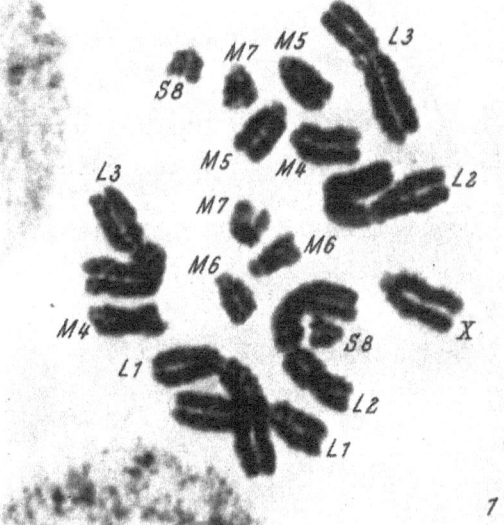

Fig. 1. C-mitosis in a spermatogonial cell of *Chorthippus brunneus*, 2n = 16 + X (orcein, ca. ×2000).

Figs. 2 and 3. Leptotene (Fig. 2, orcein, ca. ×1000) and zygotene (Fig. 3, phase contrast, unstained ca. ×1200) in *Chorthippus brunneus*. Note single threads at leptotene (arrows) and paired threads at zygotene (arrow).

In the premeiotic resting nuclei the X is positively heteropycnotic and it retains this property throughout the first phase of meiosis. With the onset of meiosis the nucleus grows rapidly in size and, although the chromosomes form a dense tangle in the main nuclear area, those toward the periphery can sometimes be identified as fine single threads (leptotene stage, Fig. 2). As the nucleus

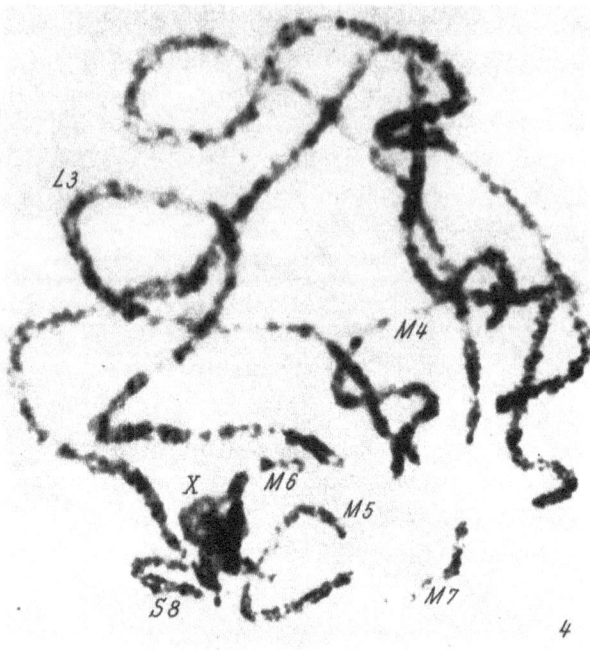

Fig. 4. Zygotene. Note bouquet arrangement of chromosome ends (*Chorthippus brunneus*, orcein-phase contrast, ca. × 1500).

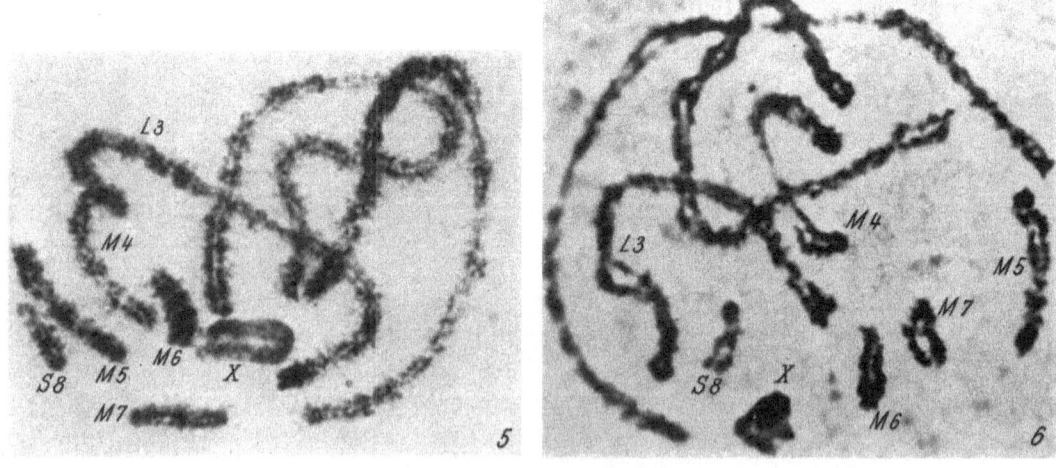

Figs. 5 and 6. Polarised pachytene in *Chorthippus brunneus* (orcein and orcein-phase contrast respectively ca. × 1500).

grows in size the chromosomes become easier to resolve because, coincident with the increase in nuclear volume, they themselves shorten in length. By the time all threads can be followed clearly, they can be seen to be double structures which possess a granular appearance due to the presence of small deeper staining chromomeres along their length. At this stage eight paired threads or bivalents

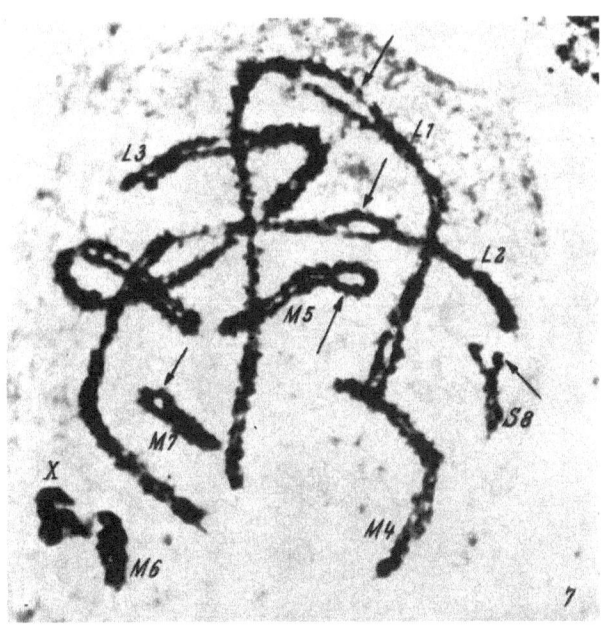

Fig. 7. Onset of diplotene in *Chorthippus brunneus*. Separation of the homologues begins at the centromeres (arrows). Note chromosomes still polarised (orcein-phase contrast, ca. ×1500).

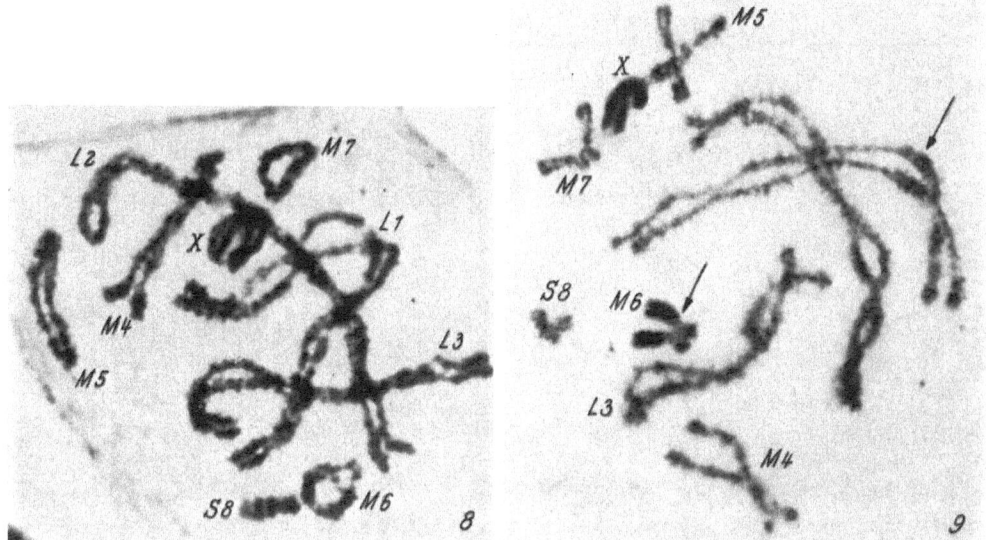

Figs. 8 and 9. Early (Fig. 8, orcein-phase contrast, ca. ×1500) and mid-diplotene (Fig. 9, orcein, ca. ×1500) in *Chorthippus brunneus*. Note that the chiasma in the precocious M_6 bivalent is in the short euchromatic segment (arrow) and that the separating homologues are now each divided into two chromatids (arrow).

can be distinguished (zygotene stage, Figs. 3 and 4). The unpaired heteropycnotic X-chromosome is bent back on itself at this time and is frequently

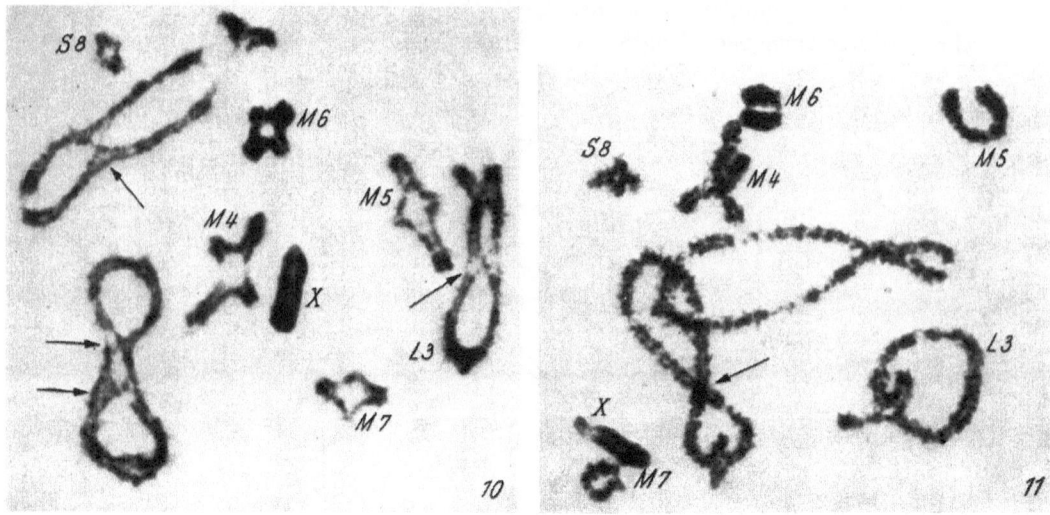

Figs. 10 and 11. Diplotene in *Chorthippus parallelus* (Fig. 10, orcein, ca. × 1500) and *Chorthippus brunneus* (Fig. 11, orcein, ca. × 1500). Note chiasmata in Fig. 10 (arrows). Note also interlocking of L 1 and L 2 bivalents in Fig. 11 and the relational coiling of homologues (arrow).

Table 1. *The Chiasma Frequencies in Individual Bivalents for 20 Cells of the Same Individual of Chorthippus brunneus* ($\bar{x} = 12.85$).

Cell No.	Bivalent Type and Chiasma Frequency								Cell Total
	L 1	L 2	L 3	M 4	M 5	M 6	M 7	S 8	
1	3	2	2	2	1	1	1	1	13
2	3	4	2	2	1	1	1	1	15
3	3	2	2	2	1	1	1	1	13
4	3	2	3	1	1	1	1	1	13
5	4	3	2	2	1	1	1	1	15
6	3	3	2	2	1	1	1	1	14
7	1	3	1	1	1	1	1	1	10
8	2	2	2	2	1	1	1	1	12
9	2	2	2	1	1	1	1	1	11
10	3	2	2	1	1	1	1	1	12
11	3	3	2	2	1	1	1	1	14
12	3	2	3	1	1	1	1	1	13
13	3	3	2	1	1	1	1	1	13
14	3	2	2	2	1	1	1	1	13
15	3	3	2	2	1	1	1	1	14
16	2	2	2	2	1	1	1	1	12
17	3	2	2	1	1	1	1	1	12
18	3	2	2	2	1	1	1	1	13
19	3	2	2	1	1	1	1	1	12
20	3	2	2	2	1	1	1	1	13
Totals	56	48	41	32	20	20	20	20	257

associated with the M 6 bivalent, a large segment of which is also condensed precociously.

We have never seen the actual stages of the pairing process in *Chorthippus* but by the time of its completion it is frequently possible to see that the chromosomes are held in a polarised orientation with the ends of the autosomes grouped in a bouquet arrangement (pachytene stage Figs. 5–6). In bouquet polarisation the simplest arrangement is that where all the chromosomes lie with their ends directed towards one centriolar field. This, however, is subject to modification by a variable number of the chromosome ends, more especially those of M 6 and S 8, lying attached to the X (see Fig. 5). Bouquet po-

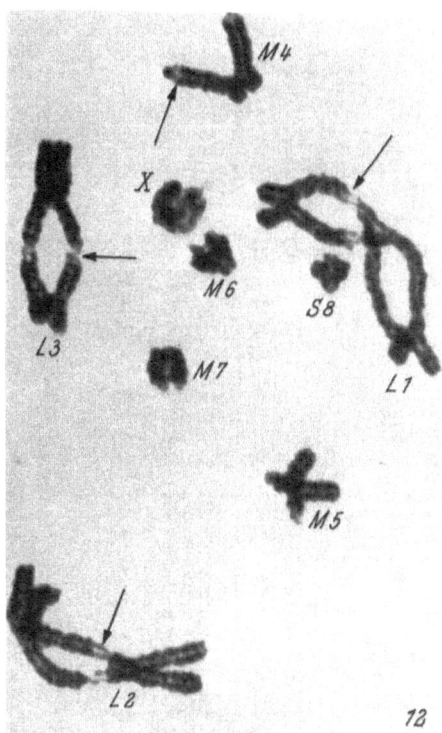
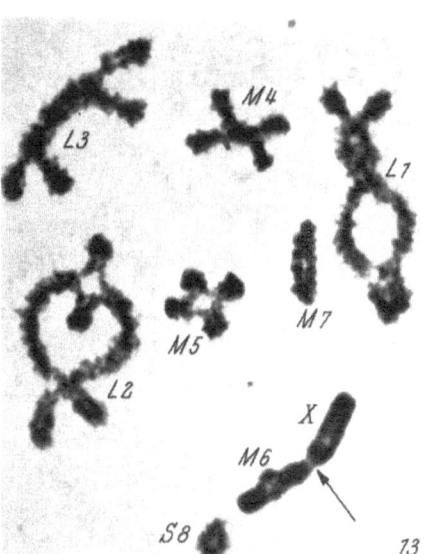

Figs. 12 and 13. Diakinesis in *Chorthippus brunneus*. Note clarity of centromeric constrictions in Fig. 12 (orcein, ca. ×1500) and heterochromatic association of X and M₆ in Fig. 13 (orcein-phase contrast, ca. ×1500).

larisation relaxes and is gradually lost as chromosome contraction proceeds (early diplotene stage, Fig. 8). Coincident with this the paired members of each bivalent start to separate from one another. This separation commonly begins at or near the centromere (Fig. 7) though subsequently other regions also seem to fall apart (Figs. 8–9). The X-chromosome gradually unbends during this stage.

As homologous chromosomes separate they can be seen to be double, each consisting of two chromatids (diplotene stage, Figs. 9–11). This is true also of the unpaired X (Figs. 8–10). Separation of homologues proceeds much more quickly in the shorter members of the complement and by the time separation is complete one finds that, at one or more regions along the length of each bivalent, pairs of non-sister chromatids exchange partners with one another at so-called chiasmata (Figs. 10 and 11). Occasionally, L-bivalents are interlocked at this stage (Fig. 11). The number of chiasmata per bivalent is related to the length of the chromosomes involved in the bivalent but the relation is not a simple one. Although the longer chromosomes have more chiasmata than the shorter ones there is no direct proportionality between chromosome length and chiasma

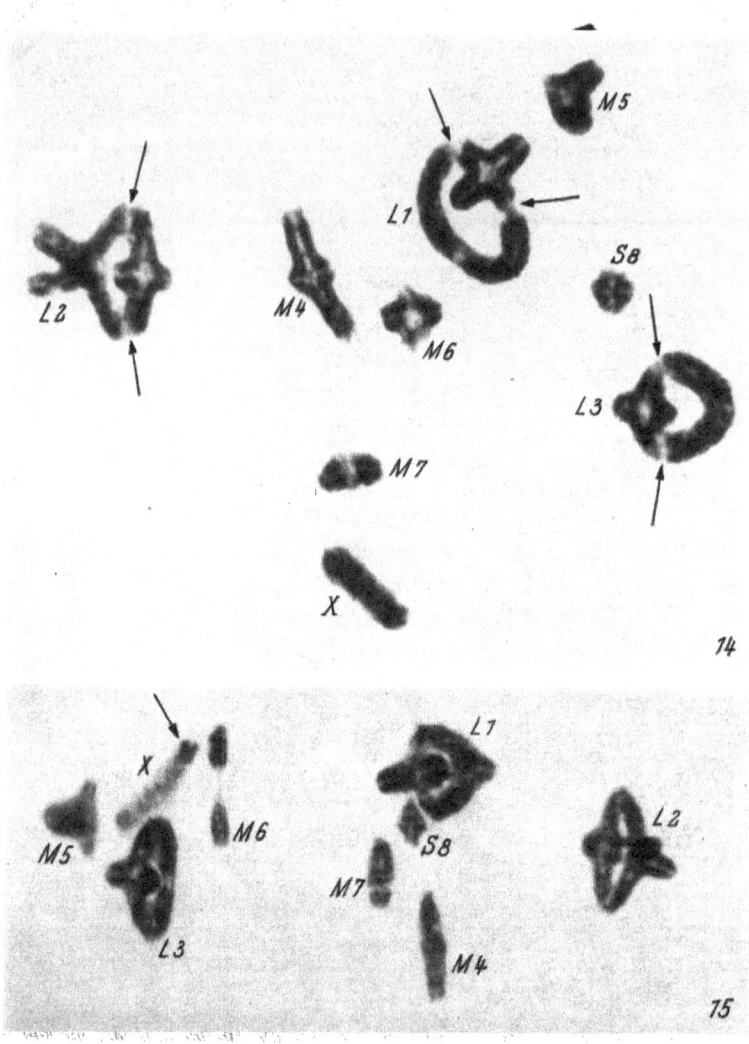

Figs. 14 and 15. Pro-metaphase-1 (Fig. 14, orcein, ca. ×1500) and metaphase-I (Fig. 15, orcein, ca. ×1500) in *Chorthippus brunneus*. Note clarity of centromeric constrictions in Fig. 14 (arrows) and negatively hetero-pycnotic X with isopycnotic centromere (arrow) in Fig. 15.

frequency. In the small and medium members of the complement only one chiasma usually forms, though the M 4 and M 5 bivalents and less commonly the other M-members sometimes form two (Figs. 8 and 11). Where a single chiasma is produced all the chromosome arms lie initially in one plane. But the arms rotate, so that they come to lie at right angles with respect to each other forming characteristic cross configurations (compare Figs. 9–11). A similar process of rotation also occurs in the segments beyond the most distal chiasma of the large metacentrics which usually form two, three or less commonly four or five chiasmata (Table 1). Where three or more chiasmata are present in one bivalent, loops form between successive chiasmata as the bivalent opens out. Initially these lie in the same plane but a rotation of the chromosome arms, equivalent to that which transforms single chiasmate bivalents into crosses, causes successive loops to lie at right angles to one another (diakinesis stage, Figs. 12 and 13). This arrangement is most evident, however, after the nuclear membrane breaks down and the centro-

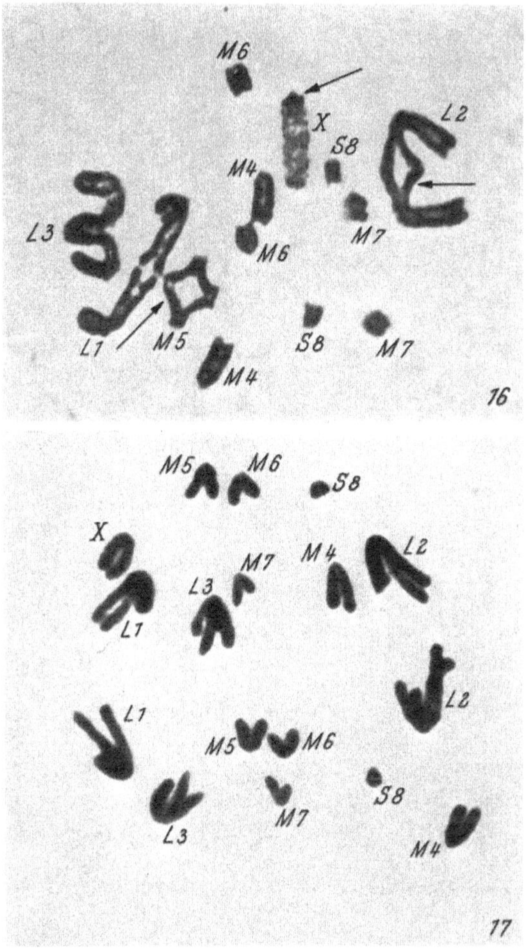

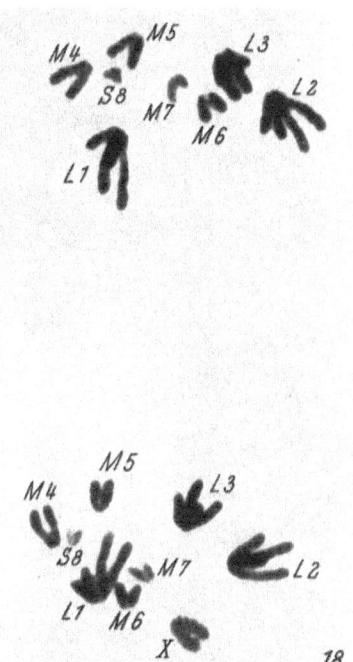

Figs. 16 and 17. Onset of anaphase-I (Fig. 16, orcein-phase contrast ca. ×1500) and mid-anaphase-I (Fig. 17, orcein ca. ×1500) in *Chorthippus brunneus*. Note lapse of attraction between homologous chromatids (arrows) and centromere of X in Fig. 16.
Fig. 18. Late-anaphase-I in *Chorthippus brunneus* (orcein, ca. ×1500).

meres become attached to the spindle which forms coincident with this breakdown, for now the centric loops become considerably extended (pro-metaphase 1, Fig. 14).

The initial points of centromere attachment are determined by chance relationship between the chromosomes and the spindle system but in the case of the autosomal bivalents this is converted into a very precise alignment. This involves each bivalent vacillating through the spindle until they end in a position of congression where the two centromeres lie equidistant above and below the equator of the spindle (metaphase-I stage, Fig. 15). The X-chromosome, being unpaired, does not congress in company with the autosomes though it too

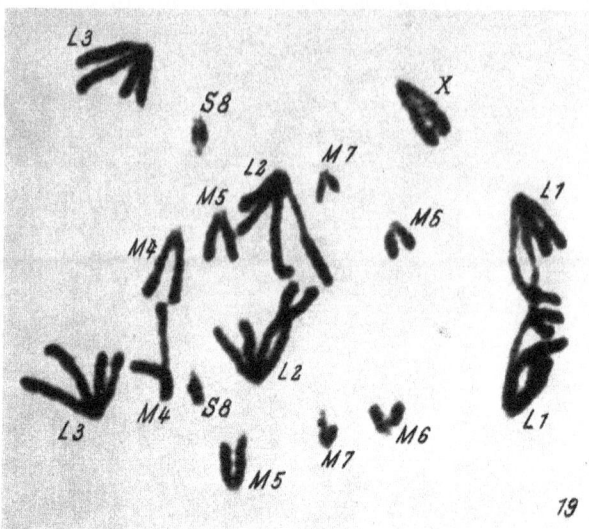

Fig. 19. Anaphase-I in *Omocestus viridulus* to show divergence of chromatids (orcein, ca. × 1500).

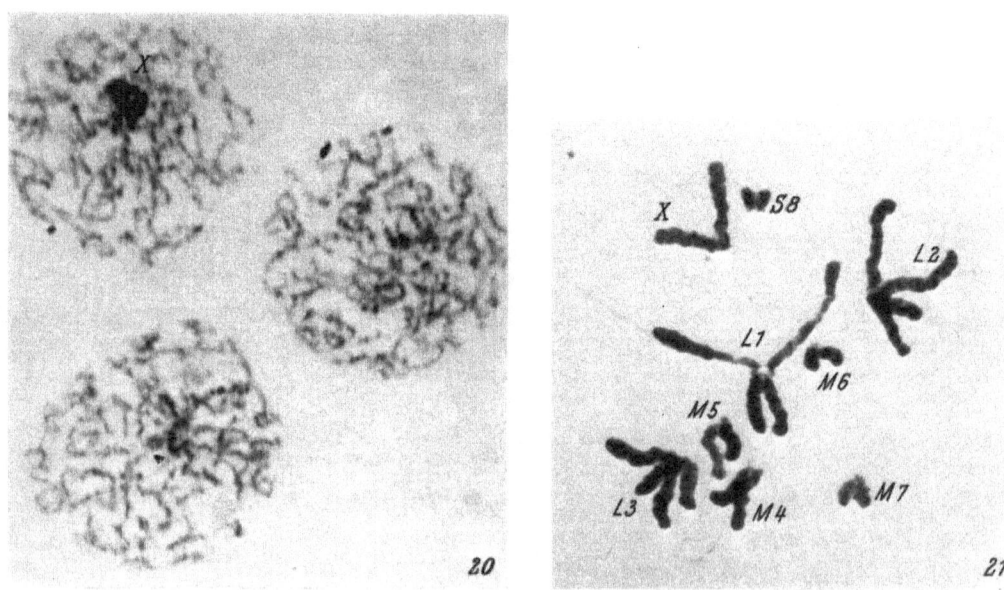

Fig. 20. Interkinesis nuclei in *Chorthippus brunneus* with and without an X-chromosome (orcein, ca. × 1500).

Fig. 21. Prophase II in *Chorthippus brunneus* with chromatids diverging. X-chromosome present (orcein, ca. × 1500).

moves through the spindle from pole to pole (see pg. 175). Following this it may remain near one of the spindle poles or else come to lie in the vicinity

of the spindle equator. By now it is frequently negatively heteropycnotic
except in the centromere region (Fig. 15 and see section VIII–I).

In the position of congression the homologous chromatids of each bivalent
suddenly lose the parallel alignment they have maintained since each homo-

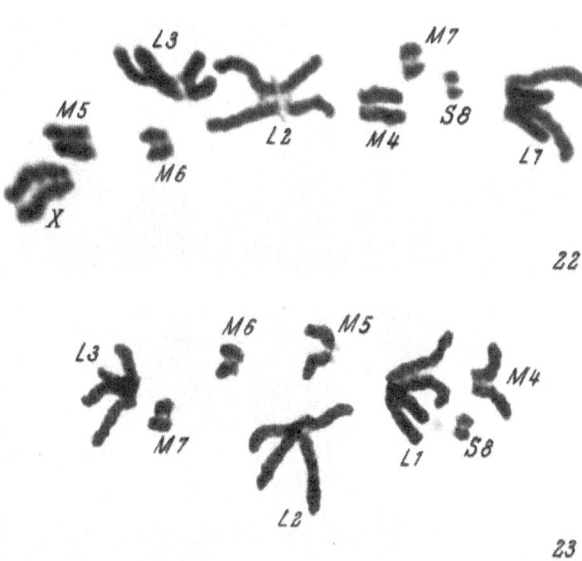

Figs. 22 and 23. Metaphase-II, side view of the equator, in *Chorthippus brunneus*. Cells with (Fig. 22—8 + X)
and without (Fig. 23—8) an X-chromosome (orcein, ca. × 1500).

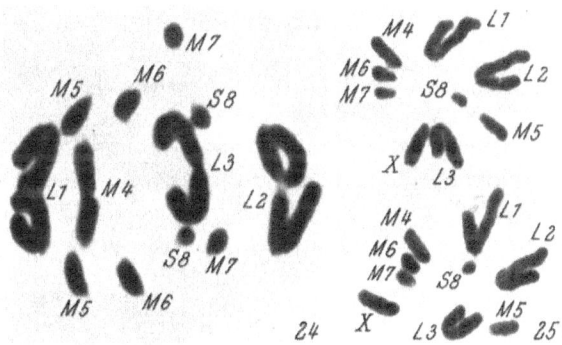

Figs. 24 and 25. Early and late-anaphase-II in *Chorthippus brunneus*, side views (orcein, ca. × 1500).

logue could be first identified as a double unit (Fig. 16). This, together with the
movement of the two centromeres of each bivalent to opposite poles, tears
chiasmata apart or else allows for their easy movement to and off the chromosome ends. The paired homologues thus disengage completely and move polewards in opposite directions (anaphase-I, Figs. 17–18). In this movement
the centromeres lead the way and the chromosome arms trail passively behind.
Each metacentric half-bivalent now appears as a four-armed structure while
the acrocentric half-bivalents appear V-shaped (Fig. 18). These events are even

easier to see in the related species *Omocestus viridulus* (Fig. 19). The two chromatids of the X-chromosome also diverge at this stage: nevertheless, they usually remain together and pass to one spindle pole (but see pg. 175). When the X-chromosome lies just off the spindle equator at first-meta-phase, it sometimes moves onto it when the half-bivalents begin their sepa-ration. And in a few such cases we have found the X to divide (see Fig. 161), the sister chromatids then passing to opposite poles. In the majority of cases, how-ever, two kinds of nuclei are produced following first-anaphase separation—one

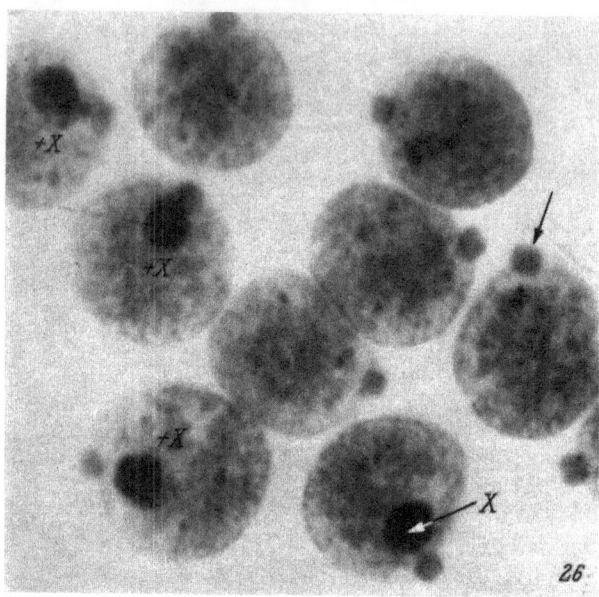

Fig. 26. Spermatid nuclei, with and without a positively heteropycnotic X-chromosome, in *Tylotropidius gra-cilipes*. Note clarity of centrioles (arrow) just outside the nuclear membrane (carmine, ca. ×1500).

with and the other without an X-chromosome (interkinesis stage, Fig. 20) and the X now becomes positively heteropycnotic for the second time in the meiotic cycle.

The end of this first sequence of chromosome movements is followed in *Chorthippus* by division of the cytoplasm so that the sister secondary spermato-cytes separate from one another and then behave independently but syn-chronously in the second stage of meiosis. When the chromosomes reappear at the onset of this second meiotic sequence they are often loosely coiled but other-wise have the same form they showed at the end of the first meiotic phase: each chromosome is still composed of two chromatids and these are widely spaced except in the vicinity of the centromere (prophase-II, Fig. 21). The chromosomes then contract by internal coiling and, when the nuclear membrane disrupts, they become attached to a spindle system by auto-orientation, in which the two sister half-centromeres of each chromosome orient to opposite poles. In this position they congress to the spindle equator (metaphase-II, Figs. 22–23). The two chromatids which comprise each chromosome are then separated in a second phase of spindle-movement. And in those secondary spermatocytes which possess an X-chromosome this now divides in company with the autosomes (anaphase-II, Figs. 24–25) unless of course it is one of the products of the rare first-division separations of the two chromatids of each X.

The net result is thus the production of four spermatid nuclei from each cell that undergoes meiosis. Two of these products have, and two are devoid of, the distinctive X-chromosome. In those cells where it is present it is once more positively heteropycnotic (Fig. 26). And meiosis is over.

As we pointed out earlier, the course of meiosis is subject to variation. Some of these variations are sporadic and accidental, others result from a change in chromosome structure or number. Some of these are within and others beyond the range of normality. But first we will consider those differences which we can suppose to be under genotypic control and which have been favoured by selection to meet the particular needs of different species (see pg. 254) and more particularly those which affect chiasma formation and the stage at which chiasmata first become visible.

II. Patterns of Variation in Chiasma Formation

1. Localisation

In *Chorthippus* chiasmata form at various points along the length of the chromosome. But even in cases like this, with so-called random chiasma formation, it has long been evident that the pattern is not completely random (see pg. 207). However, in some organisms chiasmata occur exclusively, or nearly so, in very localised regions of the chromosome. Two such types of localisation are known—proximal and distal. In the first of these, chiasmata are restricted to the neighbourhood of the centromere, while in the second they are confined to short regions near the non-centric ends of the chromosome (Table 2, Figs. 27–30). This localisation rarely appears to be an absolute nuclear condition affecting all chromosomes equally. Thus in *Stethophyma* localisation is more complete in some bivalents than others and in some individuals than others (WHITE 1936). Again in *Bryodema*, nine of the eleven acrocentric autosomal bivalents possess only one chiasma in the long arm; this forms adjacent to, or in the immediate neighbourhood of the centromere. But two chromosomes do not conform to this behaviour (WHITE 1954). The smallest bivalent invariably forms a single distal chiasma, as too does the fifth bivalent, but in this case 64 of the 166 first metaphases examined also had a proximal chiasma leading to the production of a ring bivalent. The same principle applies to plants too. Thus in *Allium fistulosum* there is an average of 1.7 non-localised chiasmata per cell.

The cause of this localisation varies. Since chiasmata can form only between the paired parts of chromosomes then theoretically their localisation could be determined directly by a strict localisation of pairing. In only one case, however, has this been shown to operate, namely *Fritillaria* (DARLINGTON 1935). Here different species differ in their pattern of pairing and hence their pattern of chiasma formation. In some species, including *F. meleagris*, pairing is interrupted while it is still far from complete. And, since pairing begins at the proximal regions, the distal and mid parts are still unpaired, or at the most only intermittently paired, when chiasmata form, so that they are localised proximally. In other species where pairing is more complete, chiasmata are either semi-localised (*F. elewesii*) or else random (*F. imperialis*).

In all other cases of localisation pairing appears to be complete. The pairing of homologues may, however, be variable not with respect to its com pleteness but rather to the site of its initiation. In these terms one can distinguish four variants:

i) Undetermined or random pairing which is initiated simultaneously at several independent points which vary at random (*Tulipa, Chorthippus* — DARLINGTON 1940).

Table 2. *Patterns of Chiasma Localisation.*

Type	Organism	
	Plant	Animal
1. Proximal	*Allium fistulosum* (LEVAN 1933) *Allium porrum* (LEVAN 1940) *Fritillaria meleagris* (DARLINGTON 1935) *Paris quadrifolia*[1] (DARLINGTON 1937) *Trillium kamtschaticum*[1]	*Bryodema tuberculata* (WHITE 1954) *Chromacris miles* (SAEZ 1930) *Schistocerca paranensis* — mutant (JOHN and HENDERSON 1962) *Stethophyma* (= *Mecostethus*) *grossus* (JANSSENS 1924, WHITE 1935)
2. Distal	*Campanula persicifolia* (DARLINGTON and GAIRDNER 1937) *Oenothera* (MARQUARDT 1938) *Paeonia californica* (WALTERS 1942) *Rhoeo discolor* (DARLINGTON 1929) *Secale cereale*[2] (DARLINGTON 1933)	*Blaberus discoidalis* (JOHN and LEWIS 1959) *Chrysochraon dispar* (KLINGSTEDT 1937) *Metrioptera brachyptera* (WHITE 1936) *Oswaldocruzia filiformis* (JOHN 1957) *Periplaneta americana* (JOHN and LEWIS 1960) *Tetrix ceperoi* (HENDERSON 1961)

[1] HAGA and MATSUURA claim that in *Paris*, as in *Trillium kamtschaticum*, the centromeres of homologous chromosomes generally remain synapsed until late metaphase-I so that a synapsed centromere-pair appears as if it were a chiasma node. There can be no doubt that this is a misinterpretation. Indeed many of the differences between the japanese and european schools concerning the nature of chiasma formation can be resolved in terms of such a misinterpretation.

[2] REES (1955) has described inbred mutants with proximal localisation (so-called P-bivalents) in rye.

ii) Procentric pairing which is initiated at the centromere, irrespective of whether the chromosome is meta- or acro-centric (*Stethophyma, Tegenaria*). This is facilitated by relic or Rabl polarisation (REVELL 1947).

iii) Proterminal pairing, which is initiated at the ends as in *Chrysochraon* (KLINGSTEDT 1937). Where the chromosomes are acrocentric, this may involve only the non-centric end (proterminal distal) as in *Tetrix* (HENDERSON 1961) or both centric and non-centric ends (proterminal proximal and distal). Proterminal pairing is facilitated by the occurrence of active prophase polarisation leading to bouquet formation.

iv) Semi-localised pairing where pairing begins simultaneously at both near-centric and near-distal ends (*Lilium candidum* — DARLINGTON 1940).

If chromosome regions which pair first are preferred in chiasma formation then this too may lead to a localisation of chiasmata. This implies, as DARLING-

TON first recognized, that a time limit is operative which acts as the causal agent in such cases. And this, in turn, emphasises the need to distinguish between effective pairing, which leads to chiasma formation and non-effective pairing, which does not.

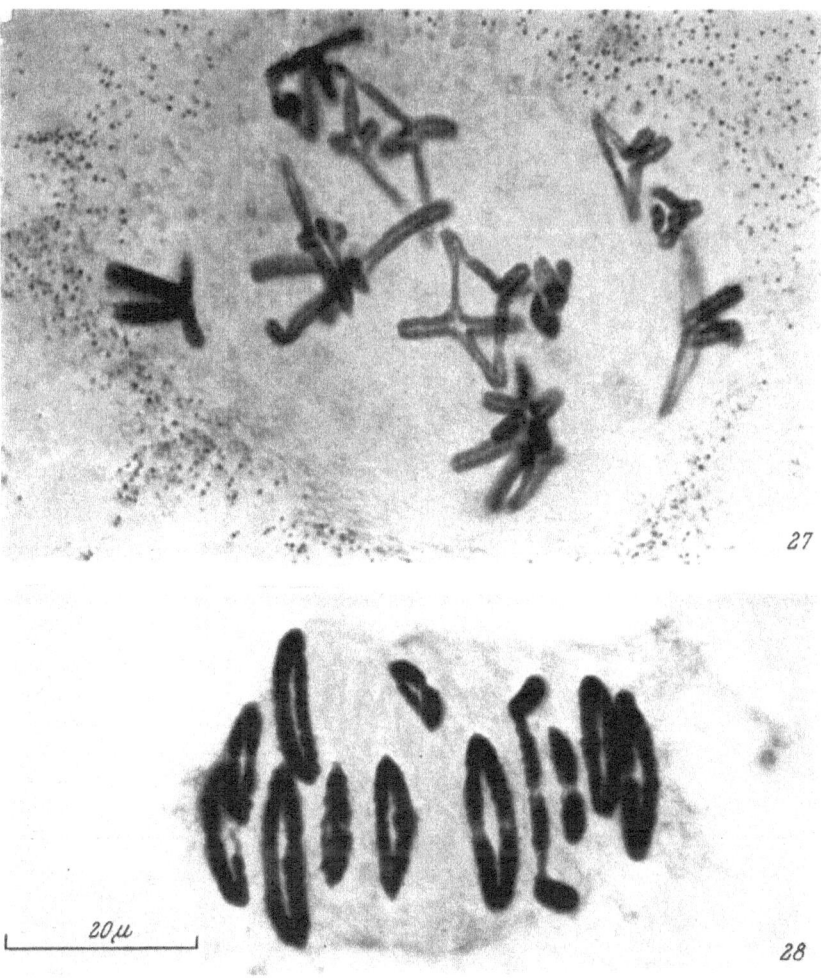

Figs. 27 and 28. Distribution of chiasmata in the newt *Triturus helveticus*, unrestricted to proximal in the female (Fig. 27) and distal in the male (Fig. 28). Photographs kindly supplied by Prof. H. G. CALLAN (see WATSON and CALLAN 1963).

A second factor which may influence pairing, and hence chiasma formation, is the presence of heteropycnotic segments. Thus in *Oenothera* (MARQUARDT 1938) and *Rhoeo* (COLEMAN 1941) proximal chromosome regions do not form chiasmata and these are positively heteropycnotic during the pairing period. The same is true in *Lycopersicon esculentum* (BARTON 1949) and *Blaberus discoidalis* (JOHN and LEWIS 1959) and here it is possible to show that these segments pair last. Two genetical studies support the cytological observation that chiasmata do not form in H-segments. In *Drosophila* one of the genetic characteristics of the centric heterochromatin is the very low frequency of crossing-over in it and this is maintained when this heterochromatin is removed from its position near the centromere (BAKER 1958, Fig. 31). Similarly SNOAD (1963) has carried out a comparative study of the behaviour of heterochromatic and euchromatic regions of X-ray induced interchanges in tomato chromosomes and has found that crossing-over does not occur in the heterochromatic regions (Fig. 32 and compare BARTON 1951).

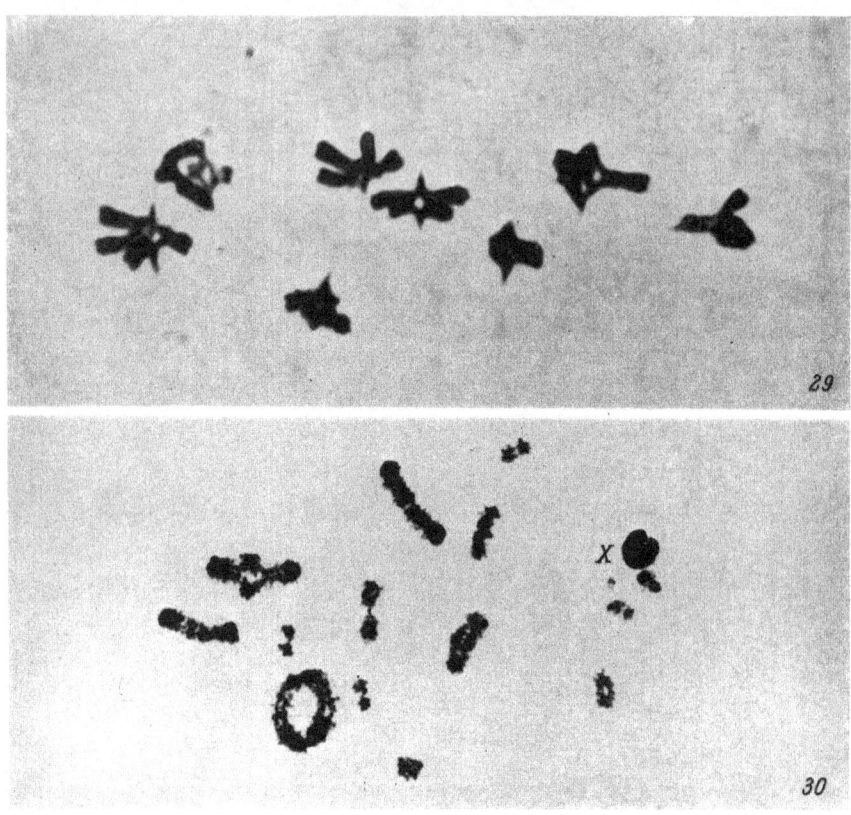

Figs. 29 and 30. Localisation of chiasmata. Proximal localisation in *Allium fistulosum* (Fig. 29, carmine-phase contrast, ca. ×1500) and distal localisation in the tettigoniid *Metrioptera brachyptera* (Fig. 30, orcein, ca. ×1500).

DYER (1963) draws attention to the interesting pattern of relationship which exists between the distribution of chiasmata on the one hand and the structure and distribution of H-segments on the other (Table 3). Thus in *Fritillaria*, *Cestrum* and *Tulbaghia* chiasmata are proximally localised. In the first two genera this distribution overlaps that of the heterochromatin and the H-segments are interrupted by small regions of euchromatin. Here DYER is of the opinion that chiasmata form in the adjacent euchromatin rather than in the H-material. In *Tulbaghia*, on the other hand, where heterochromatin and chiasmata are localised at opposite ends of the chromosome arms, the heterochromatic segments are entire. *Chorthippus* shows an interesting intermediary for in the precocious M 6-bivalent chiasmata actually form in a short euchromatic segment (Fig. 9).

One of the most comprehensive studies on the distribution of chiasmata in relation to H-material was carried out by Linnert (1954) in the genus *Salvia*. She came to three main conclusions:

(1) Species with the most concentrated heterochromatin had the lowest chiasma frequency.

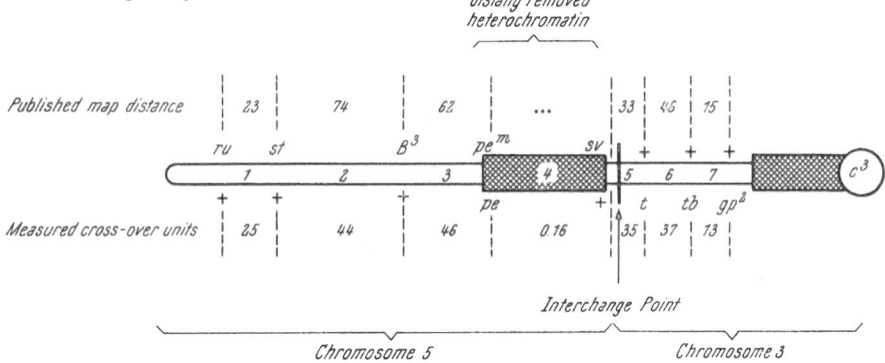

Fig. 31. Cross-over frequencies in the long chromosome of T (3,5) pe^{MSI} of *Drosophila virilis* (♀ F$_1$ produced by the cross, T (3,5) pe^{MSI}, pe, t, tb, gp^2 ♀ × T (3,5) pe^{MSI}, ru, st, B^3, pe^M sv ♂). These frequencies (measured crossover units) are compared with the map distances obtained for the normal, pre-translocation chromosomes (published map distance). Modified from Baker 1958.

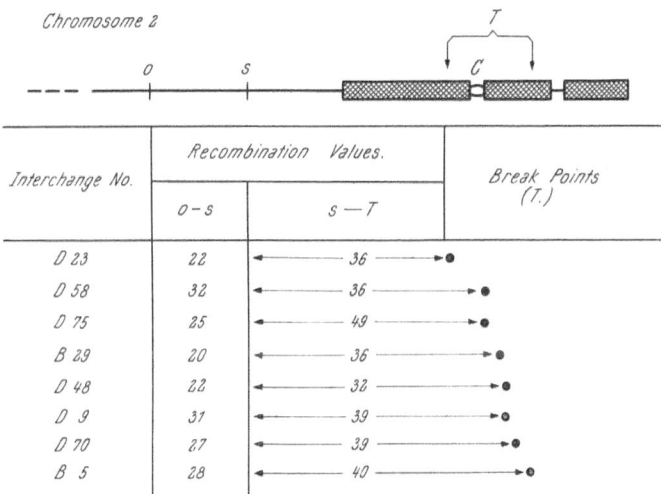

Fig. 32. Crossing-over in interchanged chromosome-2 of tomato. The variation in cross-over frequency between S and the various interchange points (T) within the centric heterochromatin is not significantly greater than that between S and O, even though the lengths of the segments concerned vary in the first and not the second set of data. This suggests that crossing-over does not occur in the heterochromatin. Thus, crossing-over between S and the varying T points is a function of the length of euchromatin between S and the centromere (unpublished data kindly supplied by Mr. B. Snoad).

(2) The concentration or dispersion of H-material had little influence on the position of the chiasmata. Thus species like *S. jurisicii* and *S. pratensis* which lack heterochromatin had predominantly terminal chiasmata while *S. austriaca*, where heterochromatin is similarly absent, had a large number of median and subterminal chiasmata. Likewise, in species with pronounced heterochromatin, *S. hormonium* had many median chiasmata while *S. argentea* had most of its chiasmata at the ends of the chromosomes (Fig. 33).

(3) In *S. nemorosa* it was possible to determine the position of the chiasmata in the individual chromosomes of the complement (Fig. 34). This showed that:

(a) Many of the chiasmata occurred at the boundary between eu- and heterochromatin (compare WHITE 1954). Perhaps such boundaries impede the terminal movement of chiasmata.

(b) The nucleolus-forming H-segment is free of chiasmata.

Table 3. *Relationship Between the Distribution of Chiasmata and Heterochromatic Segments.*
(After DYER 1963.)

Species	Distribution of		Structure of H-segments
	H-segments	Chiasmata	
(1) With overlapping distributions of Xta. and H-segments			
Trillium kamtschaticum			
Trillium smalli		proximal	
Fritillaria recurva	proximal		interuppted
Paris japonica			
Triturus cristatus		dispersed	
Hordeum vulgare	prox., median	prox., median	± interuppted
Cestrum parqui	median, distal	prox., distal	
Secale cereale	distal	distal, terminal	small
Trillium grandiflorum			± interuppted
Vicia faba	dispersed	dispersed	small
Cestrum elegans		prox., distal	± interuppted
(2) With distinct distributions of Xta. and H-segments			
Triturus vulgaris		median, distal	interuppted
Triturus palmatus		distal, terminal	
Bufo calamita	proximal		± entire
Bufo viridis		distal	
Rana esculenta			
Tulbaghia pulchella		prox., distal	
Adoxa moschatellina	terminal		entire
Paris podophylla		proximal	

2. Achiasmate Meiosis

In most meiotic cycles, as in *Chorthippus*, bivalents are held together by chiasmata and these, we shall see, are necessary for regular segregation, for in their absence the univalents that form do not show an orderly behaviour (see pg. 54). In some species, however, meiosis proceeds regularly in the absence of chiasmata (Table 4). No plants are known with such an achiasmate meiosis and most of the cases in the animal kingdom are to be found in the Arthropoda. In no case are chiasmata dispensed with in both sexes, though in some of the hermaphroditic Enchytraeids chiasmata are absent in both the male and the female line (CHRISTENSEN 1961). In the well known case of *Drosophila* and in *Phryne fenestralis* (BAUER 1946, WOLF 1950) it has been shown that the absence of chiasmata in the male sex is accompanied by an absence of crossing-over.

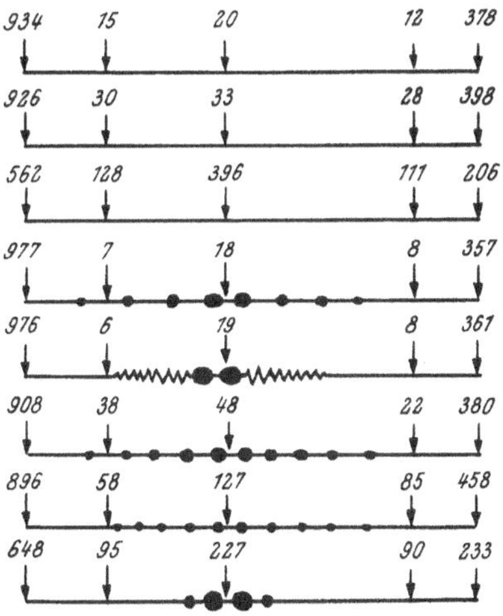

Species	Total Xta		
	1. Ends	2. Median	3. Interstitial
S. jurisicii *934 15 20 12 378*	1312	20	27
S. pratensis *926 30 33 28 398*	1324	33	58
S. austriaca *562 128 396 111 206*	786	396	239
S. sclarea *977 7 18 8 357*	1334	18	15
S. argentea *976 6 19 8 361*	1337	19	14
S. cleistogama *908 38 48 22 380*	1288	48	60
S. glutinosa *896 58 127 85 458*	1354	127	143
S. horminum *648 95 227 90 233*	881	227	185

Fig. 33. The distribution of chiasmata (× 1000) in the chromosomes of different species of the genus *Salvia*. (LINNERT, 1955.)

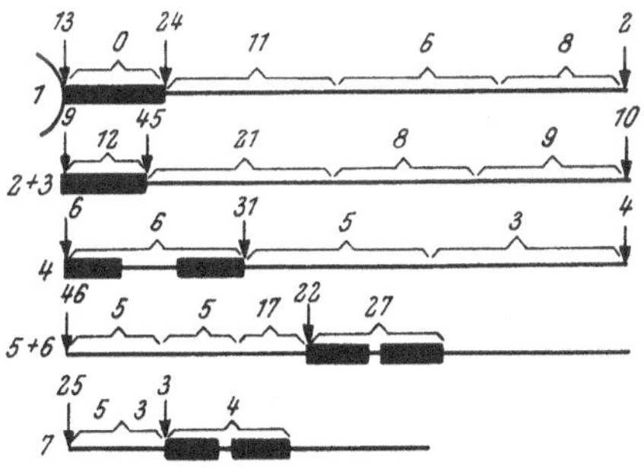

Chromosome N° and Type	Number of chiasmata				
	In Hetero-chromatin	In Eu-chromatin	At Het.-Eu. boundaries	At ends	Total
1 *13 0 24 11 6 8 2*	0	25	24	13 + 2	64
2+3 *9 12 45 21 8 9 10*	12	37	45	9 + 10	113
4 *6 6 31 5 3 4*	6	8	31	6 + 4	55
5+6 *46 5 5 17 22 27*	27	27	22	46	122
7 *25 5 3 3 4*	4	8	3	25	40
	Total chiasmata				394

Fig. 34. The individual distribution of chiasmata in the seven chromosomes of *Salvia nemorosa* (LINNERT, 19 55)

Segregation in the absence of chiasmata in one sex only thus affords a means of reducing the extent of recombination and hence variation within the species as a whole.

Most of the organisms showing achiasmate meiosis have small chromosomes which make its study difficult. Through the kindness of Professor HANS BAUER

Table 4. *Achiasmate Meiosis in Animals.*
(Based on ULLERICH 1961.)

Species	Sex	Reference
Protozoa		
1. *Heliozoa — Actinophrys sol*		BĚLÁŘ 1923 ⎱ see BAUER
2. *Coccidia — Aggregata eberthi*		BĚLÁŘ 1926 ⎰ 1946 a
3. *Foraminifera — Rubratella intermedia*		GRELL 1958 (see also LE CALVEZ 1950)
4. *Gregarinea — Stylocephalus longicollis*		GRELL 1940
5. *Flagellata — Trichonympha*		CLEVELAND 1949
Metazoa		
1. *Annelida — Enchytraeidae*	♀ (± ♂)	CHRISTENSEN 1961
2. *Mollusca — Sphaerium corneum*	♂	KEYL 1956
3. *Arthropoda*		
(a) *Acarina — Eylais setosa* and *Hydrodroma despiciens*	♂	KEYL 1957
(b) *Copepoda*		
(i) *Ectocyclops strenzkei*	♀	BEERMANN, W. 1954
(ii) *Cyclops strenuus*		BEERMANN, S. 1959
(iii) *Tigriopus californicus, japonicus* and *brevicornis*	♀	AR-RUSHDI 1963
(c) *Arachnida* *Scorpionidea — Tityus, Isometrus*	♂	PIZA 1943, 1947
(d) *Insecta*		
(i) *Orthoptera*		
a. *Manteidae — Callimantis, Acanthiothespis, Pseudiomiopteryx, Miomantis*	♂	HUGHES-SCHRADER 1943, 1950
b. *Eumastacidae — Thericleinae* Several species	♂	WHITE 1965
(ii) *Mecoptera — Panorpa communis, germanica* and *cognata*[1]	♂	ULLERICH 1961
(iii) *Lepidoptera — Cidaria* and *Galleria*	♀	SUOMALAINEN 1953 SMITH 1938
(iv) *Diptera*		
a. *Nematocera — Phryneidae*[1] *Bibionidae Scatopsidae Thaumaleidae*	All in ♂	BAUER 1946 b, WOLF 1950 ⎱ WOLF 1941 ⎰
Blepharoceridae		WOLF 1946
Mycetophilidae		LE CALVEZ 1947 FAHMY 1950
Chironomidae Metriocnemus cavicola		BAUER and BEERMANN 1952

Table 4, Contd.

	Species	Sex	Reference
Metazoa	*Tipulidae* *Tipula caesia* and *pruinosa*		Bauer and Beermann 1952, Bayreuther 1956
	Sciaridae		Metz 1933
	Cecidomyiidae		White 1954 for summary
	b. *Orthorrapha — Asilus sericeus* and *notatus*		Metz and Nonidez 1921, 1923
	c. *Cyclorrapha —* many species including *Drosophila*[1]		Darlington 1934, Cooper 1941, 1944, Ribbands 1941
	(v) *Hemiptera — Homoptera* *Protortonia primitiva* *Parlatoria oleae*		Schrader 1931 Nur 1965

[1] In these cases genetical studies have confirmed the absence of crossing-over.

we have had the opportunity of examining the process in *Tipula caesia* where this limitation of chromosome size does not apply (Figs. 35–40). Up to pachytene the course of meiosis appears to be normal but after this there is a marked change. There is no diplotene and no diakinesis, the chromosomes simply go through a contraction phase during which the paired homologues of each bivalent maintain their parallel alignment (Figs. 36–37) until the spindle forms (Fig. 38). It has often been assumed that this parallel pairing is an exaggerated form of the somatic pairing which characterises the mitotic divisions of dipteran flies. There are a number of facts which militate against this, not the least of which is that the details of somatic pairing and its time of onset appear to be quite variable within the order. Further, in meiosis of the fly *Melophagus ovinus* the autosomes of the male conjoin, not by somatic pairing, but by small segments which Cooper (1941) suggests represent non-genic organelles which he refers to as collochores.

All the cases we have mentioned so far involve organisms in which the centromere is localised. There is however one known example of an achiasmatic sequence occurring in a species with a non-localised centric system (see pg. 134). Males of *Protortonia primitiva* have $2n = 4 + X$ (Schrader 1931, Hughes-Schrader 1948) and the meiotic nucleus is divided at the onset of first prophase into four vesicles. The largest of these contains two, presumably homologous, autosomes; two medium-sized vesicles each contain one autosome while the smallest vesicle includes the X-chromosome. By mid-prophase the constituent chromatids of all these chromosomes are completely dissociated, the largest vesicle contains four condensed bodies and the others two each. The vesicles elongate and jointly form a spindle system. The chromatids then re-associate giving chains of two or four, as the case may be, arranged in the long axis of the vesicles. The association is clearly non-chiasmate and is comparable with the secondary pairing which takes place in interphase in other coccids (see section VI–2).

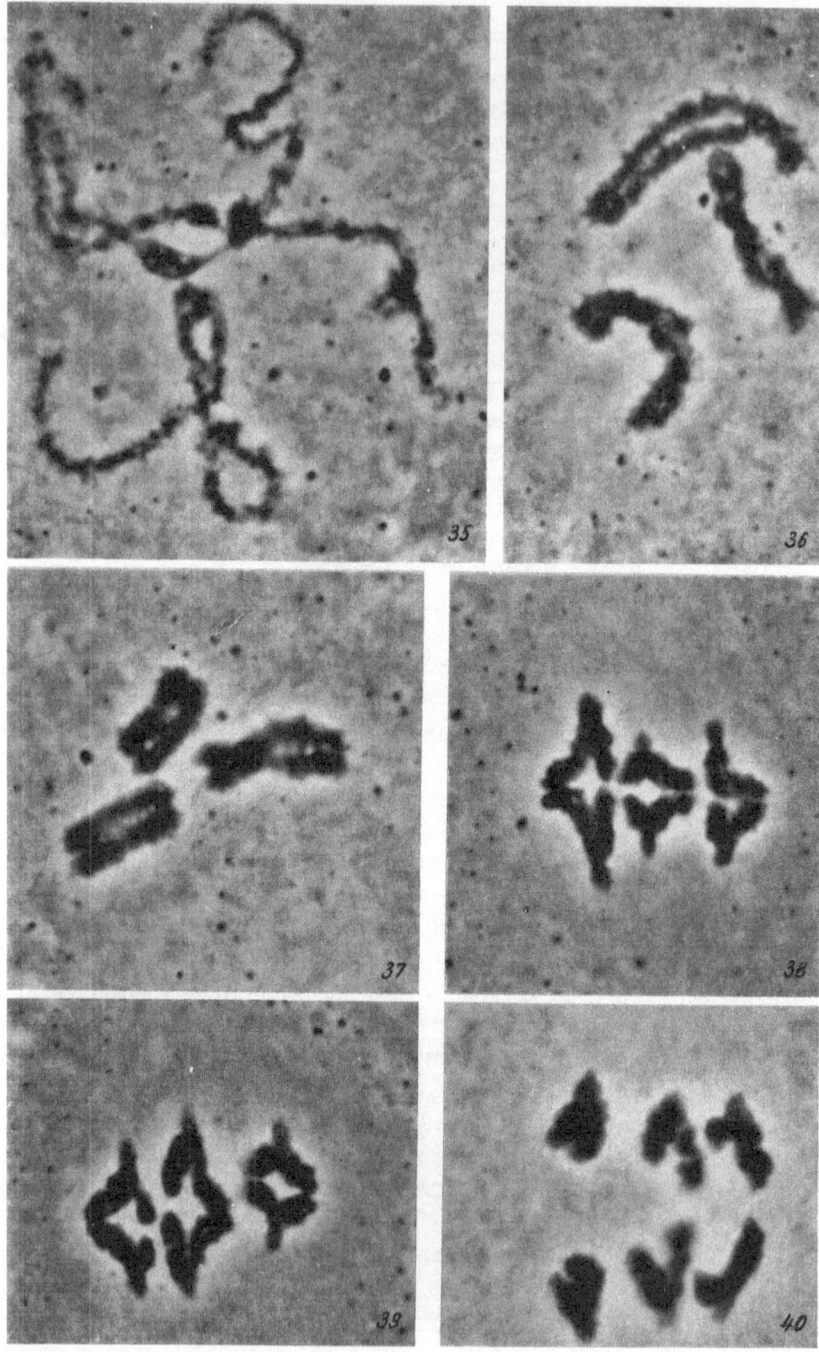

Figs. 35–40. Achiasmate meiosis in the male of *Tipula caesia* (2n = 6). Fig. 35. Pachytene.—Figs. 36 and 37. Post-pachytene contraction.—Figs. 38 and 39. First metaphase.—Fig. 40. First anaphase. Photographs kindly supplied by Professor H. BAUER (phase contrast, ca. × 2500).

3. Female Meiosis

Compared to the information available on male meiosis that for females is scant. The reason for this is largely technical. First, spermatocytes and microsporocytes can be obtained in large numbers whereas oocytes and megasporocytes are fewer in number. Second, the presence of yolk surrounding the egg nucleus makes it difficult to handle the oocyte, especially by the squash method.

Nevertheless, in those cases where female meiosis has been studied there are clear and important differences compared to the equivalent male process. The most obvious dissimilarity is in relation to the survival of the four meiotic products. In the vast majority of plants and animals all four products develop into definitive gametes on the male side whereas in the female only one of the products normally functions as a gamete. But there are more important differences which apply not only in those cases where the sexes are differentiated from one another but also between the separate sex cells of hermaphrodites. These differences fall into two main categories:

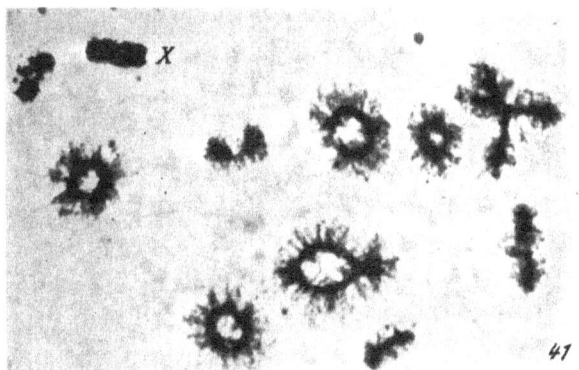

Fig. 41. Lampbrush chromosomes at diplotene in the male of *Duronia chloronata*. Photograph kindly supplied by Dr. S. A. HENDERSON.

(1) The occurrence of a diffuse diplotene—In many species the chromosomes enter a marked diffuse diplotene phase on the female side. This appears to be correlated with the growth of the nucleus which takes place at this time (CALLAN and LLOYD 1961). In female molluscs, echinoderms, fish, amphibians, reptiles and birds this leads to the development of giant lampbrush chromosomes. In female mammals it is even more extensive, producing an extreme lampbrush phase referred to by mammalian cytologists as dictyate. It reaches the most extreme form, however, in female insects where the chromosomes disappear completely from view at the end of pachytene and do not reappear until the organisation of the first metaphase plate shortly before release of the egg from the ovariole.

In the lampbrush phase of the amphibia, each chromosome consists of an apparently single row of granules—the chromomeres—connected by a fine strand. Basically these chromosomes, like others at diplotene, consist of two chromatids but these can be resolved only near the ends, in the region of a chiasma and in the loops for they arise in pairs from the chromomeres. These chromosomes can be isolated mechanically without fixation and they will survive for several days in physiological saline. Active enzymes can be applied to the lampbrush chromosomes in this isolated condition. Treatment of such chromosomes with pepsin, trypsin or RNA-ase leads to the dissolution of the matrical material which surrounds the lateral loops. But these enzymes have no effect on the basic chromosome thread itself. This is, however, disrupted, either in the main axis between successive chromomeres or in the lateral loops, by treatment with DNA-ase (see CALLAN 1963, GALL 1963 and page 41). On this basis it has been concluded that DNA runs continuously along the chromosome and that it is responsible for the longitudinal integrity. However, this conclusion is at variance with those reached from comparable studies on other types of chromosome (see LEWIS and JOHN 1963a).

There is little doubt that male diplotene chromosomes may also show a lampbrush appearance (RIS 1955, NEBEL and COULON 1962) through usually in a far less exaggerated form. In some cases however, as in the *Heteroptera*

Table 5. *Some Differential Chiasma Patterns in Male and Female Animals.*

Species	Chiasma Pattern		Reference
	Female	Male	
1. *Drosophila melanogaster*[1]	chiasmate	achiasmate	DARLINGTON 1934
2. *Bombyx mori*[1]	distal	unrestricted	MAEDA 1939
3. *Paratettix cuccullatus* and *P. toltecus*[1]	unrestricted	distal	NABOURS 1929, 1947
4. *Tigriopus californicus, japonicus* and *brevicornis*	achiasmate	chiasmate	AR-RUSHDI 1963
5. *Triturus helveticus*	unrestricted	distal	WATSON and CALLAN 1963
6. *Triturus cristatus* subspecies *cristatus, carnifex* and *karelinii*	proximal	unrestricted with high Xa freq., distal with low Xa freq.	

[1] In these species the cytological picture has been confirmed by genetical studies.

(LEWIS and SCUDDER 1958) and *Tetrix* (HENDERSON 1961) there is quite a marked diffuse diplotene (see also Fig. 41). Again, in the growing spermatocytes of *Drosophila*, what are apparently paired, loop-shaped structures similar to those of the lampbrush chromosomes of amphibian oocytes make their appearance. Their degree of development varies between species but they are frequently extensive and species are distinguished in respect of them (MEYER 1963). The lampbrush state is thus certainly not exceptional as RHOADES (1961) recently claimed. Indeed NEBEL and COULON (1962) report that they have found a lampbrush structure in the somatic chromosomes of onion and barley.

(2) Chiasma frequency and pattern—The second important difference between male and female meiosis is concerned with the frequency and pattern of chiasmata in the two sexes. A variety of different patterns are now known (Table 5). Superimposed on such differential patterns, however, one may also find differences in chiasma frequency (Table 6). In twelve of these cases, including all the hermaphrodite forms, the chiasma frequency is lower on the male side but in three of the dioecious ones it is higher. While some of these differences are probably not significant those marked with an obelisk (†) are almost certainly so.

Table 6. *Chiasma Frequencies in Male and Female Cells of the Same Species.*
WILSON tested his data statistically and found the difference between male and female
was not significant at the 5% level. Other cases cannot he tested on the data given.
But where the range of variation on the two sides does not overlap (†) significant
differences are very probable.

	Species	Mean Xa Frequency		Reference
		♀ Cells	♂ Cells	
Hermaphrodite	1. *Dendrocoelum lacteum*	20.40 (17 — 22)	11.82 † (9.3 — 13.0)	PASTOR and CALLAN 1952
	2. *Endymion nonscriptus*	18.2 (16.36 — 20.0)	17.67 (16.7 — 19.35)	WILSON 1960
	3. *Lilium longiflorum*	31.48 (26.5 — 37.2)	27.26 (24.2 — 31.8)	FOGWILL 1958
	4. *Lilium martagon*	41.0 (40.0 — 42.0)	36.3 † (36 — 36.5)	
	5. *Fritillaria meleagris*	37.8 (30.0 — 42.5)	24.8 † (23.3 — 26.2)	
	6. *Allium ursinum* *paniculatum* *flavum* *pallens* *macranthum*	14.14 16.0 18.83 19.40 58.66	13.75 14.58 14.90 15.0 42.25	VED BRAT unpublished
Dioecious	1. *Triturus helveticus*	25.0 ± 1.2	21.5 ± 1.1 — 22.4 ± 0.6	WATSON and CALLAN 1963
	2. *Triturus cristatus cristatus*	24.5 ± 1.0	36.5 ± 1.1 —† 38.5 ± 1.6	
	3. *Triturus cristatus carnifex*	23.0 ± 2.2	30.7 ± 1.7 —† 32.1 ± 2.8	
	4. *Triturus cristatus karelinii*	24.6 ± 0.55	39.5 ± 2.5 —† 42.2 ± 1.3	
	5. *Mus musculus*	40.9	34.2	SLIZYNSKI 1960[1]

[1] We are unhappy about the interpretation on which SLIZYNSKI bases his chiasma
counts so that while the relative situation will still obtain the absolute chiasma frequences
may well be lower.

As far as the higher female frequencies are concerned, FOGWILL (1958) has
suggested that it may depend on the facts that:

(i) The female nucleus is larger and this may facilitate pairing. Moreover
embryo sac mother cells have a considerably larger mass of cytoplasm con-
tributing materials to the nucleus than is present in the equivalent male cells, and

(ii) The meiotic prophase appears to be much longer in duration in female cells. Thus FOGWILL estimates it to be at least twice as long in the species she studied.

Unfortunately there is no reason to doubt that these differences apply also in those species where the female has a lower frequency than the male.

III. Structural Abnormalities at Meiosis

The variations on the meiotic theme described in the prevoius section were within the range of normality. The normal course of meiosis may also be considerably impaired as a consequence of abnormal genotypic or environmental conditions. And in this section we will consider those abnormalities which, though they do not depend on them, can lead to structural and numerical change.

1. Chromosome Breakage

Meiotic chromosomes sometimes show evidence of spontaneous chromosome breakage for free fragments are present at diplotene (Fig. 42) or first-metaphase (Fig. 43). In other cases fragments produced by breakage remain attached by chiasmata (Fig. 44). This type of breakage is important because it can lead to modified anaphase behaviour. As we shall see subsequently, bridges and fragments at anaphase can arise following crossing-over in the reverse pairing loop of paracentric inversion heterozygotes (see section IV–Ia). This was first shown by McCLINTOCK in 1931 and since that time there has been a general tendency to regard the combination of anaphase bridge and fragment as diagnostic of inversion hybridity. In other words, since A has been shown to be able to determine B, B has been taken to be indicative of A. Owing to compensation of error, it is possible to obtain true conclusions from false hypotheses by valid deduction. But not only is the above argument illogical, the conclusion is also known to be untrue. In 1953 HAGA suggested an origin by breakage and union for first anaphase bridges and such an origin was clearly shown by REES and THOMPSON (1955) in certain inbred lines of rye. This is what they found.

At metaphase-I, free fragments were present in about 1% of the PMC's, while fragments attached by chiasmata were found in a further 5%. The rarity of free fragments at this stage led REES and THOMPSON to conclude that chromosome breakage was a post-pairing event though there is no reason to believe that the pairing ability of acentric fragments is impaired. Be this as it may, about half the cells at first and second anaphase contained fragments. Only about 10% of the total breakage is detectable at first metaphase because manifestation at this stage depends on the movement apart of the broken ends. While this occurred in closed loops it did not occur in half-loops unless of course the half-loop was centric for, in this event, spindle forces are operative in addition to repulsion between homologues. Further the fragment was not always released even at first anaphase when the aberration showed post-reduction.

Measurements of fragment size indicated that, although there may have been a secondary region of localisation near the centromere, breakage was mainly pro-terminal. This is in keeping with the low manifestation at metaphase and the equational first division separation of the aberration in 70% of the cells.

Thus, these spontaneous chromosome breaks in rye appear to be localised in the same region as are the chiasmata. But breakage was not accompanied by a change in chiasma frequency. In *Paris* on the other hand breakage was procentric or interstitial and so also are the chiasmata (HAGA 1953). It should be borne in mind, however, that bridges and fragments of inversion origin too can be produced only if the inversions occur in regions of crossing-over.

In rye, breakage is accompanied by sister union in at least 95% of the acentric fragments while a minimum value of 70% was obtained for centric fragments. This too suggests breakage after pairing (see below). Bridges were almost as frequent as fragments, indicating the rarity of chromatid (B') breaks. And bridges without fragments, attributed to "splitting errors" (see pg. 42) were found to constitute about 5% of the total bridge frequency.

MARKS (1957a) concluded that the spontaneous bridges and fragments he found in *Oxalis dispar* were likely to be due to breakage (Fig. 45). In this material, however, chromatid fragments were found as well; there were also some exchanges between non-homologous chromosomes and a few cells showed extreme fragmentation. We interpret some of our own unpublished observations, mainly on acridoids, in the same way. In fact we find a syndrome of anomalies similar to that found by HAGA (1953) in *Paris* and by SMITH-WHITE and McCUSKER (1960) in *Asteroloma pinifolium*.

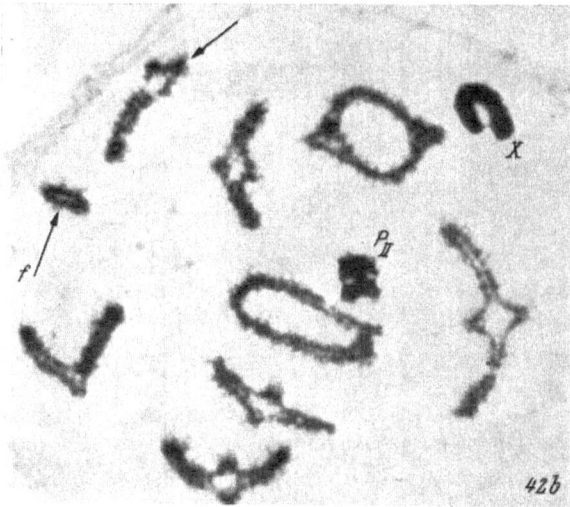

Figs. 42a and b. Diplotene in a normal cell (42a, 2n = 18 + X) and one showing an asymmetrical bivalent and a free acentric fragment produced by chromosome breakage (42b, arrows) in *Pyrgomorpha kraussi*. Note precociously condensed autosomal (PII) bivalent (orcein-phase contrast, ca. × 2000).

For some unaccountable reason it has not always been appreciated that the bridges and fragments produced following crossing-over in a paracentric inversion must be of constant size for a given inversion: fragment size does not vary with the position of the chiasma. Where, therefore, bridges and fragments are claimed to be the result of crossing-over in heterozygous paracentric inversions, a different inversion must be postulated for every size of bridge and fragment. This has not tempered the enthusiasm with which cytologists discover (or create!) inversion hybridity.

For example, PÄTAU (1948) claimed to have observed "a few inversion bridges and fragments" in the spider *Aranea reaumuri*. And an inversion origin was claimed even though "the size of the fragment seemed to vary" and "a very careful search did not give any indication of a pachytene loop." If then inversion hybridity can be claimed solely on the basis of bridges and fragments at anaphase and even in the face of conflicting evidence from pachytene, we must be at least suspicious of the claims based only on anaphase observations. And there have been many such

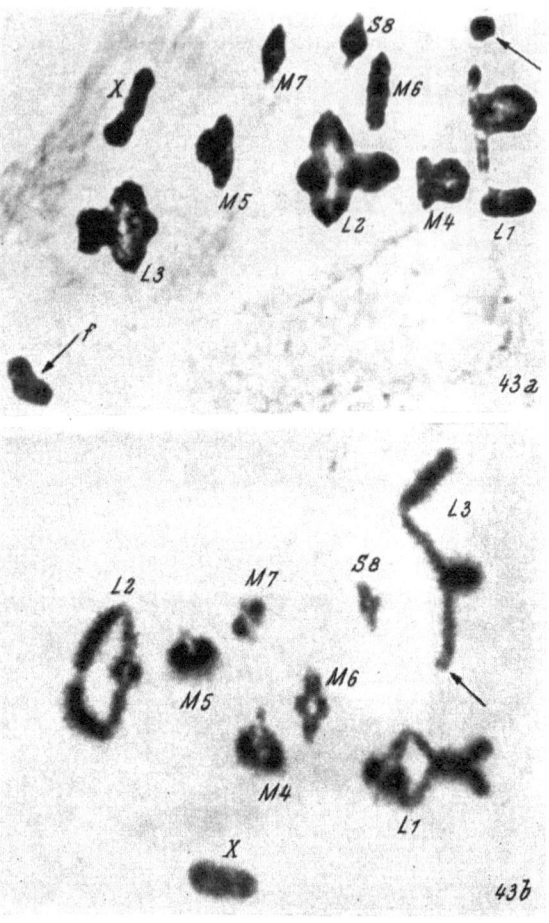

Figs. 43a and b. First metaphase cells showing asymmetrical bivalents produced by chromosome breakage in *Chorthippus brunneus* (Fig. 43a, orcein-phase contrast) and *Ch. parallelus* (Fig. 43b, orcein). In 43a a free fragment (f) is present.

claims. Thus GEITLER (1937, 1938) attributed the bridges and fragments observed by him in natural populations of *Paris quadrifolia* to crossing-over in paracentric inversion heterozygotes. And to accomodate the variation in fragment (and bridge) size, twenty-one distinct inversions were required and postulated. HAGA (loc. cit.), on the other hand, attributes the same anomaly in the same organism to breakage and sister union. HAGA, like MARKS (1957a), found some configurations which apparently resulted from breakage and fusion of chromatids belonging to different non-homologous bivalents. This led him to conclude that the breakage occurred at a stage when the chromosomes were effectively split. But differences in the unit of reunion rather than breakage may be involved.

In addition to the more-or-less easily definable breakage so far described with its comparatively simple consequences, there have been many reports of extensive chromosome shattering (Loveless 1954). In the case of oat hybrids, Holden and Mota (1956) found that the condition was correlated with an abnormal peripheral position of the nucleus in atypical bi-nucleate pollen mother cells. Similar cells were described by Jain (1958) in heat-treated *Lolium*. Darlington and Haque (1955), on the other hand, found that breakage in the pollen mother cells of *Allium ascalonicum* was associated with a delay in entering meiosis. The effect was thus correlated with others attributable to reduced preco-

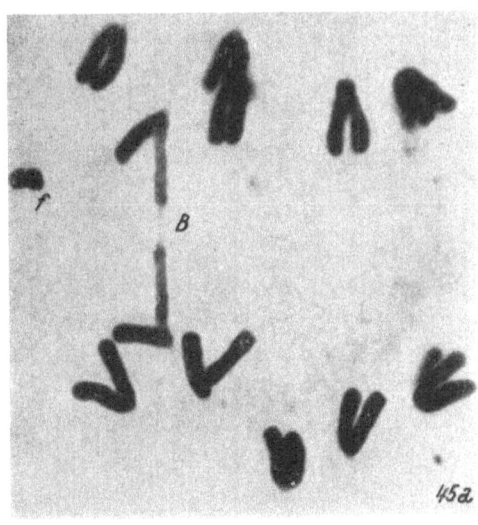

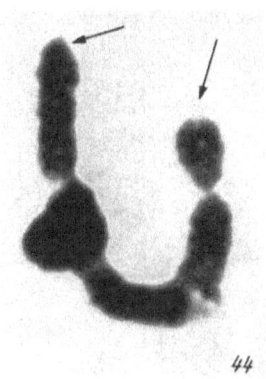

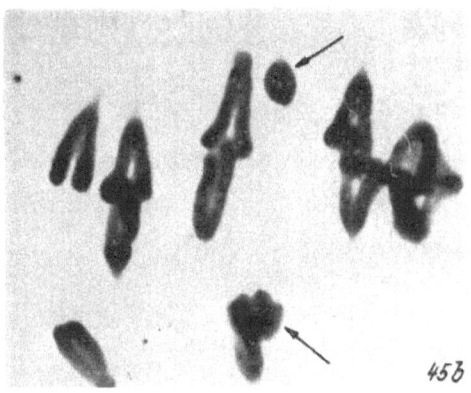

Fig. 44. An L-bivalent at metaphase-1 in *Chorthippus brunneus* showing a fragment produced by chromosome breakage which is attached to the unaffected homologue by a chiasma (orcein, ca. × 2000).

Figs. 45a and b. Spontaneous chromosome breakage at meiosis in *Oxalis dispar* showing equational separation of B″ + 2 SU giving bridge and fragment at anaphase-1 (45a) and reductional separation of same to give loop chromosome and normal chromosome with attached fragment (45b). Preparations from material kindly supplied by Mr. G. E. Marks (carmine, ca. × 1500

city, namely, failure of pairing, reduced spiralisation and increased cell size. The aberrations here also were of the chromosome (B″) type with a preponderance of sister unions. This was attributed to breakage during pachytene for, at this stage, sister chromatids are at a positional advantage.

The abnormal cells in *Allium* seemed to be confined to the ends of the pollen sacs and away from the tapetum. They could therefore be attributed to a defect in the cytoplasm. In rye, on the other hand, the condition doubtless reflects an unbalance in the genotype wrought by the enforced inbreeding. It would

Table 7. *Summary of Abnormal Chromosome Complements Obtained Following Transfer of Rana pipiens Nuclei into R. sylvatica Cytoplasm.*
(Data of HENNEN 1963.)

Note the Normal Diploid Complement of *R. pipiens* Consists of 26 Metacentrics.

Transfer Type	Developmental Type	No. of Embryos	Embryos Showing Chromosome Counts of:									Embryos Possessing Abnormal Chromosomes		
			22	24	25	26	27	28	29	30	37	Rings	Minutes	Acentrics
A. Back-transfer embryos	1. Abnormal a) arrested early (gastrulae, neurulae and early tailbuds)	18	1	4	1	1*	2	2	1	1	—	12	10	1
	b) arrested late (late tailbuds and abnormal tadpoles)	8	—	—	2	—	5	1	—	—	—	—	—	—
	2. Normal tadpole	1	—	—	—	1	—	—	—	—	—	—	—	—
B. Nuclear transfer hybrids	1. Sacrificed as donors (at mid-blastula stage)	6	Range of 25—37 from approximate counts									5	—	—
	2. Arrested as late blastulae	2	Range of 23—28 from approximate counts									2	—	—

* This Karyotype though showing a normal diploid count in fact contained a ring and a minute.

appear that unbalance resulting from wide hybridisation can have the same effect, as has been demonstrated in *Bromus* (WALTERS 1951) and, perhaps, *Allium* too (EMSWELLER and JONES 1938). In species hybrids, however, there is a greater expectation of structural hybridity and the genotypically versus structurally determined aspects of chromosome behaviour are less easily distinguished (see LEWIS and JOHN 1963a). One might, in consequence, be more likely to assume an inversion origin for bridges and fragments in species hybrids. There can be little doubt that this is their origin in many cases. But it is worth bearing in mind that, while genotypic unbalance in hybrids may result in various abnormalities including breakage, alien cytoplasm can have the same effect, for HENNEN (1963), using the technique of nuclear transfer, has shown that chromosome abnormalities arise when nuclei of *Rana pipiens* undergo replication and division against a *Rana sylvatica* cytoplasmic background. This demonstration depended on the transplantation of diploid nuclei from normal blastulae of *pipiens* into experimentally anucleate *sylvatica* eggs. After some 10–12 mitotic generations in the alien cytoplasm the nuclear descendents of the original transfer nucleus were transferred back to enucleate *pipiens* eggs. All but 1 of the 62 back-transfer embryos derived from such eggs developed abnormally and were arrested at stages varying from late blastulae to young tadpoles. Cytological analysis of 26 such abnormal back-transfer embryos showed that they had abnormal chromosome complements. And the earlier the stage of arrest the more pronounced the degree of abnormality.

These abnormalities were of two kinds (Table 7):

(a) Irregularities of chromosome number resulting from the failure of centromeres either to reproduce or else to move normally, so leading to states of aneuploidy.

(b) Irregularities of structure resulting from breakage and reunion so leading to the production of ring chromosomes, minutes and acentric fragments.

Examination of the chromosomes from 8 nuclear transfer hybrids showed that these too had similar abnormalities (Table 7). Evidently, therefore, *pipiens* chromosomes develop these abnormalities partly as a consequence of replicating at the expense of materials derived from *sylvatica* cytoplasm and partly from moving on spindles derived from *sylvatica* cytoplasm.

Presumably, breakage in endosperm cells, which is frequently extensive following wide crossing, can have a similar origin as, indeed, may hybrid inviability or sterility (see section VII).

2. Side-arm Bridges

The separation of chromatids at mitosis and that of chromosomes (first division) or chromatids (second division) at meiosis is sometimes prevented or impaired by the "adhesion" of chromatids at homologous points. These points, may be terminal or, more often, interstitial. Similar "adhesions" have been described between non-homologous loci within chromatids and between chromatids both within and between chromosomes. The terms used to describe this type of aberration include point unions, errors or affects, pseudo-chiasmata, partial chiasmata and sub-chromatid errors. Some of these terms imply an explanation, others are intentionally non-commital.

When the adhesions are interstitial they give rise to so-called two-side arm bridges (Fig. 46). Such bridges have been found to occur spontaneously in plants unbalanced genetically owing to inbreeding, hybridisation, polyploidy or aneuploidy. They have also been seen under apparently normal genetic and environmental conditions, more especially in animals (Table 8). Configurations which are superficially similar, at least, have been seen in cancer cells. They can also be induced by cold or heat shock (Table 9) and by X-irradiation (Table 10). In the mitotic cycle they can only be obtained following prophase irradiation and they are then visible at the anaphase immediately following treatment, the X_1-anaphase. In the meiotic cycle, on the other hand, irradiation of metaphase also gives the same effect (see page 39). Finally, side-arm bridges can also be induced by treatment with a variety of chemicals (Table 11).

Many of the investigations which have been carried out on these side-arm bridges have been made at mitosis rather than meiosis. But since the anomaly appears to be of the same type in the two kinds of division, the results of these investigations also will be considered here.

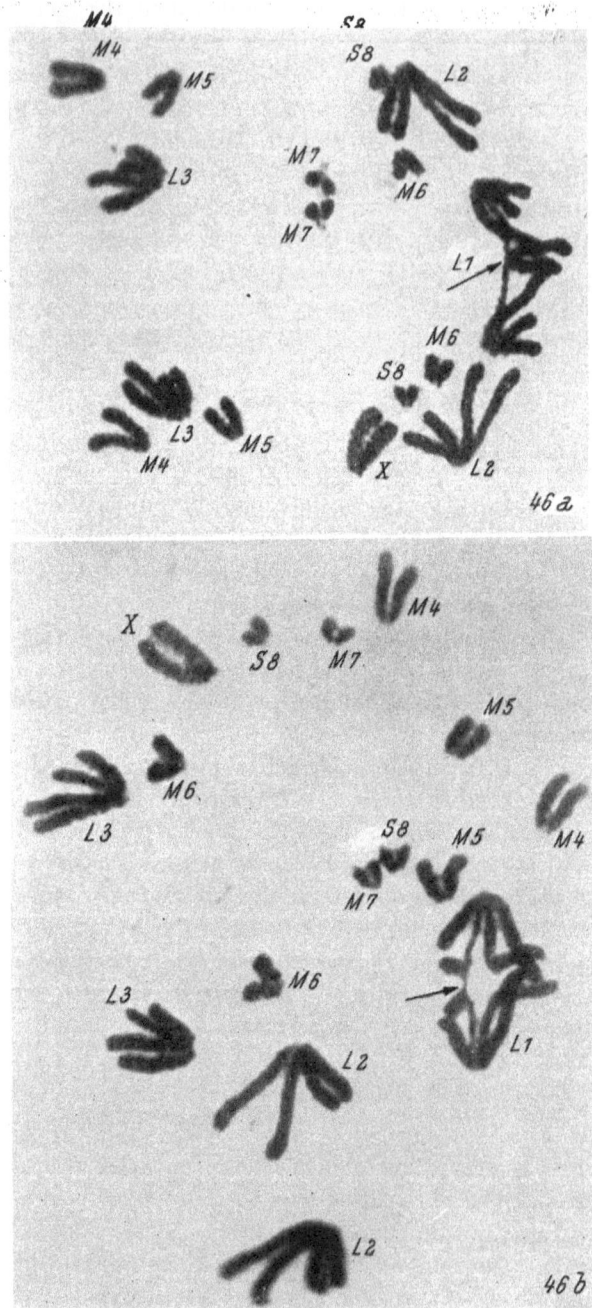

Figs. 46a and b. Side arm bridges (S.A.B.) in *Omocestus viridulus*. In (a) the side arms arise from a point (arrow) but in (b) they are separated by a bridge (arrow). Compare with Fig. 47 (orcein, ca. ×1500).

Table 8. *Spontaneous Side Arm Bridges at Mitosis and Meiosis.*

	Condition and Reference	Material and Cell Type	Comments
Mitosis	1. Developmental (MELANDER 1963a and b)	*Calliphora erythrocephala* (embryonic and pupal)	Subterminally located in the long metacentric chromosomes from 9–12th cleavage mitoses. Interstitially located in the long acrocentrics during the 12th and 13th mitoses.
		Polycelis, Dugesia, Planaria, Dendrocoelum and *Bdellocephala* (embryonic)	Formed regularly in a small fraction of anaphases in the free blastomeres. From 1–4 bridges per anaphase may form. Normally they resolve without breakage.
	2. Hybridity (KLINGSTEDT 1939)	*Chorthippus bicolor × Ch. biguttulus* (Sperm)	Few cells only affected. Attributed to prolongation of pairing.
	3. Unbalance (BROCK 1955)	*Hyacinthus* (Endosperm)	More common in nucleolar arm of the long chromosome (compare DAVIDSON 1957) the centromere of which also showed breakage (compare DARLINGTON and KEFALLINOU 1957, KIHLMAN 1955). Mainly homologous. No detectable heterochromatin.
Meiosis	1. Natural and regular (JOHN, LEWIS and HENDERSON 1960)	*Chorthippus brunneus* (SMC) (and see Table 12)	Occur in all chromosome types but predominantly in the long arms of the longest chromosomes. Maximum of 3 bridges per cell may form and from 2–40% first anaphases may be affected within an individual. Invariably between homologues.
	2. Hybridity (KLINGSTEDT 1939)	*Ch. bicolor × Ch. biguttulus* (SMC)	Attributed to prolongation of pairing (compare mitosis 2 above) but see also meiosis 1. above. We have found these bridges quite regularly in all chorthippoid populations we have examined.
	3. Hybridity (WALTERS 1957)	*Bromus trinii × B. maritimus* (PMC)	Bridges involve homologous and non-homologous chromosomes.
	4. Cultivated (DARLINGTON and KEFALLINOU 1957)	*Gasteria undulata* (PMC)	In 48 of 261 cells examined. Distal. Long arms of long chromosomes only.

Sp = Spermatogonial; SMC = sperm mother cell; PMC = pollen mother cell,

a) Properties of the Side-arm Bridges

(i) Homologous point unions—In many of the studies summarised in Tables 8 to 11 only homologous point unions were found. This was the case in:—

a) Untreated SMC's of *Chorthippus* (JOHN, LEWIS and HENDERSON 1960),

b) Cold-treated root tips of *Trillium* (DARLINGTON and LA COUR 1940, SHAW 1958),

c) X-irradiated root tips of *Scilla campanulata* (REES 1952) and *Vicia faba* (PEACOCK 1961),

d) X-irradiated PMC's of *Tradescantia reflexa* (SAX 1938) and *Trillium erectum* (WILSON et al. 1959).

Homologous unions also predominated in:

a) Untreated endosperm of *Hyacinthus* (BROCK 1955),

b) X-irradiated root tips of *Vicia faba* (DAVIDSON 1957),

c) X-irradiated PMC's of *Lilium longiflorum* (CROUSE 1961) and

d) 8-EOC-treated root tips of *Vicia faba* (DAVIDSON 1957).

Table 9. *Side-arm Bridges Induced at Mitosis by Cold Shock and at Meiosis by Heat Shock.*

Condition and Reference	Material and Cell Type	Comments
1. Cold (DARLINGTON and LA COUR 1940)	*Trillium* sp. (RT)	In heterochromatin. Homologous.
2. Cold (CALLAN 1942)	*Triton* sp. (T)	A few cases. Heterochromatin present.
3. Cold (SHAW 1958)	*Trillium* sp. (RT)	In heterochromatin. Homologous.
4. Heat (BARBER 1940)	*Fritillaria meleagris* (PMC)	Diplochromosomes observed.

RT = Root tip; T = Tail; PMC = Pollen mother cell.

In fact, the frequency of non-homologous unions is generally very low. For example, DAVIDSON (1957) recorded 3% and 0–8% of them respectively in 8-EOC treated and X-irradiated roots of *Vicia*. In only two investigations has a high frequency of non-homologous unions been claimed. ÖSTERGREN and WAKONIG (1954) found that they exceeded the homologous unions in frequency under "certain types of treatment." Many non-homologous unions were found also by CROUSE (1961) at the second division of meiosis but only following the irradiation of prophase II — metaphase II. In this case, however, it should be borne in mind that, as pointed out by CROUSE herself, the lapse of attraction between homologous chromatids, which determines the onset of first anaphase, persists to the prophase of second division or even to second metaphase (Figs. 22 and 23). Consequently, homologous chromatids at this stage have a unique spatial relationship—unique compared both with mitotic prophase and with the prophase of the first meiotic division when their association is especially intimate. At second division, therefore, homologous chromatids are at a disadvantage from the point of view of proximity which must be of importance whatever the precise cause of the point unions.

(ii) Distribution between chromosomes—In some cases point unions were observed only in long chromosomes or long chromosome arms. This was the

Table 10. *Side-arm Bridges Induced at Mitosis and Meiosis by X-irradiation.*

	Material and Cell Type	Reference and Treatment	Comments
Mitosis	1. *Tradescantia reflexa* (PG)	Sax 1938, 75–200 r	In about half the cells. Terminal and sub-terminal.
	2. *Tradescantia paludosa* (PT)	Swanson 1943	Half-chromatid lesions and exchanges.
	3. *Trillium grandiflorum* (RT)	Darlington and La Cour 1945, 45–375r	Appear after 4 hours with 45 r.
	4. *Scilla campanulata* (RT)	Rees 1952, 50r/58sec./18° C	Long chromosomes disproportionately affected. Homologous only. No detectable heterochromatin.
	5. *Scilla sibirica* (E)	La Cour and Rutishauser 1954, 54 r/18° C	Homologous and intra-chromosome non-homologous (intra and inter-chromatid). Absent near centromere.
	6. *Vicia faba* (RT)	Davidson 1957, 100 r/50 sec	Mainly in middle third of chromosome arms. Errors in S- and M-types proportional to length except that proportionately fewer errors in M following mid-prophase irradiation. 0–8% non-homologous.
	7. *Hyacinthus* (RT)	Davidson 1957, 100 r/50sec	Present in L-, M- and S-chromosomes but disproportionately in L-types.
	8. *Trillium erectum* (PG)	Wilson *et al.* 1959, 25 r	See text (p. 44)
Meiosis	1. *Tradescantia reflexa*	Sax 1938, 75–2000 r	Homologous. Interstitial and terminal.
	2. *Uvularia perfoliata*	Darlington and La Cour 1952, 90 r/20° C	Long chromosomes only.
	3. *Tradescantia subcaulis*	Haque 1952, 18 r/23° C	Observed at A II after irradiation of diakinesis.
	4. *Lilium longiflorum*	Crouse 1954, 1961 5–30 r	See text. Frequency not affected by chloramphenicol treatment.
	5. *Lilium longiflorum*	Mitra 1958, 15–60 r	The only error obtained from irradiation 2, 3 and 6 hours before A I. Homologous and non-homologous, X and U-shaped.
	6. *Trillium erectum*	Wilson *et al.* 1959, 25 r	Homologous only. Random along length. Tri-radial seen at PGM-I. Y- and E-types not observed.
	7. *Vicia faba*	Peacock 1961, 36 r/60 sec	Homologous only. Y- and E-types observed (See Fig. 52).

case, for instance, in untreated PMC's of *Gasteria undulata* (DARLINGTON and KEFALLINOU 1957) which has four pairs of long and three pairs of short chromosomes, the latter being less than a quarter the size of the former.

Table 11. *Side-arm Bridges Induced by Treatment with Chemicals.*

Chemical and Reference	Material and Cell Type	Comment
1. Mustard gas (DARLINGTON and KOLLER 1947)	*Tradescantia bracteata* (PGM)	Homologous only. Centric errors observed.
2. Coumarin (ÖSTERGREN 1948)	*Allium cepa* (RT)	Usually homologous. Random along chromosome. Increase in frequency with concentration.
3. Phenols (LEVAN and TJIO 1948)	*Allium cepa* (RT)	"Rings" seen occasionally. Gaps common
4. Acenaphthene (D'AMATO 1950)	*Allium cepa* (RT)	Interpreted as whole break (1 hit) with union at one point.
5. α-Naphthalene derivatives (AVANZI 1950)	*Allium cepa* (RT)	Interpreted as "fusion of protein fibres in *eroded* chromosomes".
6. Ethyl urethane (OEHLKERS and MARQUARDT 1950)	*Paeonia* (PMC)	Lesions and exchanges.
7. Ethyl urethane (DEUFEL 1951)	*Vicia faba* (RT)	Lesions and exchanges.
8. Penicillin (LEVAN and TJIO 1951)	*Allium cepa* (RT)	Erratic frequencies of this and other radiomimetic effects.
9. Coumarin (ÖSTERGREN and WAKONIG 1954)	*Allium cepa* (RT)	Nearly all chromosome effects at X_2. See text.
10. 8-Ethoxycaffein (KIHLMAN 1955)	*Allium cepa* (RT)	Misdivision observed.
	Vicia faba (RT)	M-chromosome affected disproportionately to length but no difference between nucleolar and non-nucleolar arms. 3% non-homologous.

In other cases a disproportionately high frequency has been found in the long or longer chromosomes of complements with a range of chromosome size (Table 12). Thus:

a) In *Scilla campanulata* the over-all length ratio of long: short is 3:1 but the error-ratio of X-ray induced point unions was found to be 27:1 (P = 0.01, REES 1952).

b) Point errors occurred in long, medium and short chromosomes following the X-irradiation of *Hyacinthus* roots but the long chromosomes were disproportionately affected (DAVIDSON 1957), and

c) Following the treatment of *Vicia* roots with 8-EOC, the long, nucleolar, chromosome was affected disproportionately (DAVIDSON 1957).

This latter investigator obtained curious and partly opposite results following X-irradiation of the same material. Irradiation of mid-prophase resulted in relatively fewer point errors in the long chromosomes but treatment of other prophase stages gave error-frequencies which were a function of length.

(iii) Distribution within chromosomes—Non-random distributions of point errors within chromosomes have frequently been reported. Thus:

a) In cold-treated *Trillium* roots the error is confined to those heterochromatic segments which are revealed by the cold treatment. This and other properties, e. g., their occurrence only at homologous points, suggests that coldinduced, side-arm bridges may be a somewhat different class of error from the others though they give superficially the same aberration. Thus both spontaneous and X-ray induced point unions have been described in species where heterochromatin is not revealed by cold treatment, e. g., *Scilla campanulata* and *Hyacinthus*.

b) BROCK (1955) claimed that the spontaneous point unions he observed in the endosperm of hyacinths occurred more frequently in the nucleolar arm of the long chromosome. But, in *Vicia* the nucleolar and non-nucleolar arms of the long M-chromosomes, which are of approximately equal length, did not show a difference after either X-ray or 8-EOC treatment (DAVIDSON 1957).

c) Many have commented on the absence or uncommonness of point unions in the vicinity of the centromere. This was found in the case of:

i. X-ray induced unions in *Scilla sibirica* endosperm (LaCour and RUTISHAUSER 1954).

ii. Spontaneous point unions at first anaphase in *Gasteria* (DARLINGTON and KEFALLINOU 1957) and *Chorthippus* (JOHN, LEWIS and HENDERSON 1960).

iii. Spontaneous and X-ray induced unions in *Scilla campanulata* (REES 1952).

d) Distal localisation of the error is common but DAVIDSON (1957) found the middle third of the chromosome arm to be the most susceptible both in the long, metacentric, nucleolar chromosome and the longer arm of the short, acrocentric chromosomes.

e) There have, however, been a few reports of the occurrence of point unions at random along the length of the chromosomes. For example in:

i. Coumarin-induced errors in *Allium cepa* roots (ÖSTERGREN 1948), and

ii. X-ray induced errors in *Trillium* PMC's (WILSON *et al.* 1959).

(iv) Change in distribution with stage irradiated—DAVIDSON (1957) found that the point unions induced by the irradiation of late prophase occur nearer the ends of the chromosomes than those induced by irradiation at earlier prophase stages. More terminal adhesions were seen with late treatment also but non-homologous unions did not occur.

(v) Frequency distribution and dose response—While the frequency distribution for point unions per cell seems to fit a Poisson distribution, that per chromosome does not. With regard to the former, DAVIDSON (1957) found a fit with Poisson for each of his four prophase fixations made at different times. But the four were not homogeneous. With regard to the latter LaCOUR and RUTISHAUSER (1954) found that the observed frequency of two point unions per chromosome was less than half that expected on a Poisson series (P < 0.01).

(vi) Susceptibility of various stages—Point unions are induced at mitosis only by prophase irradiation. And early prophase was found to be the most susceptible stage in both X-irradiated root tips of *Vicia faba* (DAVIDSON 1957) and X-irradiated endosperm of *Scilla sibirica* (LaCOUR and RUTISHAUSER 1954). This is also the stage most susceptible to 8-EOC treatment (KIHLMAN 1955).

In contrast it would appear that metaphase irradiation also will induce point unions in the meiotic cycle both at first and second division (CROUSE 1961). But point unions are not induced by irradiation before late pachytene. The yield of point unions obtained following treatments over the range of susceptible stages in mitosis and meiosis are also quite different (Fig. 47). But if the two peaks are held to correspond then, although that of meiosis occurs later in relation to the onset of division, they more-or-less coincide relative to the respective times of DNA-synthesis and chromatid formation. Finally, over a range of low doses (CROUSE 1954, 1961) the yield has been found to be linear with dose (Fig. 48) but whether this aberration shows an oxygen effect has not been determined.

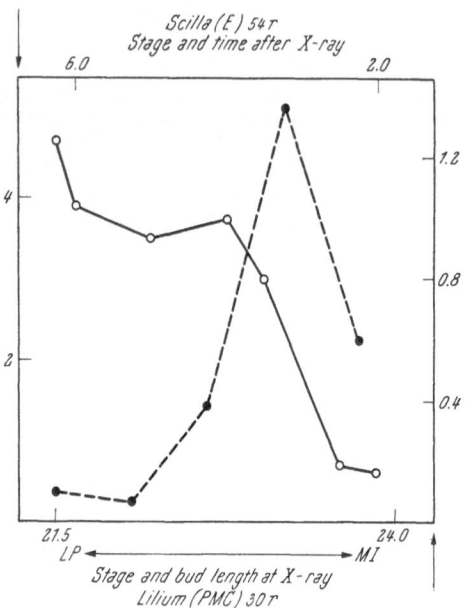

Fig. 47. Comparison of the sensitivity of different parts of prophase to X-rays with regard to the induction of side arm bridges at the following anaphase. Mitotic data (—○—) from LaCOUR and RUTISHAUSER (1954), meiotic data (---●---) from CROUSE (1961). Since endosperm (E) was used in the mitotic study there is no gradual approach to a peak such as would inevitably obtain in the case of a non-synchronised population. Vertical axes refer to the S.A.B.'s per cell.

(vii) The nature of the side-arm bridges—In handling chromosomes, the investigator is usually concerned with a structure composed of two visibly distinct chromatids or else with the equivalent of a single chromatid; in fact, with chromosomes after or prior to their reproduction in interphase and prior or after their division in the mechanically active phase. This simple two-level system can account for, or rather is revealed by the results of both breeding experiments and cytological studies. Such properties as pairing, crossing-over and segregation can be considered in terms of it. So also can most, at least, of the effects of mutagenic agents. Thus the results obtained following the irradiation of nuclei in interphase, or their treatment at this time with radiomimetic compounds like 8-EOC, conform to and confirm the simple, two-level, chromosome/chromatid system. So also do the consequences in the X_2-mitosis of irradiation during the X_1-sequence.

There are, however, certain observations which suggest that while the chromatid (or unreproduced chromosome) behaves as a lateral unit in respect of its mechanical and hereditary work, it has, in fact, a multiple structure. It is no longer doubted that the chromosomes are multi-stranded or polytene in tissues like the salivary gland cells of dipteran larvae. It is equally clear, however, that this condition depends on a multiplication of the original chromosome (or some of its components at least) and that it cannot be accounted for simply in terms of the individualisation of existing sub-units. But a multiple structure for chromosomes in the germ-line is suggested by various lines of evidence of unequal value (see LEWIS and JOHN 1963a). These include:

a) Claims to have seen two or even four sub-chromatid units with the light microscope both at mitosis and meiosis (NEBEL and RUTTLE 1939).

b) The mechanical autonomy shown by half-chromatids at meiosis in certain coccid bugs all of which have non-localised centromeres (see section VI–2).

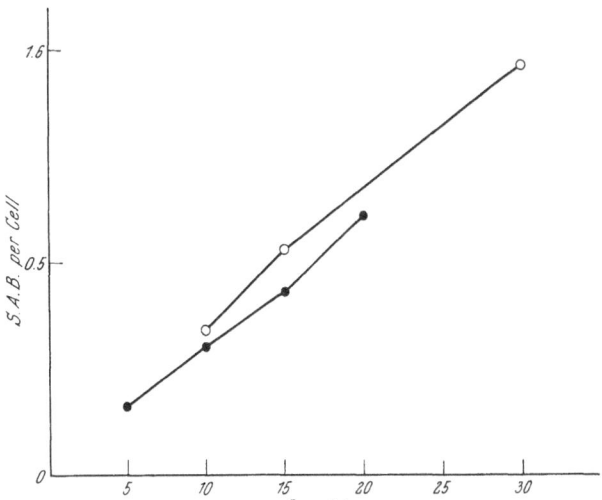

Fig. 48. The relationship between dose and induction of side arm bridges (S. A. B.) at meiosis (data of CROUSE 1954, open circles, and 1961, solid circles).

c) Claims that a chromosome may give rise to two chromatids without making DNA.

d) The claim that certain chromosomes divide at both meiotic divisions.

e) The large differences in the amount of DNA per nucleus in related species.

f) Large size-differences between the chromosomes of related species.

g) A 2 : 1 ratio with regard to DNA value in related species having the same chromosome number.

h) A 2 : 1 ratio of chromosome numbers in con-generic species having the same DNA value.

i) A 2 : 1 ratio of chromosome numbers in related species having a 1 : 2 ratio with regard to DNA values.

j) The delayed effects of certain mutagenic treatments and the mosaics that sometimes develop following the treatment of gametes, and

k) The evidence from electron microscopy.

So far as the observational evidence is concerned, it may well be that the appearance of doubleness is an artifact or an illusion. But a four or more (ie. two or more double helices) stranded structure may be compatible with an essentially two-stranded behaviour (see below).

It is also conceivable that the components of a double helix begin to move apart during the later stages of division for, although synthesis is an interphase event for the most part, there is no compelling reason for believing that the two strands of DNA uncoil only immediately prior to synthesis. Indeed, such an uncoiling may be the basis of the two types of DNA which have been found in growing rat tissues (BENDICH 1952). There is, however, little to support this suggestion and, in itself, it does not account for the 4-partite structure which some have claimed for telophase chromosomes.

The most formidable arguement against the multi-stranded concept comes from a study of lampbrush chromosomes (see page 24) and it renders unlikely any model which postulates a large number of threads per chromosome. Thus, the loop axis—the delicate thread on which the ribonucleoprotein matrix is accumulated and the equivalent of a chromatid—has been described as no more than 60–80 Å units in diameter while the main axis—the equivalent of a chromosome—has a diameter of 100–200 Å units. Threads of these diameters could accomodate only 2–4 twin helices of DNA at most in the case of the loops and a correspondingly higher number in the main axis. And the chromosomes of newts are amongst the largest known. It should be borne in mind, however, that the main axis is extended to about 40–50 times its normal length so that the arrangement of the macromolecules may be an unusual one and of temporary duration.

The question of lateral multiplicity in lampbrush chromosomes has been studied by GALL (1963) who examined the kinetics of the breakage caused by DNA-ase treatment. Where:—

(i) The sub-units are attacked equally and independently,

(ii) A visible break requires the scission of all sub-units at the same transverse level, and

(iii) The rate of enzyme activity remains constant with time, then the number of visible breaks (b) is given by:—

$$b = k_1 t^n \equiv \log b = n \log t + k_2$$

where k is a proportionality constant,
t is time, and
n is the number of independently attacked sub-units.
Plotting $\log b$ against $\log t$ should give a straight line of slope n.

The data for the fragmentation of a group of giant loops which occur near the middle of chromosome 10 gave $n_{23} = 2.6 \pm 0.2$. The comparable value for inter-chromomeric breaks was $n_{19} = 4.8 \pm 0.4$. The ratio between them indicates that there are twice as many sub-units in the inter-chromomeric segments ($4.8/2.6 = 1.8$).

GALL concluded that the absolute values were two and four although these fall, respectively, three and two standard errors away from the experimentally-determined means.

The nature of the sub-units cannot be deduced from the breakage kinetics the interpretation of which must take into account the nature of the action of the enzyme and that of the molecule it attacks. In solution, the two polynucleotide strands of duplex DNA are broken independently and at random by DNA-ase and complete fracture results only if the two bonded columns are broken within an interval of a few base pairs. If the action on chromosomal DNA *in situ* is the same, then the sub-units correspond to single polynucleotide chains.

Of course, evidence for multistrandedness does not necessarily mean that the lateral multiplicity can be halved, for example, to give two structures of unimpaired mechanical and genetical function. But the evidence mentioned above in which DNA estimations were involved does support this possibility.

The terms suggested to describe the side-arm bridges are out-numbered by the explanations that have been offered for them. These include:

a) Interlocking of plectonemic coils.

b) Failure of reproduction (DARLINGTON and LACOUR 1940).

c) Fusion of broken protein fibres in "eroded" chromosomes (AVANZI 1950).

d) Chromosome breakage with union at one point (D'AMATO 1950).

e) Matrical stickiness with whole chromatid exchange (ÖSTERGREN and WAKONIG 1953).

f) Two-plane splitting (DARLINGTON 1949).

g) Sub-chromatid exchange (NEBEL 1937).

Some of these (b, f and probably a) could explain only homologous unions which, as we have seen, are the commonest and sometimes the only type obtained. In fact the second explanation (b) was offered for the point unions observed in heterochromatin following cold shock and where all the unions were homologous. One of the explanations (g) implies a multiple structure for the chromatid though not necessarily a genetically divisible one. A multistranded chromatid structure has been proposed also to account for so-called partial lesions without exchange. These have been described in untreated species hybrids (WALTERS 1955) and in material treated with either X-rays or ultra-violet light (see EVANS 1962). In fact most modern workers attribute the production of side arm bridge configurations to sub-chromatid exchange, more specifically in most cases to half-chromatid exchange.

b) Sub-chromatid Exchange

Fig. 49 shows that sub-chromatid exchanges between both sister and non-sister chromatids could give rise to side arm bridges at either first or second anaphase of meiosis (Figs. 50–51). Thus exchange between non-sister chromatids gives a bridge at anaphase-I in the absence of a chiasma in the intercept between the sub-chromatid exchange and the centromere. But a chiasma in this region will give a second division bridge if it involves one (but not both or neither) of the chromatids concerned also in the sub-chromatid exchange.

Exchange between sister chromatids gives a bridge at second division unless a chiasma forms in the critical intercept. Neglecting chiasma formation between sister chromatids, such a chiasma must involve one or other of the chromatids concerned in the sub-chromatid exchange and it will result in a bridge at first division. This also follows three-strand double crossing-over but two- and four-strand double cross overs give a second division bridge. And so on.

It is clear, therefore, that intra-chromosome as opposed to inter-homologue, sub-chromatid exchange cannot be distinguished simply by looking at the anaphase configurations obtained. Further, the frequency of first versus second division side-arm bridges expected on either type of exchange cannot be calculated without information regarding the positions of the sub-chromatid ex-

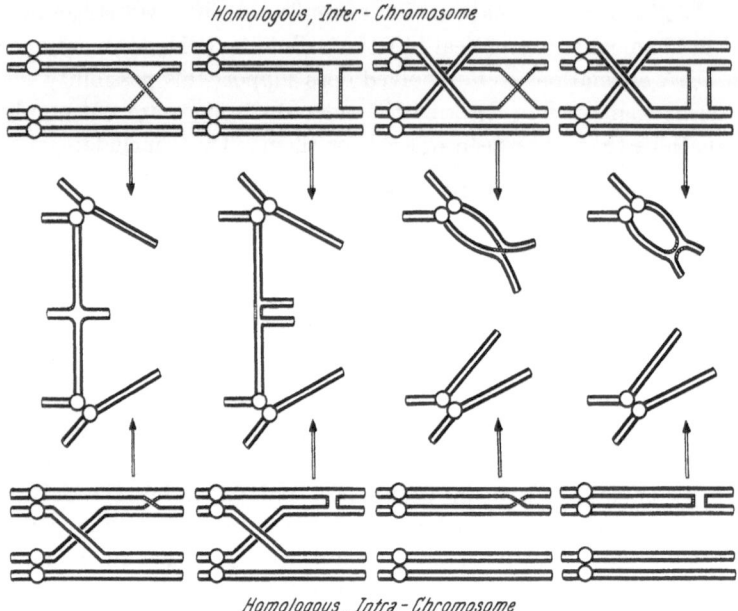

Homologous, Inter-Chromosome

Homologous, Intra-Chromosome

Fig. 49. An explanation of side arm bridges in terms of half-chromatid exchange in dinemic chromosomes.

changes relative to the centromere, the frequency of chiasmata in the critical intercept, and the chromatids involved in them.

Homologous unions at mitosis in haploid pollen grains must, of course, involve chromatids of the same chromosome. WILSON *et al.* (1959) concluded that the homologous unions at meiosis (the only type observed in their study) also involved intra-chromosome, sub-chromatid exchange. The evidence was as follows:—

(a) Outside of the regions affected by chiasmata, sister chromatids in *Trillium* coil in the same direction at any given point

(b) The direction of the internal coiling in one pair of chromatids is at random relative to that of the other pair in the same bivalent, and

(c) The coiling of the two chromatids in the side-arms of the side-arm bridges was always in the same direction. It was concluded, therefore, that the chromatids in the side-arms were sisters and, consequently, that the sub-chromatid exchanges involved parts of only sister chromatids.

Even if this evidence is accepted it is a curious situation for the coiling of the two side-arms would "match perfectly" only if chiasmata distal to the point of union were excluded. In this work, however, a random distribution of point unions was claimed, some of them giving side-arms "virtually a whole arm in length."

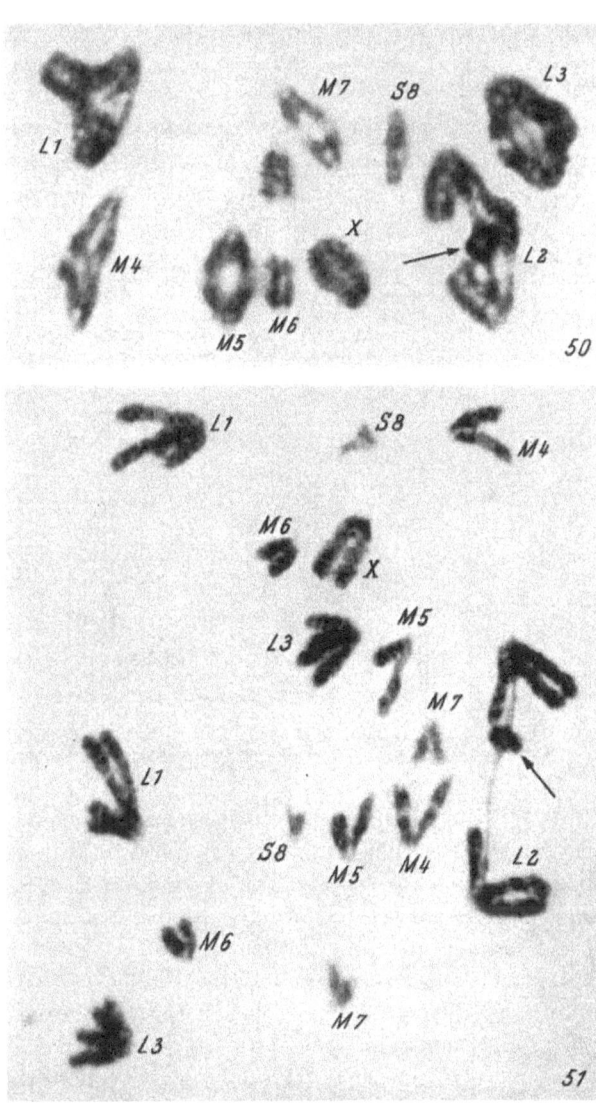

(i) The target area —Assuming that the broken ends of half-chromatids behaved independently with regard to rejoining, WILSON et al. (1959) worked out the kinds of configuration which could be obtained following the breakage of two, three and four half-chromatids in a chromosome or half-bivalent. Those which can be obtained following the breakage of two half-chromatids (either one from each sister or both from the same chromatid) can, of course, be obtained when more than two are broken. But the reverse is not the case. The two configurations which can be obtained only when at least three half-chromatids are broken are illustrated in Fig. 52.

All the configurations observed by WILSON et al. could be explained on the basis of breakage in only two of the four half-chromatids in a chromosome or half-bivalent but only if the two half-chromatids were members of different chromatids.

Fig. 50 and 51. Side arm chromatid bridges with chiasma-like structure (arrows) at first-metaphase (Fig. 50) and first anaphase (Fig. 51) of *Chorthippus parallelus* (orcein, ca. ×1500).

PEACOCK (1961), on the other hand, claims to have observed the associations indicative of either 3- or 4-strand, sub-chromatid breakage, i. e., the configurations illustrated in Fig. 52. Consequently, in contrast to WILSON et al. (1959) he concluded that all four half-chromatids are broken and that sub-chromatid exchange (as opposed to chromatid or chromosome exchange) depends not on a difference in the unit of breakage but on that of union. It is true that the

Y-shaped chromatid opposite a deficiency could be the result of secondary breakage in the bridge close to the side-arms. The E-shaped configuration, on the other hand, cannot be explained in this way.

(ii) Test of the sub-chromatid exchange hypothesis for the origin of side-arm bridges—Even if the chromatid has a multiple structural and/or genetic organisation, it does not follow that any of the side-arm bridges need be the result of sub-chromatid exchange. In other words, the various lines of evidence mentioned earlier in support of a multi-stranded structure are not necessarily directly relevant to the interpretation of the side-arm bridges. The only direct evidence in favour of these being due to sub-chromatid exchange depends entirely on the claims of the cytologist to be able to resolve the sub-chromatid elements.

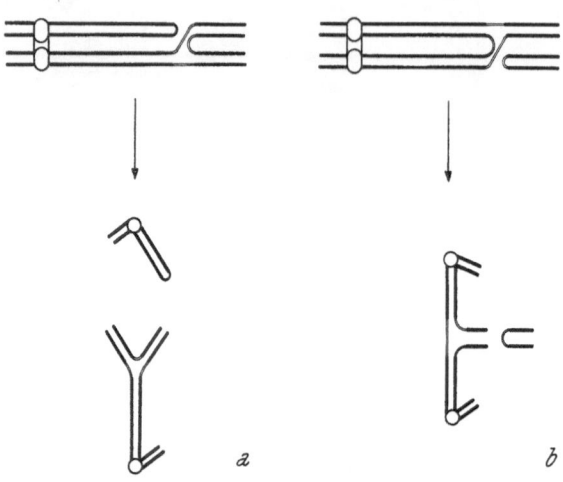

Fig. 52. Side arm configurations obtainable only by breakage in three or four half-chromatids. Left, Y-shaped; right, E-shaped chromosomes.

An indirect test of the hypothesis has been suggested however; but it depends on an assumption regarding the relationship in replication of the sub-chromatid elements. It is assumed that the half-chromatid, let us say, of X_1 is related to the chromatid of X_2 in the same way as is the chromatid of X_1 to the chromosome of X_2 (Fig. 53). It is assumed, in other words, that half-chromatids are segregated into distinct chromatids at replication in the same way as chromatids are segregated to separate nuclei at anaphase. The assumption has been generally accepted without question as has the test which depends on it, namely that some, at least, of the sub-chromatid errors of the X_1-division should be revealed as full-chromatid errors at the X_2-division.

To see whether this assumption is justified, let us consider those conclusions regarding chromosome replication reached on the basis of autoradiographic studies. So far, studies of this kind have been successful only in connection with mitosis but, presumably, the mechanism is essentially the same in meiotic cycles.

It is generally accepted that the replication of the DNA molecule is semi-conservative; complementary helices separating and each then specifying a new complementary chain. The results of MESELSON and STAHL (1958) are consistent with this conclusion though even here it is worth noting that if their primer molecules consisted of say, two double helices, the results they obtained could be explained on the basis of conservatism*.

In a series of investigations TAYLOR (1957, 1958, 1959) obtained results which indicated that replication at the chromosome level was semi-conservative. And

---

* See note 1 Appendix.

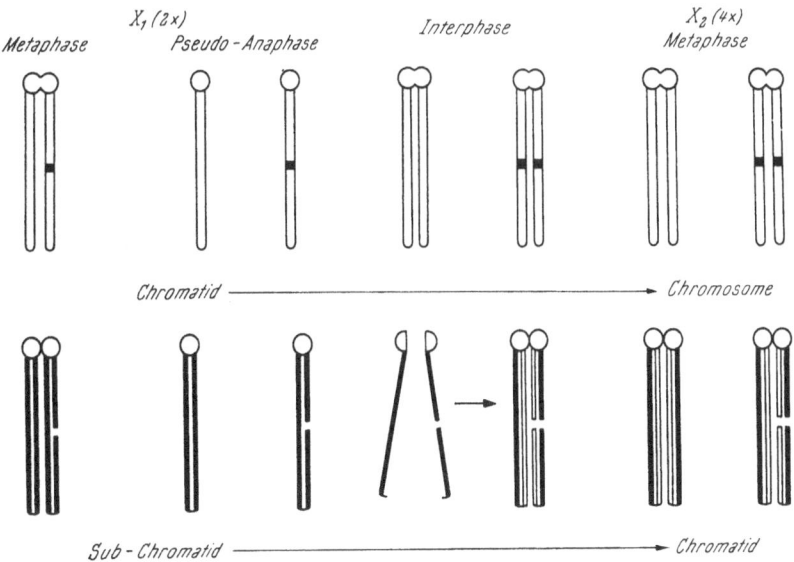

Fig. 53. That the sub-chromatid errors of one division should appear as chromatid errors in the next, assumes that sub-chromatids are related to chromatids in the same way as are chromatids to chromosomes (see text).

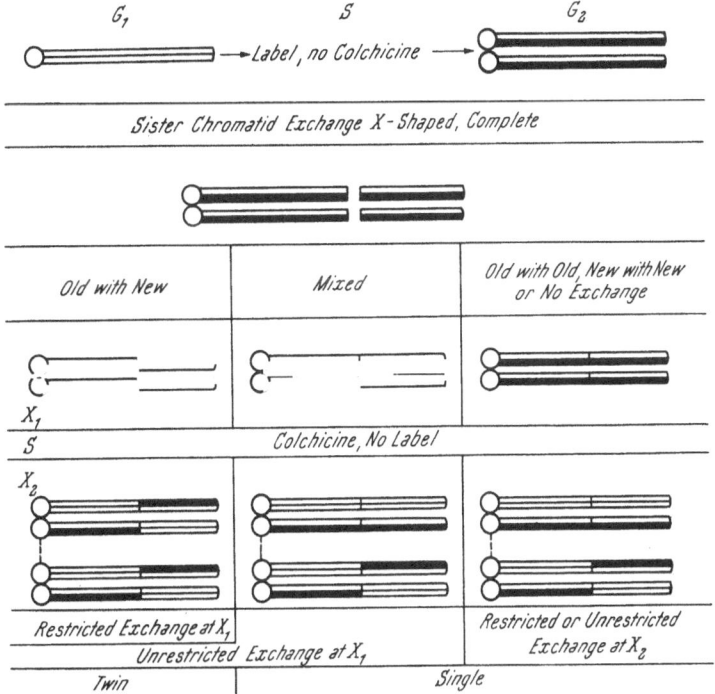

Fig. 54. Restricted (old with new) exchange at X_1 will give twin exchanges at X_2. Unrestricted exchange at X_1 will give the same result, go undetected or else be detectable as a single exchange at X_2 (when the two chromatids behave differently). But exchange at X_2, restricted or unrestricted, will appear as a single exchange. On the basis of various assumptions and a comparison of the frequency of single to twin exchanges observed at X_2, Taylor concluded that exchange was restricted.

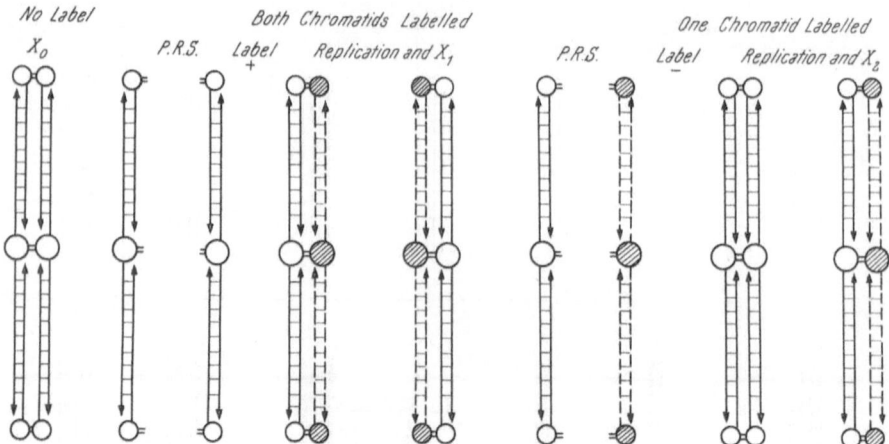

Fig. 55. Labelling of both chromatids at X_1 and only one chromatid at X_2 is consistent with conservative replication at the DNA level if the chromosome is dinemic. But this model does not impose any restriction on sister-chromatid exchange. A restriction is imposed, however, if there is a polarity within the links and sister-chromatid exchange involves these links (see Fig. 56). P. R. S. = Pre-replication split.

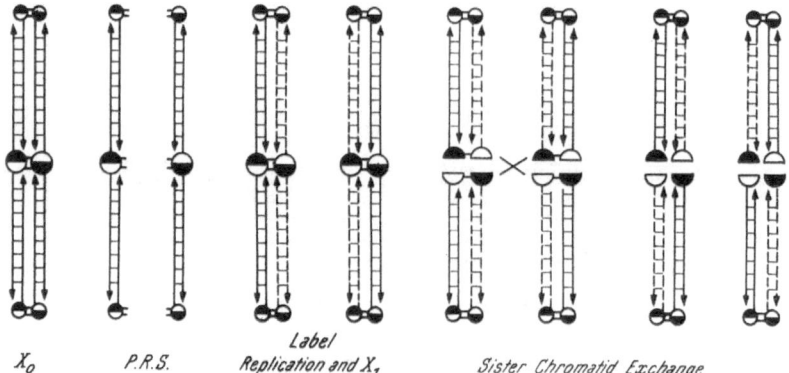

Fig. 56. See text and Fig. 55.

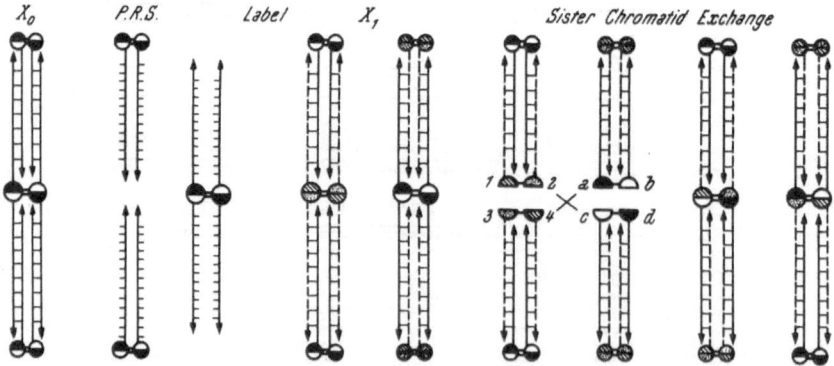

Fig. 57. A dinemic model with semi-conservative DNA replication giving iso-labelling at X_1, label segregation at X_2 and spurious restricted sister-chromatid exchange (see text and compare with Fig. 58 left).

these results, together with his evidence for restricted sister-chromatid exchange
(see Fig. 54) led him to conclude that the anaphase chromatid consists laterally
of a single DNA double helix, i. e., it is mononemic[1].

Now, although the polarity implicit in restricted sister-chromatid exchange,
like the exchange itself, is revealed by the pattern of DNA labelling it does not
necessarily follow that the restriction is imposed by the known polarity of single
DNA helices (see Figs. 55 and 56). We cannot enter into a detailed discussion
here but in Fig. 57 we have illustrated a dinemic model (based on the mononemic

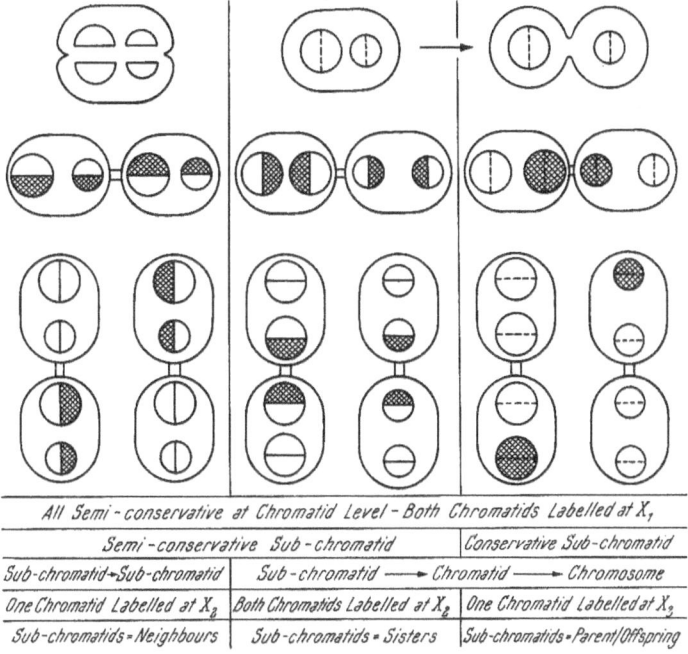

All Semi-conservative at Chromatid Level – Both Chromatids Labelled at X₁		
Semi-conservative Sub-chromatid		Conservative Sub-chromatid
Sub-chromatid→Sub-chromatid	Sub-chromatid ──► Chromatid ──► Chromosome	
One Chromatid Labelled at X₂	Both Chromatids Labelled at X₂	One Chromatid Labelled at X₂
Sub-chromatids = Neighbours	Sub-chromatids = Sisters	Sub-chromatids = Parent/Offspring

Fig. 58. Possible relationships between the individualisation of chromatids and the plane of pre-replication
splitting (see text).

model of FREESE 1958) which satisfies TAYLOR's findings. We have illustrated
sister chromatid exchanges as occurring in the links. This will lead to the appearance of restriction (old DNA with new) when no restriction really exists. Un-
coupling at the DNA-link junctions will have the same effect. But in this event,
and in the case of breakage within the DNA, restriction is imposed by the polarity
of the single helices.

However, the results obtained by LaCOUR and PELC (1958, 1959) and some
of those obtained by PEACOCK (1963) do not conform with those of TAYLOR.
The claim made by LaCOUR and PELC of label-segregation at X₁ was not con-

---

[1] We would suggest mono-nemic, di-nemic etc. for single-stranded, double-
stranded etc. chromosomes instead of uni- and bi-nemic which mix Latin and Greek
roots (see PEACOCK 1963). Since polynemic, with its uniformity of Greek roots is
established, then the uniformity should be preserved in the corresponding terms.
The only alternative would be to coin multinemic which, along with macromolecule,
would, at least, ensure a consistent confusion.

of both sister chromatids in corresponding segments at X_2 has been found by
TAYLOR, by LaCOUR and PELC, and by PEACOCK. And this is not expected on
TAYLOR's basic scheme. It was attributed by him to interchromosome exchange
and, more specifically to exchange between non-sister, homologous chromatids
firmed either by WOODS and SCHAIRER (1958) or by PEACOCK. But the labelling

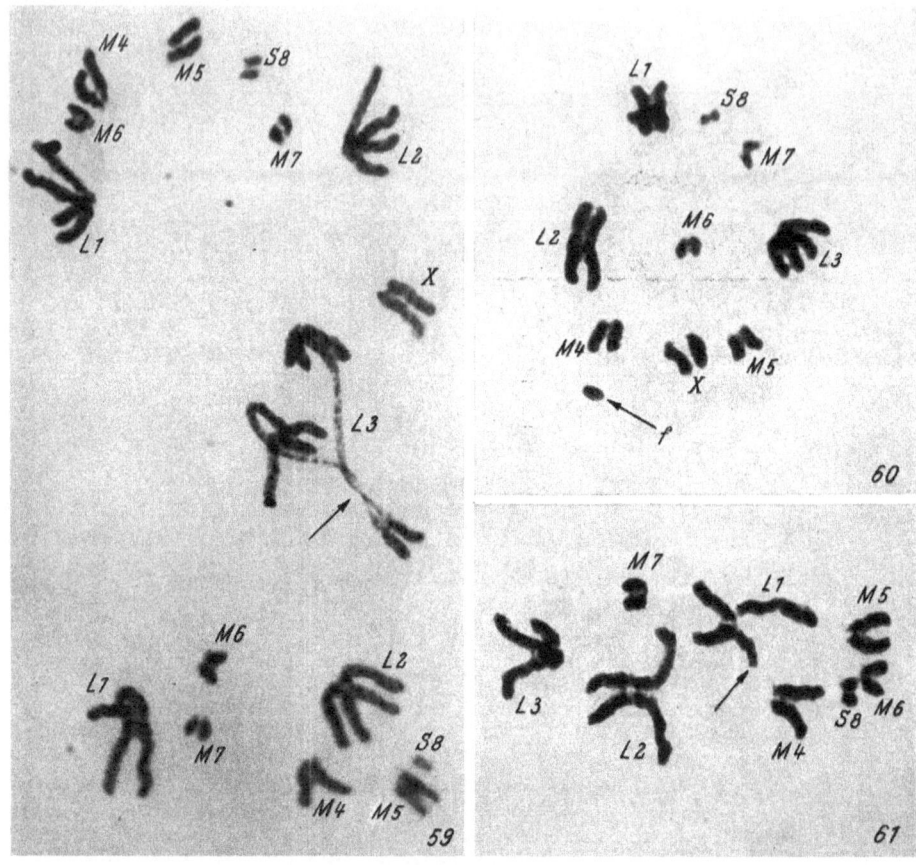

Figs. 59—61. The consequences of first-division side arm bridge formation at second division in *Chorthippus
brunneus*. Fig. 59—collapsed side arm bridge. Note the relational twisting of the bridge chromatids which has
resulted from the resolution of the major coils of first division. Fig. 60—Supernumerary acentric fragment (f)
in an (8 + X) cell. Fig. 61—Asymmetrical "half-bivalent" in an (8 + 0) cell (all orcein preparations, ca. ×1500).
Figs 60 and 61 could represent complementary products of breakage in a side arm bridge at first anaphase,
though both have been seen in the same cell.

since he observed them in tetraploid cells induced by colchicine. PEACOCK,
however, rejects this explanation because he found the condition in the absence
of colchicine treatment.

In some cases, PEACOCK found X_2 chromosomes in which both chromatids
were labelled along their entire length; in other cases short segments, usually
terminal ones, were "iso-labelled" while label-segregation was found in the
remaining parts of the chromosome. PEACOCK's explanation is based on a dinemic
model. However, even on a multistranded model, uniform labelling of both
chromatids at X_2 in some cases, and uniform segregation in others can be acco-

modated only if one assumes a variable relationship between the plane of separation of the "half" chromatids and that of their replication split (see Fig. 58 and Lewis and John 1963a). Where some parts of a chromosome show iso-labelling and others label-segregation, we must either assume that this variable relationship can vary within a chromosome or else invoke, as Peacock does, sub-chromatid exchange involving one of the double helices in the dinemic model.

Table 12. *Frequency Distribution of Spontaneous Two-Side-*

Species and Population		Bridge Frequency								
		1 Bridge per Cell								Total
		L 1	L 2	L 3	M 4	M 5	M 6	S 8	Unid.[1]	Total
Chorthippus brunneus	Birmingham	25	16	14	9	...	...	1	16	81
	Cardiff	3	2	1	1	...	...	1	...	8
	Pembroke	3	3	1	3	...	...	...	4	14
Chorthippus parallelus	Aberystwyth	9	9	4	2	...	...	...	4	28
	Cardiff	7	3	1	3	1	1	...	1	17
	Cheltenham	9	6	1	6	...	...	...	9	31
	Oxford	7	4	2	2	...	...	...	23	38
Omocestus viridulus	Birmingham	12	8	2	2	1	...	...	...	25
Totals		75	51	26	28	2	1	2	57	242

[1] Unidentified.

This would necessitate an average of more than one such exchange per cell and, on the pattern Peacock proposes (see his Fig. 8), the sub-chromatid error is expected to replicate as such.

Thus, we have now reached a position where sub-chromatid exchange is invoked to account for the labelling results when they might have provided a test of sub-chromatid exchange! Here, then, we can conveniently return to the earlier-mentioned test of sub-chromatid exchange.

In Fig. 58 we have illustrated some possible relationships between the individualisation of chromatids and the replication of their genetic components. The two inner circles can be taken to represent DNA double helices, the size difference between them illustrating an existing sub-chromatid aberration. On the left, the plane of splitting of the DNA helices is taken to coincide with the plane of separation of the two chromatids. DNA replication is semi-conservative and this scheme is equivalent to that in Fig. 57. Under these circumstances, sub-chromatid aberrations are expected to replicate as sub-chromatid aberrations, unless of course some "healing" or "extending" process is postulated.

DNA replication is taken to be semiconservative in the centre scheme also but the two original double helices are segregated, i. e. the plane of pre-replication split does not coincide with that which individualises chromatids. We have indicated the two new chromatids of the X_1 prior to the replication of their constituents but this is not necessarily the order of events (see, for example, Fig. 42. LEWIS and JOHN 1963a). In this case, sub-chromatid errors give rise

Arm Bridges at Anaphase-I in 65 Wild Grasshoppers.

per Bivalent Type											Total	
2 Bridges per Cell						3 Bridges per Cell					Individuals Analysed per Population	A I-cells without Bridges
L1 + L2	L1 + L3	L2 + L3	L1 + M4	L3 + M4	Total	L1+ L2+ L3	L1+ L2+ M4	L1+ L3+ M4	L3+ M4+ M5	Total		
5	...	...	1	1	7	...	...	...	...	0	22	165
...	...	...	...	...	0	...	...	...	...	0	4	31
1	...	...	...	...	1	...	...	...	...	0	5	63
1	...	1	3	1	6	1	...	...	...	1	10	171
1	...	3	...	1	5	...	1	...	...	1	4	59
...	1	...	...	1	2	...	1	...	...	1	8	99
1	1	2	...	...	4	...	...	...	...	...	8	92
2	1	1	1	...	5	2	...	1	1	4	4	60
11	3	7	5	4	30	3	2	1	1	7	65	740

to chromatid and chromosome aberrations in successive cycles. They should also give iso-labelling of chromatids at X_2.

On the right (Fig. 58) DNA replication is conservative or if, in this case, each circle is taken to represent a single DNA helix, it can be regarded as a mononemic structure. In this event, a sub-chromatid error would involve only one of the helices in duplex DNA. In terms of labelling this scheme would have the same consequences as the first, but with regard to the reproduction of the sub-chromatid error it would be similar to the second. Unless further aberrations are invoked, complete iso-labelling at X_2 cannot occur on the first scheme and segregation of the label cannot occur on the second. And if one is to conclude that both schemes are possible then, clearly there is no fixed expectation for the fate of sub-chromatid errors.

Thus PEACOCK's (1963) claim to have found the "expected" chromatid aberrations at X_2 may be more of an embarassment than ÖSTERGREN and WAKONIG's (1954) failure to find them. If, however, the segmental iso-labelling of X_2-chromosomes is due to sub-chromatid exchange, then as PEACOCK points out, the mean grain-count per unit length should be less in iso-labelled regions

4*

than in labelled segments showing label segregation. His observations suggest that this may be so but quantitative comparisons of this type are fraught with error.

So far as we are aware, PEACOCK (1961) has made the only firm claim to have observed chromatid exchanges in the X_2 attributable to sub-chromatid aberrations at the X_1-division. In all other cases the "expected" chromatid effects have not been obtained. For example, although 18% of the PMC's in *Gasteria undulata* showed side-arm bridges or terminal adhesions at first division, nothing was seen at pollen grain mitosis which could be attributed to them (DARLINGTON and KEFALLINOU 1957). Further, in view of the high incidence of point errors we have observed in grasshoppers (Table 12), it seems to us very unlikely that many of them could be due to an aberration, sub-chromatid or otherwise, involving genetic components of the chromosomes. Thus although we have found chromatid effects at second division, which can be interpreted as the consequences of side-arm bridges at first (Figs. 59–61), these are rare.

In conclusion, while the evidence for multistrandedness, at least in some species, is accumulating, that for sub-chromatid exchange is not compelling except in a very few cases. It would appear therefore that, at one extreme, impediments to anaphase separation depend on true and complete chromatid bridges resulting from breakage with sister union, crossing-over in relatively inverted segments or inverted crossing-over. At the other extreme, nothing more than a delay or a reluctance on the part of chiasmata to terminalise may be involved. And between these extremes, while sub-chromatid exchange involving genetic components of the chromosome may be involved in some cases, impeded anaphase movement generally results from an anomaly involving non-genetic entities. Consequently, a repair process is possible and the component concerned may be of the same nature as the H-linkers which have been postulated in chromosome models like those proposed by TAYLOR (1963).

3. Univalence and Mis-division of the Centromere

Chromosomes which fail to establish a chiasmate association with their homologues are subject to errors of segregation unless the meiotic system has been specially adjusted to meet their needs (see section VIII). But while chiasmata can be dispensed with in certain cases, some means of maintaining pairing until, or at least securing it at, first metaphase is usually necessary.

Certain genetic systems are characterised by the presence of particular chromosomes in the hemizygous condition. These, then, have no partner. And if this is their standard condition (hereditary univalents) some means of ensuring their segregation must be evolved. Hereditary univalents typify certain sex-chromosome systems (see section VIII-1) but are not confined to them; they occur also, for example, in *Rosa canina* and *Leucopogon juniperinum*. The regular conduct of hereditary univalents is in marked contrast to the variable behaviour usually found in relation to so-called spontaneous univalents. This provides cogent evidence that special adjustments of the meiotic system are necessary to ensure the regular disjunction of the unpaired.

Of course, the occurrence of spontaneous univalents may be determined by heredity, but while ancestry may be responsible for the unpaired chromosome, posterity can have no expectation of them. Univalents at first metaphase can be determined by a variety of causes having two kinds of effect. First, zygotene pairing may fail (asynapsis—Fig. 62). Since this is qualitatively specific and quantitatively limited it may fail for structural or numerical reasons. But abnormal genotypic or environmental conditions may give the same effect. Second, though pairing may succeed, subsequent chiasma formation may fail (desynapsis or asynartesis—Fig. 63) and here too genotypic or environmental anomalies may be involved. Further, what can be called false-univalents may be released at first anaphase following the linear or indifferent orientation of multiple or multivalent associations, especially those of three chromosomes (see sections IV-1c and V-3a). They may be produced also from bivalents which fail to co-orient or congress (see below).

At first metaphase when the bivalents congress to an equatorial arrangement true univalents do not; rather they tend to remain distributed over the spindle. This distribution has frequently been said to be a random one but as ÖSTERGREN and VIGFUSSON (1953) showed, the position of univalents at first metaphase is influenced by at least two factors:

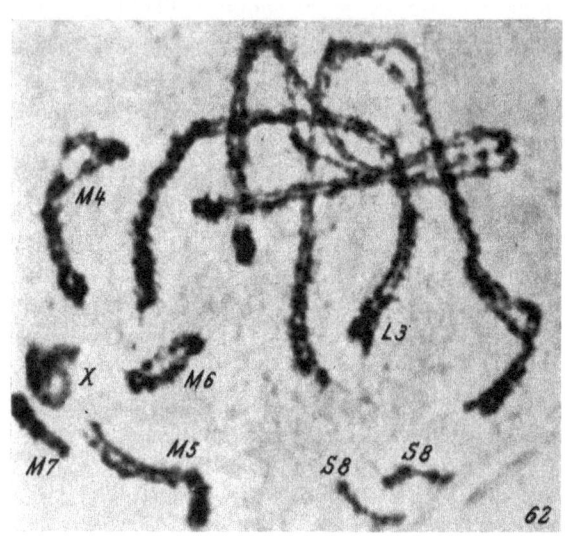

Fig. 62. Pachytene in *Chorthippus brunneus* showing proximity but failure of pairing of the S8 chromosomes (orcein-phase contrast, ca. ×1500).

a) When there are few to no bivalents the univalents tend to a more polar distribution while in cells having many bivalents they are more equatorial in arrangement (ÖSTERGREN 1949). HENDERSON (1962) came to the same conclusion with regard to the behaviour of heat-induced univalents in *Schistocerca gregaria* (Figs. 64 and 65).

b) In some cases univalents are not arranged independently of one another at first metaphase but tend to show a mutual dependence of position and univalents belonging to the same pair are preferentially located on spindle arcs close to one another (Fig. 63). This ÖSTERGREN and VIGFUSSON (1953) took to be a relic of a side-to-side alignment of homologues at first prophase since such an arrangement seems to be of quite common occurrence in the case of chromosomes which have been wholly or partially associated during pachytene but failed to form chiasma (see PRAAKEN 1943, page 489 and see also section V-3d). Indeed this position correlation may be so pronounced as to lead to the production of pseudo-bivalents. This was found, for example, in one plant of *Secale cereale* which was partially asynaptic (Fig. 66 and see section V-1). Univalents associated in such pseudo-or quasi-bivalents co-orient at first metaphase just as do

ordinary bivalents so that the members segregate regularly. On the other hand
HENDERSON (1962) has some evidence that a side by side association of univalents
is not always indicative of homology (see for example Fig. 65).

With the qualification just stated, the mode of origin of univalents does not
appear to influence their anaphase behaviour which may be of four kinds:

(a) They may be included in one or other of the daughter nuclei produced
by first division. This is generally held to be a random segregation (but see
p. 247). When univalents segregate reductionally at first division they are ex-

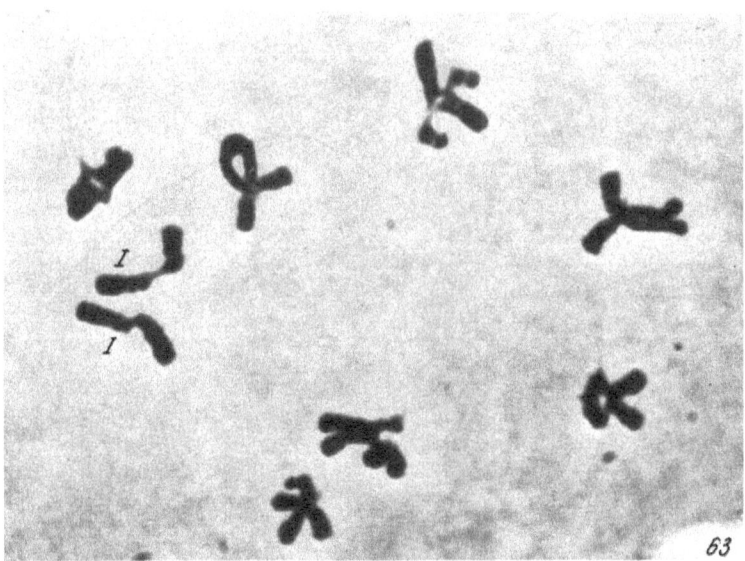

Fig. 63. Metaphase-1 in *Allium fistulosum* with two univalent chromosomes (I). Proximity of the unassociated
chromosomes, even though the spindle forces are clearly operative, suggests desynapsis rather than failure of
pairing but compare with Fig. 65 (carmine-phase contrast, ca. × 1500).

pected to behave at second division in the same way as half-bivalents to which
they are structurally equivalent (but see below, KOLLER 1938).

(b) They may fail to move, or be moved, at anaphase and so be excluded
from the main nuclei at first telophase. In this event they form micronuclei
which may include one or more univalents and which may be synchronised with
the main nuclei at second division. Indeed, if the behaviour of micronuclei at
mitosis is anything to go by, the chromosomes contained in micronuclei may
join up with those of the main nucleus on the second metaphase spindle.

(c) Supplementary components of the longitudinally multiple centromere
may orient towards opposite spindle poles at first metaphase so that the segment
between them becomes attenuated (Fig. 67). This region may even break so
that the two arms of a chromosome pass to opposite poles. This transverse
breakage of the centromere is known as mis-division. It gives rise to telocentric
chromosomes and sister union between the broken ends can give rise to iso-
chromosomes in which the two arms are mirror-images of each other (DARLING-
TON 1939b). It would appear, however, that telocentrics formed in this way can

undergo some "healing" process at the broken end and thus become stable as telocentrics (MARKS 1957).

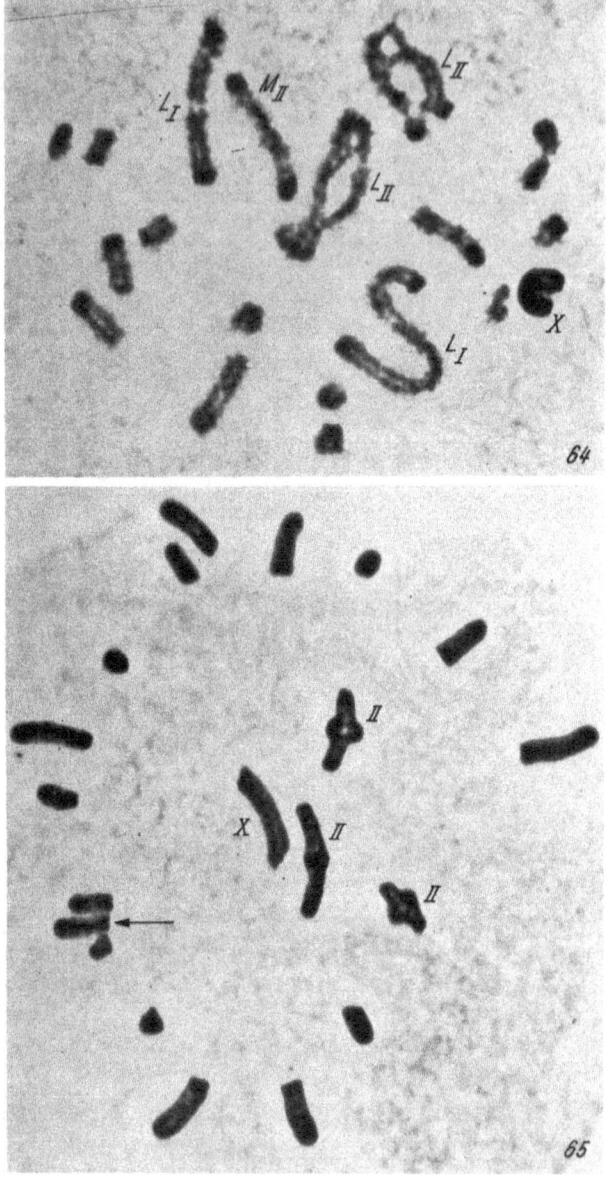

Figs. 64 and 65. Univalence induced by heat-shock (6 days at 40° C) in *Schistocerca gregaria* (orcein-phase contrast, ca. ×1500). Figs. 64—Diplotene (3 II + 16 I + X). Fig. 65—First metaphase. Three bivalents and the univalent X occupy a central position and the 16 autosomal univalents are peripheral. Note side by side association of three non-homologous univalents (arrow).

(d) Univalents may divide equationally at first division. This kind of division does not occur until the half-bivalents have undergone anaphase separation.

Consequently, the half-univalents (chromatids) may not reach the polar regions prior to the re-formation of nuclear membranes. Thus, this behaviour too can lead to the production of micronuclei. Clearly, a univalent which divided at first division cannot do so again at the second. But except for this proviso, the behaviour of half-univalents at second division parallels that of univalents at first division. Since normal "auto-orientation" is impossible, mis-division is expected to be even more likely and this is generally the case (see below).

The factors which determine this variable behaviour are not clear, for different univalents, even homologous ones, in the same cell may differ in their behaviour as may the same univalent in different cells. Certain chromosomes, however, do seem more prone to a particular type of behaviour (see below).

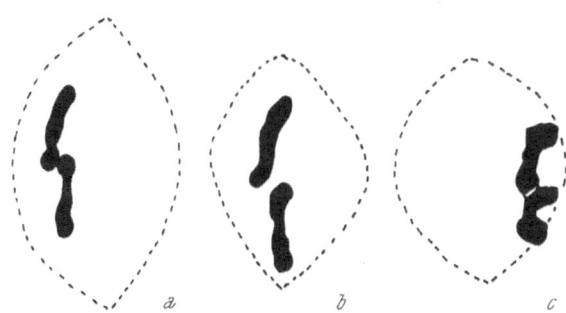

Fig. 66. Quasi-bivalents in a partially asynaptic plant of *Secale cereale* (2n = 14 + 2 B; Östergren and Vigfusson 1953, ca. ×3,300). Compare with Figs. 105–109,

Most of the variations in univalent behaviour mentioned above were described by Morrison (1953b) in his study of pentaploid, or near pentaploid, hybrids between species of wheat. The results for one such plant are summarised in Table 13. The difference between the frequency of univalents per cell at first-metaphase and that of laggards at first-anaphase (and, more particularly, telophase) is a measure of the amount of pre-reduction shown by the unpaired chromosomes, while lagging at second division is largely a reflection of equational separation at first division. The number of tetrads containing micronuclei and the mean number of micronuclei per tetrad exceeded the corresponding values for dyads. Further, while 4.3% of the first division cells contained one or more mis-dividing univalents, there was at least 10% mis-division at anaphase II.

The circumstances determining univalence at first anaphase in *Fritillaria kamtschatkensis* (2n = 2x = 24) were different from those in wheat hybrids (Darlington 1939b). Here only 15 univalents (and three "trivalents") were found in about 500 cells at first metaphase. But one cell in ten had one, two or four lagging chromosomes at first anaphase—more and not less as in wheat. Thus, a large proportion of the anaphase univalents were false. In fact, they were derived from bivalents which were seen to show failure of orientation and non-congression at first metaphase. Now, of the 109 lagging chromosomes observed at first anaphase, only two divided equationally; the others showed mis-division. In addition to simple transverse breakage across the centromere, configurations were found indicative of a combination of transverse breakage and longitudinal division.

Sister cells were observed at second division in which both the complementary products of mis-division at first anaphase behaved normally. But this relationship did not always obtain. Unequal division within the centromere may be responsible for complementary products of unequal potency. This would indicate that the centromere is more than duplex. MARKS (1957 b) however, has offered a different explanation. In *Fritillaria*, the products of delayed and partial mis-division were trapped by the wall and gave rise to micro-nuclei. And mis-division of half-univalents was observed at second anaphase.

In *Gasteria undulata* also, laggards are found at anaphase even though bivalent formation is complete (DARLINGTON and KEFALLINOU 1957). Here, however, mis-division, in some cases, took place after an equational division of the centromere. Thus, the type of mis-division shown by half-univalents at second division was seen here at first anaphase and, sometimes, in only one of the four chromatids of the collapsed bivalent. This centric error was confined to the long chromosomes. It will be recalled that the long arms of only these chromosomes are also subject to point errors (see section III-2). There was some indication, however, that point and centric errors had an alternative character between cells. A similar observation was made earlier by UPCOTT (1937) in a triploid tulip variety but here true univalents and false ones derived from trivalents were involved.

A further observation of interest was made in *Gasteria*. Iso-chromosomes were observed at second division following mis-division at the first. It would appear therefore, that union of broken ends can occur during the interphase between the two meiotic divisions although DNA synthesis is not involved at this time. Indeed if conditions such as those described on pg. 222 have been interpreted aright, union may

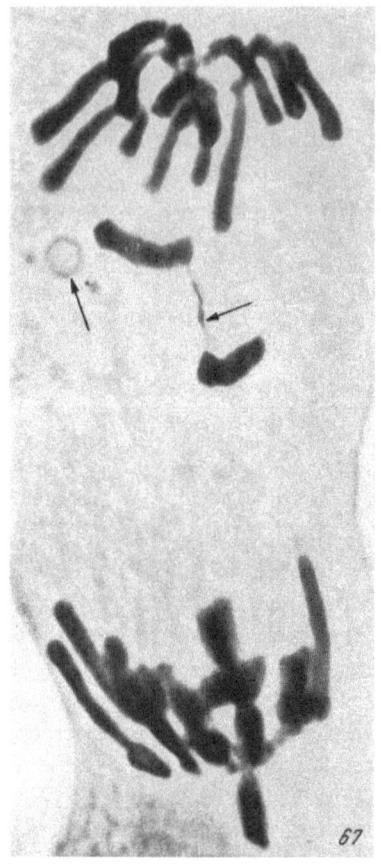

Fig. 67. Mis-division of the centromere of a univalent chromosome in triploid *Scilla*. Note attenuated centric region and movement apart of opposite arms of the univalent. Note also the persistent nucleolus near the end of the long arm of the univalent (carmine-phase contrast, ca. × 1500).

occur during the division sequence itself. There is, however, every reason for supposing that reorganisation if not synthesis of material does take place during pachytene.

An equivalent condition was described by MORRISON (1953 a) in monosomic wheats ($2n = 6x - 1 = 41$) where iso-chromosomes, formed during interphase following mis-division at first anaphase, sometimes showed mis-division at second anaphase to restore the telocentric condition. But iso-chromosomes could be re-created prior to pollen grain mitosis. Alternatively, fusion between telocentrics derived from mis-division in non-homologous chromosomes could result in interchange. Such an event was inferred from the multiple associations found in one of the progeny of a monosomic plant (MORRISON 1953c). In this material too mis-division at second anaphase was more frequent than at the first (Table 14) and the products of mis-division contributed to the formation of micro-nuclei. Some of the micro-nuclei produced at first division became mechanically active at the second, the chromosomes (or chromatids) which they contributed behaving as laggards. But peripherally-placed micro-nuclei were delayed (compare bi-

Table 13. *Metaphase Pairing and Anaphase Behaviour in a Triticum vavilovi ×*
T. dicoccum hybrid—Plant C (2 n = 5 x).
Note the T II Data Refer to Half-tetrads. (Data of MORRISON 1953b.)

Metaphase-I Associations and % Cells			Mean Univalents per Cell	Total Cells
14 II + 7 I	13 II + 9 I	12 II + 11 I		
78.0	18.0	4.0	7.96	50

No. of Laggards	A I	T I	T II
0	1.7	0	0
1	3.3	2.5	3.8
2	0	10.0	8.8
3	10.0	15.0	19.5
4	16.7	12.5	19.5
5	18.3	25.0	23.9
6	28.3	17.5	15.7
7	18.3	10.0	6.9
8	1.7	5.0	1.9
9	1.7	2.5	0
Mean per Cell	5.2	4.8	4.4
Total Cells	60	40	159

nucleate PMC's, see pg. 295) and while they entered second division, they were obliged to re-form micro-nuclei at the end of it. However, in older pollen grains (those having vacuolated cytoplasm) few micro-nuclei are found so, presumably, many are absorbed by the cytoplasm. This "digestion" may be associated with a failure to form a complete nuclear membrane.

The observations made by GILES (1943) on the origin of iso-chromosomes in a species of *Gasteria*, probably *G. maculata*, differ from those made by DARLINGTON and KEFALLINOU (1957) on a member of the same genus and, indeed, from all other accounts dealing with mis-division. GILES claimed that:

(a) Univalence was not involved even at early anaphase.

(b) Only one pair of chromosomes—the long, sub-mediocentric, nucleolar-organising chromosome—behaved abnormally and did so even though its metaphase behaviour was normal (He does, however, picture one cell where both members of this bivalent proceeded to the same pole).

(c) Telocentrics are not normally produced but isochromatids are produced directly, and

(d) There was no new mis-division at second anaphase.

A summary of GILES' results and conclusions are given in Fig. 68 and Tables 15 and 16. Early anaphase is not a common stage but even if congression is normal, as GILES claims, an early anaphase behaviour similar to that described by DAR-LINGTON and KEFALLINOU is not excluded. Be this as it may the late anaphase configurations described by GILES are unique. Separate half-centromeres produced by equational or mis-division are not found. However, what appear superficially, at least, to be normal centromeres may carry morphologically different chromatids (asymmetrical half-bivalents, Type II, Fig. 68) or both long and short arms may show pre-reduction (Type III, Fig. 68).

Table 14. *Frequency of Misdivision in Ten Wheat Monosomics.*
(Data of MORRISON 1963a.)

Monosomic	Misdivision of Univalents			
	T I		T II	
	No. of Cells Examined	% with Misdivision	No. of Cells Examined	% with Misdivision
I	64	9.4	60	33.3
III	...	...	52	27.0
V	87	14.9	23	17.4
VIII	152	13.8	46	13.0
IX	68	13.2	73	21.9
X	120	10.0	28	25.0
XI	255	13.3	135	11.1
XII	103	5.8	...	...
XIII	112	11.6	106	14.2
XXI	53	17.0	70	11.4
Total	1014	12.1	593	17.7

The possibility that Type Ia arises following equational division of the centromeres and Type Id from their reductional separation cannot be excluded. In the latter event the two arms, long and short, could be held together by the cross-bonding of longitudinally-oriented spindle fibres arising from the separate half-centromeres produced by mis-division. In any event, the "fusions" are only temporary ones for the plane of centromere division at second anaphase was held to coincide with the plane of fusion at first division. It will be noted, however, that the anaphase-II consequences from Type II and from Type III are individually the same no matter how the centromere divides. Further, these "fusions" do not involve the "chromonemata." In fact the scheme proposed by GILES suggests that the production of iso-chromatids is determined prior to anaphase separation by conditions within the centromere. This means that breakage is not a consequence of tension (as it presumably is in other cases and, presumably, must be in the case of the mis-division of half univalents) and that breakage is followed by sister-union within the centromere. In this it resembles the type of breakage considered earlier in section III-1. This, in turn, means that the "anaphase fusion" which occurs following the manifestation of the prior, primary, error and which determines the morphology of the chromosome at late anaphase, does not involve the "centromeric fibrils." This makes one rather less reluctant to accept the explanation of Type II exchange which invokes fusion between a normal half-centromere produced by longitudinal division and one produced by transverse breakage. GILES' explanation of this type is reminiscent of the fusion between broken and unbroken chromosome ends which has been claimed both after irradiation and spontaneously (HAGA 1953). This type can, however, be produced by a combination of transverse and longitudinal division involving both centromeres. It is also the configuration expected follow-

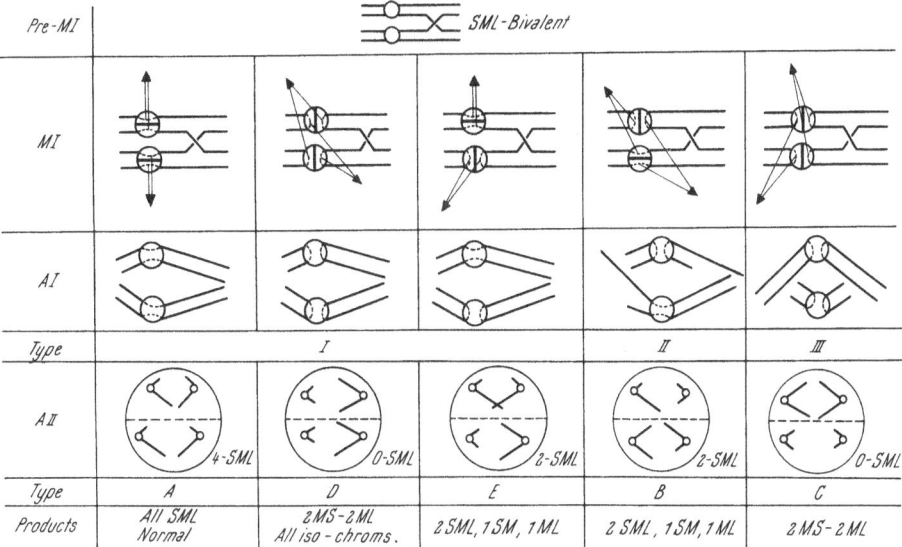

Fig. 68. Summary of anaphase behaviour observed in the SML bivalent of a species of *Gasteria* by GILES (1943).

ing single or three-strand double crossing-over in a heterozygous pericentric inversion while four-strand double crossing-over in such an inversion would give Type III.

KOLLER (1938) in one of the earliest papers describing mis-division made some curious and, to our knowledge, unique observations on some desynaptic individuals in *Pisum*. Mis-division was observed at second anaphase and KOLLER claimed that "the chromosomes showing anomalous behaviour appear to be univalents which *failed to divide during the first meiotic metaphase*" (our italics), i. e. chromosomes as opposed to chromatids were involved. The situation in *Pisum* was thus the converse of that found in *Gasteria undulata* in that the type of mis-division normally associated with the first anaphase occurred at the second. In *Pisum* the chromosomes showed a combination of transverse and longitudinal division, the former preceding or following the latter. An equational division of unpaired chromosomes was observed at first division but this was frequently preceded by extreme repulsion between homologous arms. The spindle abnormalities which are commonly associated with univalent formation were also found.

We see therefore that spontaneous univalents are variable in their behaviour. And, while the behaviour of a given univalent is not constant, there are more-or-less well defined differences between the patterns of univalent behaviour in

Table 15. *The Frequency (%) of the Three Main Types of Anaphase-I and the Five Kinds of Anaphase-II Behaviour Observed in Gasteria.* See Text and Fig. 69. These data (from GILES 1943) are in substantial but not complete agreement with each other.

Anaphase-I Types				Total Cells
I		II	III	
48.8		41.8	9.4	430
19.1	5.4	15.2	53.2	7.1	184
A	D	E	B	C	Total Cells
	Anaphase-II Types				

Table 16. *Frequency (%) of Meiotic Products at Second Anaphase, First Pollen Grain Mitosis and the Mature Pollen Grain Stage in Gasteria.* (Data of GILES 1943.)

Stage	Standard	Long-iso [1]	Short-Iso [2]	Total Cells
A II	53.4	23.3	23.3	184
P. G. M. 1	67.8	32.2	0	389
P. G.	42	24	34	100's
	Large, Full	Medium, Full	Small, Empty	

[1] Pollen grains containing the long-isochromosome are duplicated for the long arm and deficient for the short arm of the nucleolar chromosome. They are delayed in entering mitosis as shown by the higher relative frequency of this type of cell in older anthers showing PGM 1.

[2] The complementary product, deficient for the long arm and thus the nucleolus, is inviable. Presumably the other types contribute to the shrunken class of grains.

different types. Unpaired chromosomes do not, however, always behave as erratically as those in the species described above. Earlier we made a distinction between spontaneous and hereditary univalents and contrasted the variable behaviour of the former with the almost invariably constant conduct of the latter. But in *Clarkia unguiculata*, HARLAN LEWIS (1951) found that supernumerary univalents showed invariable pre-reduction. In this case, however, about 30% of the plants in his sample carried one or more extra chromosomes.

Elsewhere we have argued that plants with supernumerary chromosomes would be favoured for the same reasons and under the same circumstances as the interchange hybrids of the same species (LEWIS and JOHN 1963a). Therefore, we are dealing here with a type of chromosome which falls somewhere between the two classes we distinguished earlier. They also fall between what have been referred to as A- and B-chromosomes (see page 196).

One final feature of univalent behaviour which also appears to be variable is the influence of univalents on the spindle and cell cleavage. In at least four animals the presence of univalents in the first division spindle results in failure of cleavage due to the persistence of the spindle in the metaphase state. This has been found in *Mantis religiosa* (CALLAN and JACOBS 1957), *Schistocera paranensis* (JOHN and HENDERSON 1962), *Schistocerca gregaria* (HENDERSON 1962) and *Orthodera* (WHITE 1962). On the other hand large numbers of univalents do not prevent successful cleavage at both first or second division in interspecific grasshopper (HELWIG 1955) and newt (WHITE 1946) hybrids.

4. Meiotic Mosaics

The normal course of meiosis is sometimes disturbed. In some cases the abnormalities can be attributed to disturbances in the external environment. Alternatively, genetic peculiarities, structural or genotypic, may be involved. There are cases, however, in which both normal meiosis and abnormal meiosis characterise certain cells of the same gonad. The consistency of the anomaly in such cases rules out genetic differences as the cause of breakdown and, since normal and abnormal meiosis occur together, differences only in the immediate, cytoplasmic environment can obtain—differences which must be under the control of the genotype.

Thus, in the grasshopper *Melanoplus differentialis* as in many other insects, the sperm mother cells are grouped into well-defined cysts within each of the many testis follicles. Cells within a cyst are well synchronised in respect of both DNA synthesis and stage of division, indicating a uniform normality of the cytoplasm. LIMA-DE-FARIA and NORDQVIST (1962) found that at the onset of meiosis and after DNA synthesis, all the nuclei in some cysts broke down and gave rise to spherical, feulgen-positive bodies of varying size indicating a uniform abnormality of the cytoplasm. From one to three cysts per follicle behaved in this way.

This type of behaviour has been seen often under pathological conditions. But since it is a regular one, the above authors regard the condition in *Melanoplus* as an adaptation by which large amounts of DNA are made available at a "convenient time of development." And, though the mechanism is a different one, its role is compared with both that of the feulgen-positive body formed in contact with the sex chromosomes in oogonial nuclei of *Tipula oleracea* and that of the disproportionate DNA synthesised at certain puffs in the salivary glands of Diptera. However, the role of the DNA in these two cases may differ. Thus, it is not clear whether the authors regard the released DNA in *Melanoplus* simply as a precursor pool of value only after its breakdown into units too small to be highly specific, or whether it conveys information directly and functions in much the same way as DNA-specified RNA. In the former event it is not easy to see how the recipient cells can make better use of the DNA than the donor cells which synthesised it, especially if the recipients, like the donors, are sperm mother cells. Thus the adaptive significance of the breakdown is not obvious but cells towards the middle of the follicle may prove a more efficient source of supply than previously-differentiated follicular cells at the periphery.

Cleavage polyembryony is common in many gymnosperms and often to an extreme degree. In pines and cedars, for example, eight embryos may result from a single fertilisation—all of them alike. But only one normally survives to maturity. Here then, as in *Melanoplus*, breakdown is regular and genetically non-selective; there is no question of the differential survival of the unlike as there is following embryo-competition in types like *Citrus*, for example, with versatile reproduction. Its role then is presumably nutritive, the extra embryos making a contribution comparable with that of the endosperm of angiosperms. And we regard the condition in *Melanoplus* in the same way.

The abnormal cysts in *Melanoplus* do not give rise to sperm. But in many insects abnormal sperm are regularly produced following an abnormal meiosis. The meiotic abnormalities are, however, consistent for a given type and they frequently occur in predictable parts of the testis.

Thus in the family *Pentatomidae* (*Hemiptera*: *Heteroptera*) the testis may possess a harlequin lobe in which the spermatocytes are very small in size. In those species where the testis is 7-lobed it is always lobe 5 which is harlequin and in 19 of the 21 species possessing such a lobe which have been studied cyto-logically, meiosis is aberrant leading to the production of spermatids carrying numerically unbalanced chromosome complements (SCHRADER 1960). This depends on irregularities in meiotic separation which in most cases spring from deviations affecting meiotic pairing. Thus, in *Alitocoris*, *Mecistorhinus* and *Loxa* synapsis may be eliminated altogether so that only univalents are present at pachytene. In other cases, as in *Moncus*, complete pairing occurs but some or all of the bivalents produced undergo desynapsis.

A different situation obtains in types like *Brachystethus rubromaculatus*. The testis here is 4-lobed and the cells of the fourth lobe (the one furthest from the exit of the sperm duct) behave abnormally. The first sign of abnormality comes at metaphase when the autosomal bivalents form an aggregate which is moved laterally outwards from the spindle equator (cf. page 178). The autosomes do, however, retain connections with the poles and they move *en bloc* to one of them at first anaphase. The autosomal aggregate is again retained at second division and, again, it behaves reductionally. The sex chromosomes on the other hand behave in the manner typical of the *Heteroptera*, i. e., they show post-reduction. Consequently, four main types of sperm are produced, namely those with X only, Y only, X + 2A or Y + 2A. The only deviation is that the autosomes may form two clumps, which behave in the same way as the single one but do so independently of each other.

The abnormal meiosis which has been found in the harlequin lobe of the tribe *Discocephalini* closely resembles that in *Brachystethus* with this difference. One large autosome succeeds in freeing itself from the autosomal clump at early anaphase. It and the other autosomes proceed to opposite poles. The sex chromosomes again behave normally.   At second division, however, the autosomal aggregate is irregularly pulled into two groups while the large autosome, which shares the smaller secondary spermatocytes with an X and a Y chromatid, moves undivided to one pole—usually that to which the X proceeds!   As in *Loxa*, this behaviour characterises one of the seven testis-lobes and it has been found in *Agactitus dromedarius*, *Architas pudens*, *Mecistorhinus panamensis*, *M. sepulcharis*, *M. tripterus*, *Neodine macraspis* and *Ablaptus amazonus*.   It also obtains in *Dinocoris rufitarsus*, the only pentatomid known to possess 8 lobes in its testis.   The available evidence suggests that these aberrant meioses depend on chemical conditions within the harlequin lobe.   Thus the different dimensions attained by the small cells of this lobe appear to result from different patterns of synthetic activity in RNA and protein production (see page 269).   And these differences appear to lead to an impedance of anaphase movement by producing alterations in the surface properties of the autosomes.

POLLISTER and POLLISTER (1943) investigated a quite different type of anomaly in viviparid snails.   In these, two kinds of spermatogonia could not be distinguished but there were two kinds of spermatocytes which differed in size and behaviour.   The abnormal ones were larger owing to an increased quantity of cytoplasm.   They showed also another property associated with reduced precocity, namely failure of pairing.   Not only are bivalents absent at diakinesis but each chromosome exists as separate chromatids.   Thus, in the place of the haploid number of bivalents seen in normal cells, the tetraploid number of chromatids is present.   There is no well-defined metaphase, indeed only a few chromatids seem to establish spindle attachments, but two approximately equal groups of chromatids are present at early anaphase.   Later, however, the majority return to the equatorial region.   In the case of *Viviparus malleatus* two chromatids remain at each pole and thirty-two move, or are moved, to the equator.   The second division in this species is unequal.   One of the two chromatids which remained at the pole after the first division is included in the same spermatid as those which moved to the equator.   These, however, have formed individual vesicles in the mean time and subsequently they degenerate.   The other chromatid which remained at the pole at first anaphase forms a nucleus in a smaller spermatid which has, however, received nearly all the mitochondria.   A sperm is organised from each of the two second division products.

Now the centriospheres of the anomalous spermatocytes are very large and at diakinesis they break up to give many centrosomes of normal appearance and behaviour.   There was some evidence to indicate that the increase in size of the original centriospheres resulted from an accretion of material which originated from the nucleus during early prophase.   Further, it was found that the supernumerary centrosomes equalled in number the chromatids which returned to the equator at first anaphase.   Consequently it was concluded that the anomalous chromatids had lost their centromeres and that these had moved out of the nucleus to appear and behave as centrosomes.   And since they have many of them, the resulting sperm were multi-flagellate.

Two main kinds of spermatozoa are produced in most genera of prosobranchiate snails.   One is normal (eupyrene) having a standard haploid complement and a single centriole which produces a flagellum.   The other is abnormal, larger with many centrioles (each of which produces a flagellum) and the chromosome complement is either greatly reduced (oligopyrene) or else completely lacking (apyrene).

As ANKEL (1930) first showed, the variously abnormal sperm can be arranged in a series. The least modified are only slightly larger than normal. But while they appear to be otherwise normal in structure, they are unable to activate the eggs. Increasing atypy is shown both by a greater increase in size and by the degeneration during meiosis of greater numbers of chromosomes.

In the *Viviparidae*, as we have seen, only part of the chromosome complement degenerates. But in all the more extreme cases (eg. *Janthina*) degeneration of the whole complement occurs. Where it does obtain, the atypical sperm are invariably multiflagellate owing to the presence of supernumerary centrioles. This correlation has been used by the POLLISTER's to support their view that extra centrioles arise by the decentration of chromosomes which subsequently degenerate.

A rather different meiotic system has been described in *Bithynia tentaculata* by KEYL (1955). The spermatogonia are again all of one kind and at pre-meiosis some of the chromosomes show somatic pairing. Spermatocytes, however, are of two types. In normal spermatocytes the somatic pairing is replaced by meiotic pairing though the bivalents produced show no evidence of chiasma formation. In atypical spermatocytes those chromosomes which were paired somatically at the onset of meiosis persist as paired chromosomes which behave like ordinary meiotic bivalents but the remaining univalent chromosomes are eliminated. The sperms produced in such atypical lines are thus abnormal.

The anomalies described above form a special category in that, far from being sporadic, they are regular features of the species concerned. Further, they are frequently found only in certain parts, often very well defined parts, of the testis and the kind of anomaly is peculiar to the species or species-group. Thus, however anomalous the condition, we must consider it to be under genotypic control and consequently to be the product of selection. It is equally clear that none of the mechanisms described above produce sperm which are functional in heredity and their adaptive significance is not obvious. Perhaps these sperm facilitate the penetration of the egg by normal sperm. On the other hand polyspermy is not an uncommon event and, if by their breakdown the abnormal sperm contribute materials to the fertilised egg, a means is provided whereby the male can contribute to and share the burden of nourishing the young. This would improve the economy of dioecism as an outbreeding system for it is in many ways a wasteful one. Thus in *Brachystethus* each harlequin sperm transmits some four times as much nucleoprotein as a normal sperm so that the nucleoproteins transmitted to the egg by such supernumerary sperm are available to the developing embryo.

There are cases, however, in which an abnormal meiosis appears to produce functional sperm. For example, each testis of the house centipede *Scutigera forceps* consists of a tube like microtestis which terminates in a club-shaped macrotestis. The sperm produced in the former are 200 μ long; those in the latter are 70 times longer and about 140 times as large. The two kinds of sperm appear to be equally mobile and they seem to reach the sperm receptacles of females with equal facility. The difference between the two sperm-types is presaged early in the spermatogonial cell generations. The microtestis has one more of these and consequently it produces twice as many sperm (ANSLEY 1954).

The 36 chromosomes in the large sperm mother cells undergo a normal meiosis. But in the microtestis there is no pairing and the first meiotic division is equational. In fact it resembles the spermatogonial divisions (with which it has been confused) except in that its products do not reach full size before dividing again. The details of this second meiotic division are far from clear, but in some obscure way it effects a reduction in the chromosome number.

In the earlier sections we have given some account of the variations on the basic system which depend on "spontaneous" differences in genotypic and/or environmental conditions. But though some of them led to chromosome mutations these were not the cause of the anomalies. We will now consider the course of meiosis in those cells which possess, initially, structurally abnormal chromosomes, are hybrid for structural or numerical chromosome differences or in which homologues are not represented by two chromosomes.

IV. Structural Variants

1. Meiosis in Structural Hybrids

In a diploid chromosome complement the homologous members are structurally equivalent throughout their entire length and since pairing at meiosis is both qualitatively and quantitatively limited to structurally homologous pairs only bivalents are formed. Thus in the basic meiotic system structural homology is one of the conditions for regular pairing and hence regular segregation. Such structural homology of course is not a requirement for regular mitotic separation since here each chromosome behaves quite independently of all others whether these are homologous with it or no. But at meiosis the situation is quite different for with a change in structural homology there are likely to be problems in pairing and, following any modification in the pattern of pairing, further complications are introduced by chiasma formation.

Individuals in which the chromosomes are not precisely structurally equivalent arise from time to time in all organisms as a result of spontaneous chromosome breakage with or without reunion. One might expect such breakage to occur when the chromosomes replicate or when they are placed under exceptional strain as they may sometimes be at anaphase or as they are believed to be during pachytene (see section II-part 2). In fact, breakage appears to take place at all three of these stages—at interphase, pachytene (see section III-1) and at anaphase (see section II-2). The possibilities for reunion appear, however, to be more restricted. (But see pg. 222).

Spontaneous breakage is a rare event and both homologues are never affected in the same way so that initially only one of a pair of homologous chromosomes is structurally modified. This may, of course, subsequently become structurally homozygous but this is a secondary event. Structural heterozygosity, as such, does not affect the course of mitosis because this does not involve pairing; mitosis is affected only if the product of change is, itself, abnormal (see section IV-2). But meiosis is frequently affected.

The simplest kind of structural change that can occur involves an alteration of the sequence of the genetic material within the chromosome itself and the most direct form of this is the inversion.

a) The Inversion Heterozygote

Homologous pairing is possible in an inversion heterozygote only if the inverted segment or its progenitor forms a loop at pachytene. But pairing in the inversion need not be by homology. Indeed the very first cytological study on chromosome pairing in an inversion heterozygote (McClintock 1931, 1933) showed quite unambiguously that such reverse-loop pairing need not characterise all inversions. Thus, in maize plants heterozygous for a long paracentric inversion (*In 2a, S 6–L 6*) most PMC's did show reverse loops but in short inversions in the same species there was a pronounced non-homologous association leading to straight pairing. And in both cases there was some asynapsis of the affected arms.

Again in a type heterozygous for a pericentric inversion in chromosome 4, reverse loop synapsis was seen by McClintock only in four out of four hundred sporocytes examined. The remainder showed non-homologous association in the inverted regions. Similarly White (1962) has consistently failed to find evidence for reverse loop pairing in pericentric inversions involving grasshopper chromosomes and is of the opinion that they pair straight as indeed Coleman (1948) has demonstrated in *Trimerotropis gracilis*. Where pairing is by non-homology, chiasmata cannot form and there are no further complications to meiosis (but see page 154). Pericentric inversion heterozygotes may, therefore, not be under as great a handicap as has sometimes been supposed.

On the other hand Hoover (1938) found considerable evidence for reverse loop synapsis in salivary gland X-chromosomes of *Drosophila melanogaster*. These inversions included from 150 to 922 of the 1,024 bands known in the X and showed 67–68% reverse loop formation. If this is a valid indication of what occurs at meiosis in *Drosophila* then one must conclude that reverse loop pairing takes place fairly regularly here.

Where reverse loops are formed, chiasmata may arise within them and, when they do, secondary structural change can follow, though the precise consequences differ for para- as opposed to pericentric inversions. In both, duplication-deficiency chromatids are produced by single or 3- and 4-strand double crossing-over but with paracentric inversions these are associated with the production of dicentric chromatid bridges and acentric fragments whereas with pericentric inversions they are not. Finally, in paracentrics more complicated chiasma patterns can give rise to ring-chromatids which then bridge at second anaphase. Of course where, as in *Drosophila*, the chiasma frequency is not high enough or the naturally occurring inversions long enough, the occurrence of multiple cross-overs will be rare.

It has been standard practice for many years to accept the presence of dicentric bridges with accompanying acentric fragments as unconditional evidence for paracentric inversion heterozygosity. We have no hesitation in deprecating this practice and the conclusions it has led many authors to, for as we have already

seen (see section III-1), bridges and fragments can arise from chromosome breakage at meiosis.

The fate of the dicentric ties appears to differ in different organisms. In some
cases they may originate a breakage-fusion-bridge cycle (see section 92). In
others they simply lead to the production of unbalanced gametes while in still
others they appear to be selectively eliminated from forming functional gametes
in either the female or less commonly the male germ line. Thus egg counts from

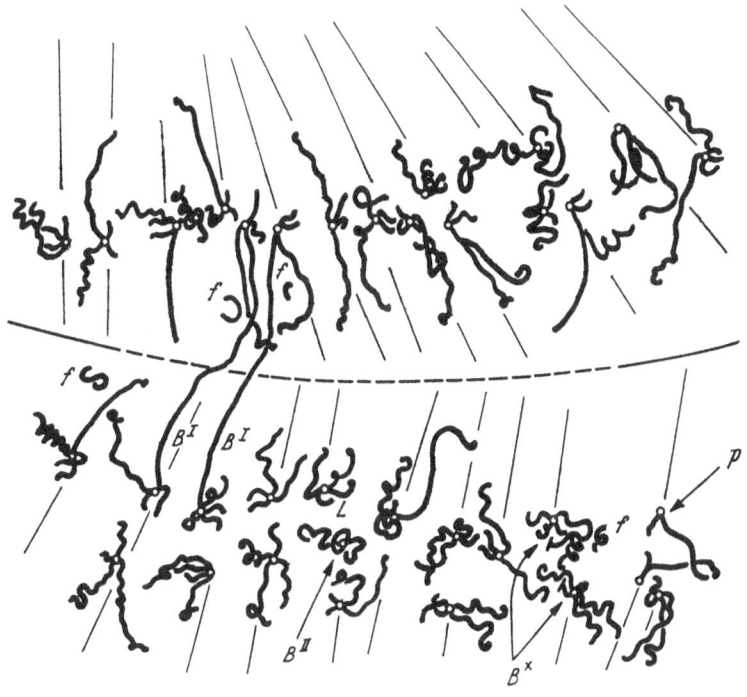

Fig. 69. The consequences of crossing-over in relatively inverted segments in a triploid × diploid cross of the
garden tulip, *Tulipa gesneriana*. There are four acentric fragments (f). Two of the corresponding dicentrics have
given first division bridges (B^I) which are persistent. A third cross-over chromatid has formed a loop which will
bridge at second division (B^{II}) while the fourth has undergone non-disjunction in a trivalent (B^X). Note also
precocious separation of the homologous chromatids (p) of one chromosome. (DARLINGTON and LA COUR 1941).

female *Drosophila* heterozygous for paracentric inversions and crossed to normal
males indicated high hatchability, contrary to expectation (GERSHENSON 1935,
STONE and THOMAS 1935). This was taken to imply simply that single cross-
overs in such inversions were rare. Experiments with overlapping inversions
showed, however, that single cross-overs did occur and this led STURTEVANT and
BEADLE (1935) to propose that in many cases the dicentric and acentric strands
were left in the polar nuclei which are destined to degenerate, so that the definitive
egg nucleus receives predominantly monocentric strands. This, they suggested,
was due to the fact that, in the linear quartet of nuclei resulting from meiosis,
dicentric bridges produce a tie at first anaphase which orients the cross-over
chromatids so that they pass preferentially to the inner meiotic products and do
not reach the end cells of the linear tetrad. Cytological evidence for such
a directed orientation was subsequently reported in *Tulipa* and *Lilium* (DARLING-
TON and LaCour 1941; see Fig. 69) and in *Sciara* (CARSON 1946). More recently

NOVITSKI (1952) has shown in *Drosophila* that this process of directed orientation holds for 4-strand doubles only in X-chromosomes which are essentially telo-centric: it is not effective for 4-strand doubles when the X-chromosomes are made 2-armed by the translocation of the long or short arm of the Y-chromo-some. The mechanism is effective, however, in single cross-overs and 3-strand doubles in both telocentric and 2-armed translocated X-chromosomes.

In maize too some inversion heterozygotes, e. g., *In 4a* and *In 2c*, show no greater ovule abortion than that obtaining in normal plants (Table 17) and here, presumably, bridges produced by inversion cross-overs act as a chromatid tie. *In 3a*, however, shows 12% ovule abortion and RHOADES and DEMPSEY (1953) noted here that the dicentric bridge did not persist at first anaphase but broke

Table 17. *The Behaviour of Some Maize Inversions.* (After BURNHAM 1962.)

| Inversion | A I Behaviour (PMC) | | | Total Cells | % Abnormal A I's | % Aborted Pollen | | % Ovule Abortion |
	Chromatid Bridge	Chromosome Bridge	Fragments Only			Expected	Observed	
In 4a	222	17	14	536	47.2	25.2	28.2	4.0
In 2c	50	0	2	86	60.0	30.2	16.5[1]	Normal
In 3a	614	19	87	1645	43.8	22.5	18.6[1]	12.0

[1] Presumably many "unbalanced" complements give apparently normal grains.

early. Clearly in *Zea* the bridge is not always effective in orienting non-cross over chromatids towards the end cells of the megaspore quartet. Similarly in the one barley inversion for which there is relevant information, an X-ray induced paracentric in chromosome-5, ovule abortion is high, ca. 30% (DAS 1951, KASHA 1961). Here too then the chromatid tie does not operate.

Chromatid ties also function on the male side in dipteran flies where by persisting into the second division they serve to hold the two sister secondary spermatocytes together so that ultimately they form one giant spermatid which is incapable of effecting fertilisation. This mechanism was first discovered by WOLF (1941) in *Dicranomyia trinotata* but was described in greater detail by BEERMANN (1956) in *Chironomus tentans*. Like the persistent tie on the female side in *Drosophila* and *Sciara* it protects the organism from some of the genetic consequences expected from inversion hybridity in the sense that it reduces the frequency of unbalanced gametes among the functional ones that are produced.

One final effect which inversions have on meiosis when in the heterozygous state is to increase the cross-over frequency not only in other members of the complement which lack inversions but also in the distal non-inverted regions of the inversion chromosome itself. This effect was predicted by STURTEVANT as early as 1919 although the nature of inversions was not revealed until 1926 by the same author.

Different inversions differ in the intensity of their effect (LEVITAN 1958) but interchromosomal effects on crossing-over have been reported for paracentric inversions (MORGAN, BRIDGES and SCHULTZ 1933, KOMAI and TAKAHU 1942,

Steinberg and Fraser 1941, Schultz and Redfield 1951, Redfield 1955, 1957, Levine and Levine 1955 and Ramel 1962) and pericentric inversions (Alexander 1952) of *Drosophila melanogaster, virilis, robusta* and *pseudoobscura*. In a recent study Suzuki (1963) measured the influence of eight different X-chromosome inversions on crossing-over in chromosome-3. All eight brought about increases as heterozygotes but two of them (sc^9 and dl — 49) failed to do so when homozygous. Significantly these were the only two inversions which did not involve movement of proximal heterochromatin to the distal end (sc^4, sc^8, rst^3 and w^{M4}) or else the reverse movement of the distal end to the proximal heterochromatin (y^4, y^{3p}—see Fig. 70). It thus appears that the presence of heterochromatin adjacent to the distal tip is responsible for the interchromosomal effects of the inversions. This is further supported by the fact that the translocation of the short arm of a Y-chromosome to the distal tip of the X (Fragment I — $Y^s \cdot X$) produced a marked increase in both the heterozygous and the homozygous state whereas the attachment of the Y to the centric end ($X \cdot Y^s$) had no effect in either state.

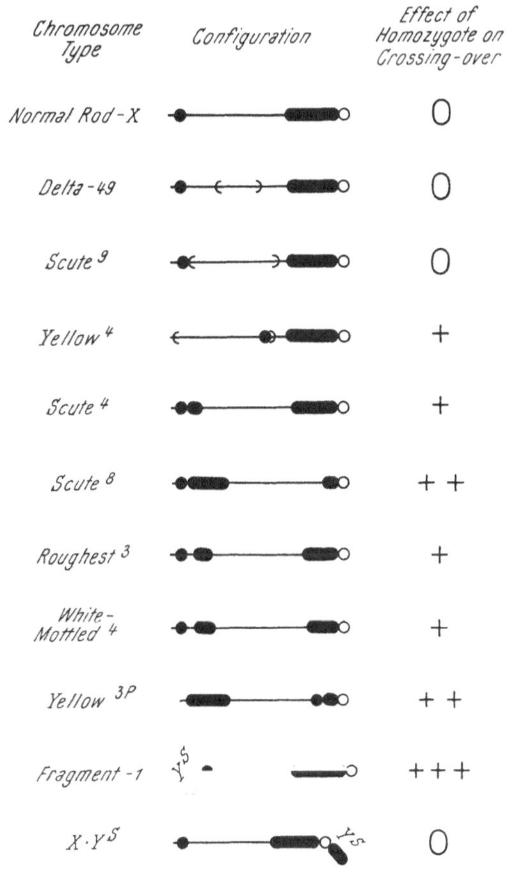

White and Morley (1956) have presented cytological findings on chiasma frequency in grasshoppers which point in the same direction as the intrachromosomal effects found in *Drosophila*. They showed that in three species — *Trimerotropis sparsa, T. gracilis* and *Moraba scurra*—pericentric rearrangements effectively suppress chiasma formation in the region heterozygous for the rearrangement presumably, as we have seen, by bringing about false pairing. Accompanying this there is a pronounced increase in the chiasma frequency of the regions distal to the rearrangement (Table 18).

Inversion heterozygosity thus reduces the frequency of recombinants in respect of the inverted region in two ways. First, by interfering with the process of homologous pairing. (This effect is likely to extend to the non-inverted segments adjacent to the inversion when the inverted segments pair homologously.)

Chromosome Type	Configuration	Effect of Homozygote on Crossing-over
Normal Rod-X		O
Delta-49		O
Scute ^{9}		O
Yellow ^{4}		+
Scute ^{4}		+
Scute ^{8}		+ +
Roughest ^{3}		+
White-Mottled ^{4}		+
Yellow ^{3P}		+ +
Fragment -1		+ + +
$X \cdot Y^S$		O

Fig. 70. The X-inversions of *Drosophila melanogaster* studied by Suzuki (1963) and their effect, in the homozygous condition, on crossing-over in chromosome 3.

Second, by the inviability or by the elimination or preferential segregation of the cross-over products. These reductions are generally accompanied by an increase in recombination elsewhere in the complement. Doubtless, this too is related to the effect of inversions on pairing.

In *Drosophila*, the segregation as well as the cross-over frequency of non-homologous chromosomes may be affected. Thus, as COOPER, ZIMMERING and

Table 18. *Mean Proximal and Distal Chiasma Frequencies in Individual Bivalents of Three Grasshopper Species with Pericentric Inversions.*
(Data of WHITE and MORLEY 1955.)

Species and Rearrangement	Bivalent		Chiasma Distribution and Frequency	Total Bivs. Analysed
	Type	Code		
1. *Trimerotropis sparsa* (Sevier — Sev) Acro → Meta-centric	Basic homozygote	St/St	0.88 1.10	1118
	Inversion heterozygote	St/Sev	0.001 1.37	2234
	Inversion homozygote	Sev/Sev	0.80 1.11	473
2. *Trimerotropis gracilis* (Humbolt — Ht) Acro → Metacentric	Basic homozygote	St/St	0.50 0.74	1204
	Inversion heterozygote	St/Ht	0.00 1.005	615
3. *Moraba scurra* (Blundell — Bl) Meta → Acro-centric	Basic homozygote	St/St	0.81 0.84	700
	Inversion heterozygate	St/Bl	0.00 1.00	800
	Inversion homozygote	Bl/Bl	0.76 0.63	1200

KRIVSCHENKO (1955) first showed, X-chromosomes with a normal cross-over frequency rarely fail to segregate. But in the presence of autosomal inversions which increase the cross-over frequency in the X the frequency of non-disjunction of the X at first anaphase also increases (MORGAN and STURTEVANT 1944, see also section IV-2, part 11).

b) The Inverted Duplication

We have seen how, at first anaphase of meiosis, dicentric chromatids may be broken by tension or else cut-through by the developing cell wall. Where the point of severance is not at the centre of the bridge the products of breakage will include a terminal inverted duplication whose length will depend on the precise point of breakage: the nearer the break to the opposite centromere the longer the inverted segment. Provided that the broken chromosome ends are stable or will "heal" as they do in maize (McClintock 1938, 1939, 1941) such inverted duplications may be included in functional gametes.

One of the most thorough studies of inverted duplications was carried out by Frankel (1949a and b) who found two such duplications in the derivatives of a varietal cross in *Triticum aestivum*. One of these was long (*LD*), repeating most of the arm to which it was attached, while the other was much (10–12 times) shorter (*SD*). By suitable crosses, plants were obtained with the full range of long and short duplications and normal chromosomes (*N*) as homozygotes,

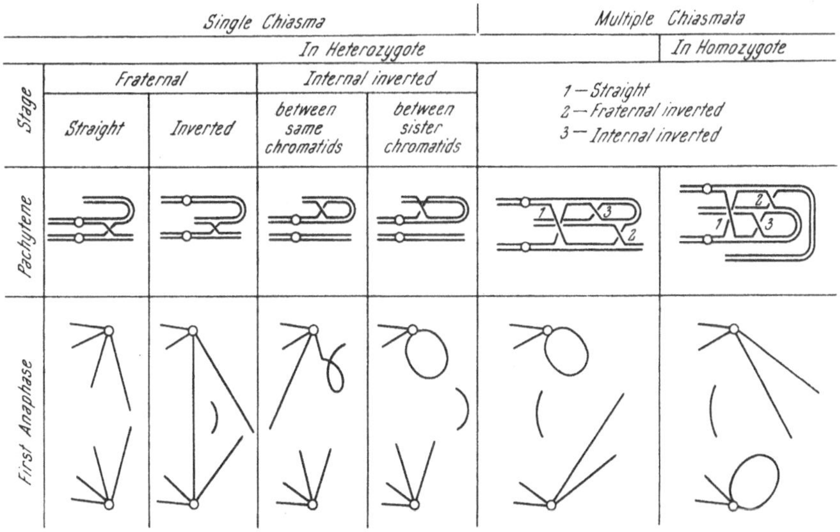

Fig. 71. Pairing behaviour and its consequences in wheat plants hetero and homo-zygous for an inverted dupli-cation (Frankel 1949).

heterozygotes and monosomics, viz. *LD/LD*, *LD/N*, *LD* monosomic, *LD/SD*, *SD/SD*, *SD/N*, *SD* monosomic, *N/N*, *N* monosomic. And all nine combinations were viable.

Chromosomes with inverted duplications possess peculiar pairing properties for there are two homologous segments within each chromosome carrying a dupli-cation. Internal pairing can therefore take place between these segments especially in duplication monosomics, or in those cases where there is no pairing between homologues over the segment present twice in the duplication. As a result of this, three possible kinds of chiasmate relationships may exist (Fig. 71):

(1) Those between sectors common to sister chromosomes (straight)

(2) Those between duplicated sectors within the same chromosome (internal inverted)

(3) Those between the duplicated sector and the sister chromosome (fraternal inverted)

Frankel's principal conclusions were:

(1) Inside pairing has a natural advantage even in the short duplication

(2) Internal chiasma formation occurs in both kinds of duplication. It is localised near the union between the duplicated segments and hence chiasma

frequency is only loosely correlated with the length of the duplication.. Moreover it is more frequent between sister chromatids than within the same
chromatid; there is chiasma interference across the bend (see section II–1 pt. II).

(3) Fraternal chiasma formation is severely restricted, or absent, in the
heterozygous duplicate/normal condition but its frequency increases progres-
sively in the sequence heterozygous LD/SD < homozygous LD/LD < homozygous SD/SD

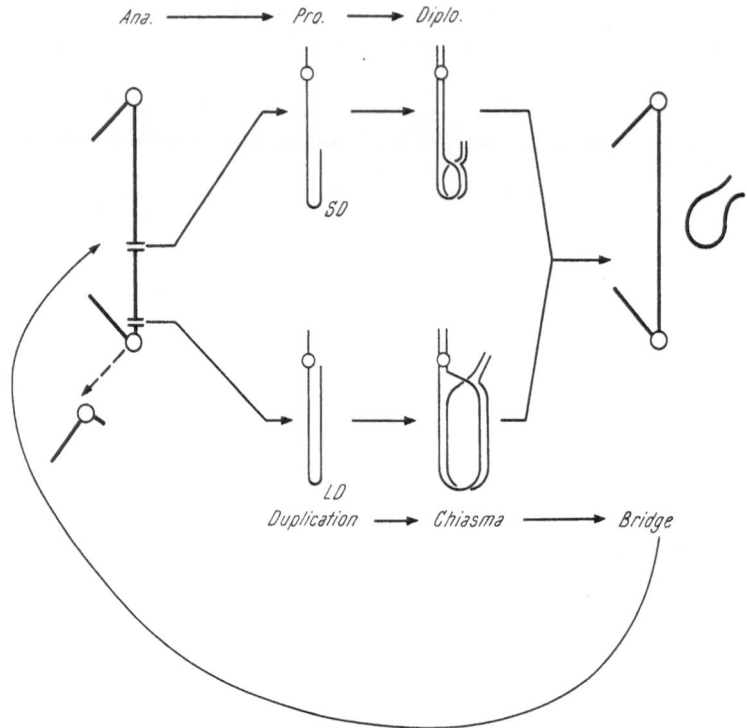

Fig. 72. The chiasma cycle which leads to a persistent inverted duplication (FRANKEL 1949).

(4) Fraternal inverted pairing occurs only between equal arms. Duplications,
one or both of which are paired internally, block fraternal pairing. When not
paired internally duplications hinder fraternal pairing without blocking it entirely;
but one duplication alone blocks pairing with the arm without the duplication.

Now the majority of chiasmata in an inverted duplication yield bridges and
fragments, a process which may result in the elimination of the duplication.
Alternatively these new bridges may in turn break unevenly and so produce
new inverted duplications in a manner reminiscent of a breakage-fusion-bridge
cycle (Fig. 72). The actual transmission of broken chromosomes is summarised
in Table 19. Two important points emerge from this table. First, the presence
of LD/N, LD/SD, SD/SD, SD/N, N/N and SD/O plants among the progeny of
LD/LD selfing proves that gametes from broken-bridge chromatids are functional
and that bridge-breakage must produce both SD and N gametes. Second, the
presence of LD/N and LD/SD plants in SD/N selfed proves that a long duplication may arise from bridge-breakage. In *Triticum* then, as in maize, chromosomes

broken in a meiotic anaphase are found in functional gametes. But, contrary to the observations in maize, broken chromosome ends fail to fuse; there is no sister union of chromatids broken in meiosis and healing is apparently rapid and permanent.

The persistence of an inverted duplication will depend on the rate of duplication crossing-over, internal as well as fraternal, and on the selective value of the duplication. The inverted duplications in wheat show a very high frequency of crossing-over but have no selective advantage and for this reason fail to persist.

Table 19. *The Progeny from Crossing and Selfing Wheat Plants Homo- or Heterozygous for Long (LD) and Short (SD) Inverted Duplications.*
(Data of FRANKEL 1949b).

Parents	Progeny									Total
	Disomics						Monosomics			
	LD/LD	LD/N	LD/SD	SD/SD	SD/N	N/N	LD/0	SD/0	N/0	
1. N/N × LD/LD	...	18	...	...	3	2	...	...	...	23
2. N/N × LD/N	...	35	...	...	4	50	...	...	1	90
3. N/N × LD/SD	...	3	...	...	5	...	...	...	...	8
4. LD/LD × SD/SD	...	1	3	1	...	1	...	...	...	6
5. LD/LD ⎫	58	18	17	1	6	2	1	1	...	104
6. LD/N ⎪	21	68	1	...	11	55	...	...	1	157
7. LD/SD ⎬ selfed	4	6	14	7	7	1	...	...	...	39
8. SD/SD ⎪	...	...	...	20	10	1	...	1	1	33
9. SN/N ⎪	...	1	2	4	42	35	...	...	...	84
10. N/N ⎭	...	...	...	...	...	35	...	...	...	35

c) The Interchange Heterozygote

Occasionally two non-homologous chromosomes undergo simultaneous breakage and then rejoin reciprocally. Two categories of reunion are theoretically possible. Either it is symmetrical producing two new monocentrics or else it is asymmetrical giving rise to a dicentric and an acentric (see section IV-2a). We shall concern ourselves here with symmetrical interchanges for they are far commoner and give rise to a true interchange of non-homologous segments— a reciprocal translocation. In an individual heterozygous for a single interchange four chromosomes are present which share partial homology but no two of these correspond exactly. The homologues, however, are such that all four chromosomes can associate together at zygotene and form a pairing-cross (Figs. 73 and 74). Theoretically the centre of the cross should indicate where the homology changes. In certain maize interchanges, however, the position of the centre of the cross varies widely as a result of an association of non-homologous segments. In others it appears to be less variable though even here there is sometimes considerable asynapsis near the centre of the cross (BURNHAM 1962).

There are three kinds of segment in this pairing cross and chiasma formation may occur in any of them. First there are the interchanged segments, second there are the non-interchanged arms and thirdly there are the segments between the centromere and the breakage point. These latter are referred to as the interstitial segments and are distinguished from the other regions which collectively are referred to as pairing segments. Of course if interchange takes place at the centromere, as it may do following mis-division, no interstitial segment will occur (see for example MORRISON 1954a). The consequences of chiasma formation in these two kinds of segment differ, indeed this is the main reason for distinguishing them. Consider first the situation where a single chiasma

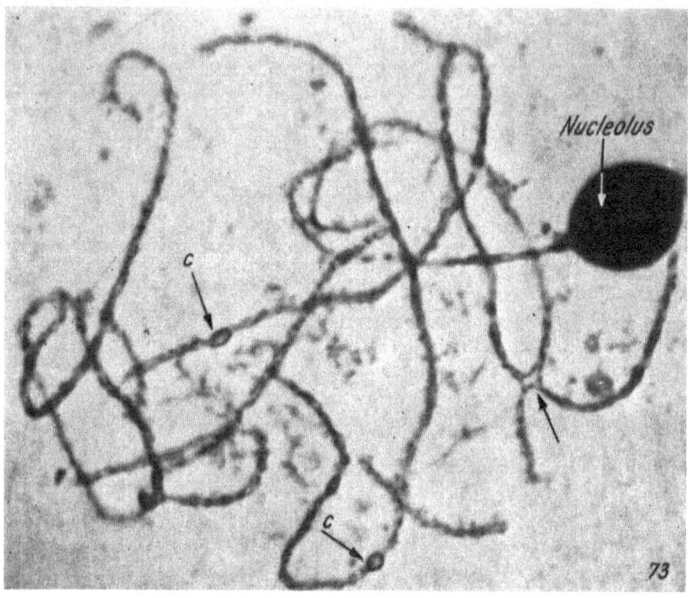

Fig. 73. Pachytene pairing cross in a single interchange hybrid of maize (arrow). Note also nucleolus and distinct centric regions (carmine-phase contrast). Photograph kindly supplied by Dr. HUBERT REES.

forms in each of the 4 pairing segments and the pachytene cross opens out at diplotene-diakinesis to form a ring-of-four. At first pro-metaphase this multiple can orientate in a variety of positions.

Numerically equal disjunction involving the orientation of all four chromosomes can occur in three ways (Fig. 75). Open ring configurations can orient in such a way that one pair of adjacent homologous centromeres proceed to one pole and the other pair to the opposite pole or adjacent non-homologous centromeres can move together, two to each pole. Both of these give qualitative chromosome non-disjunction. But a twist in the ring determined by the co-orientation of all adjacent centromeres leads to the movement of adjacent homologous centromeres to opposite poles and qualitative as well as quantitative chromosome disjunction. These three types are called adjacent homologous, adjacent non-homologous and alternate orientation respectively, the terms indicating which centromeres proceed to the same pole (Figs. 76–79).

Some have maintained that, in the absence of directed orientation, these three types should occur with equal frequency while others have held that alternate orientation should be as common as the joint frequency of the two adjacent arrangements. Indeed, on the assumption that homologous centromeres segregate in preference to non-homologous centromeres (which we do not accept), the expected frequency of alternate orientation has been held to approach that of adjacent non-homologous orientation while the adjacent homologous arrangement is rare. However, in a ring composed of equal sized, iso-brachial chromosomes associated at metaphase by terminal chiasmata (or any chiasmata equidistant from the centromeres) there is no basis for distinguishing, or therefore showing preference between, adjacent homologous and adjacent non-homologous orientation though these can be distinguished under certain other conditions. Further the ease with which the twist, necessitated by alternate orientation, can form will depend on the size of the ring, the rigidity of the chromosomes and other variables.

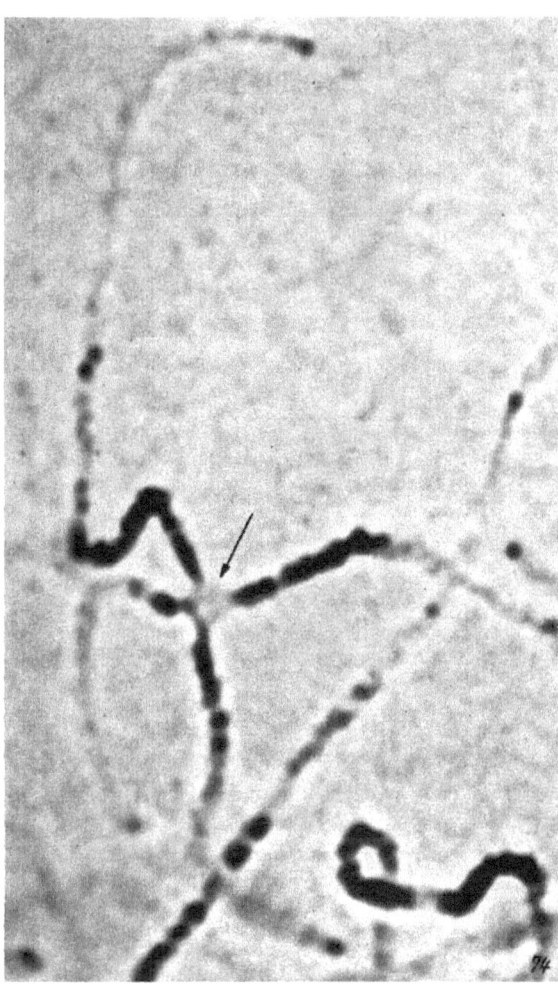

Fig. 74. Pachytene pairing-cross in an X-ray induced translocation of tomato. The interchange points are in the centric heterochromatin (× 4000). Photograph kindly supplied by Mr. BRIAN SNOAD.

Where chiasmata form in only three of the four possible sets of pairing segments a chain-of-four multiple results and such chains are expected to have different properties of orientation. Thus, mechanically speaking (LEWIS and JOHN 1963b), a chain in which the "missing" chiasma is in a pairing segment between two non-homologous centromeres (i. e., a non-centric arm of the pachytene cross) and showing adjacent non-homologous orientation is mechanically comparable to one showing adjacent homologous orientation if its "missing" chiasma is in the pairing segment between homologous centromeres (i. e., in a centric arm of the pachytene cross). There is little basis, therefore, for making generalised assumptions as to the expected frequencies of these various types of orientation because even simple rings can be "expected" to behave in a random manner only under particular structural and genotypic conditions. However, this much must be said. The expectation based on randomness depends on one's concept of randomness (see below). In our view of random behaviour, zig-zag orientation should represent a half of the numerically-equal separations and not one third as claimed by LA CHANCE et al. (1964) and by JONES (1964) for example.

We base this expectation on the fact that, for each type of equally-likely adjacent orientation, "there is a corresponding alternate arrangement" (BURNHAM 1956, p. 425). In other words, the zig-zag orientations are actually of two types. In one the twist occurs between the associated arms which join non-homologous centromeres (the alternate equivalent of adjacent non-homologous orientation) while in the other it occurs in the arms joining homologous centromeres (the equivalent of adjacent homologous orientation). Thus:—

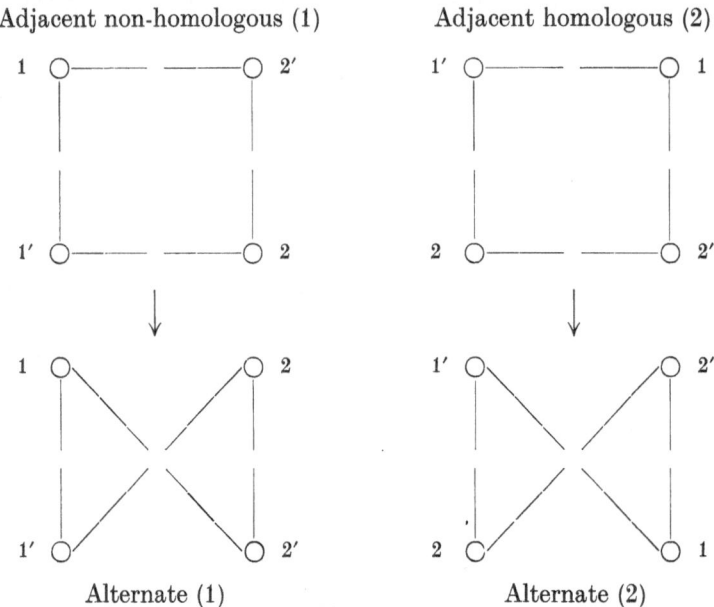

This 1 : 1 expectation of Adjacent : Alternate is not, of course, based on total randomness but on the random behaviour of two pairs of co-oriented adjacent centromeres. This is "realistic randomness."

If, on the other hand, one considers that any two of the four centromeres can go to one pole, leaving the remaining two to proceed to the opposite pole, and that all possibilities are equally likely, then, adjacent homologous, adjacent non-homologous and alternate arrangements are expected to be equally frequent among the numerically-equal segregations. This is complete randomness. It is also unrealistic because it gives the expectation on the basis of univalence. But in multiples, as in bivalents and multivalents, separation has a basis in co-orientation.

This means that one must consider, not the possibilities in relation to variously-constituted pairs of centromeres moving together to the same pole but, rather, those in relation to the segregation of centromere-pairs showing co-orientation. And, of the latter, one must ignore the possibility of co-orientation between two centromeres which alternate in the ring because this does not determine the co-orientation of the remaining pair which are similarly alternate. On the contrary it prohibits this event; it gives discordant orientation which is not expected to give numerically-equal separation. Herein lies the basis of the difference between the two "expectations." Therfore, the more realistic expectation on the basis of randomness is 1 Alternate : 1 Adjacent and, *mutatis mutandis*, this is also the expectation on the basis of random bivalent formation where, again, co-orientation is the basic of disjunction (see also page 127 and cf. JONES 1964; RICKARDS 1964).

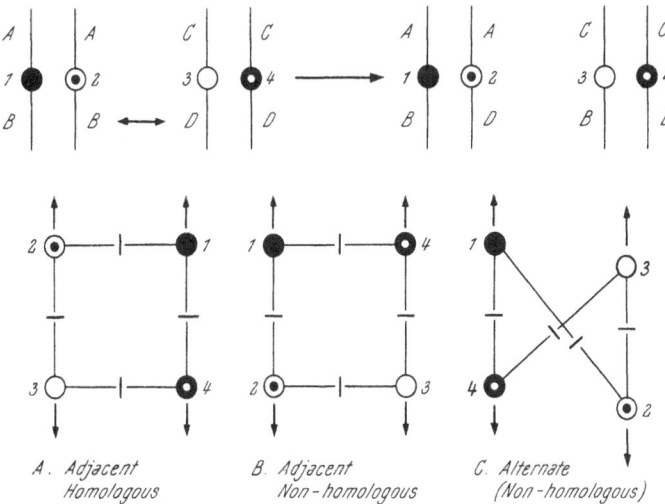

A. Adjacent   B. Adjacent   C. Alternate
  Homologous   Non-homologous   (Non-homologous)

Fig. 75. Three types of orientation giving numerically equal segregation in a ring of four chromosomes.

Following either kind of adjacent orientation all 4 meiotic products are un-balanced because they carry complementary duplication-deficiency chromatids. Where orientation is alternate, on the other hand, all four cells are balanced though two of them carry the basic, and the other two the interchanged arrange-ment of segments. Occasionally only two of the four centromeres orient giving a discordant ring (Fig. 79) whose subsequent behaviour also leads to complete unbalance. The proportion of unbalanced gametes may be even further increased by the cross of four failing to form or else to maintain itself as a multiple of four and giving either a chain of three and a univalent or else two bivalents. It is for such reasons that many floating interchanges fail to survive in nature (LEWIS and JOHN 1963a).

Chiasma formation in the interstitial segment has quite different conse-quences. If a chiasma forms in one of the interstitial segments then adjacent homologous orientation either does not occur (McCLINTOCK 1945, BURNHAM 1950, JOHN and HEWITT 1963) or else its frequency is reduced. But, as described above in the case of chain multiples, the position of the interstitial chiasma in relation to the free arms has a powerful if not unconditional influence on the behaviour of the multiple (LEWIS and JOHN 1963b). Because crossing-over in the interstitial segments inevitably means chromatid non-disjunction half the meiotic products are automatically unbalanced, whether the multiple adopts adjacent non-homologous or alternate orientation. The only difference, in fact, is in the kinds of chromatid carried in the balanced gametes. Following adjacent non-homologous orientation they carry cross-over chromatids whereas following alternate orientation the chromatids are non-crossover. Crossing over in both

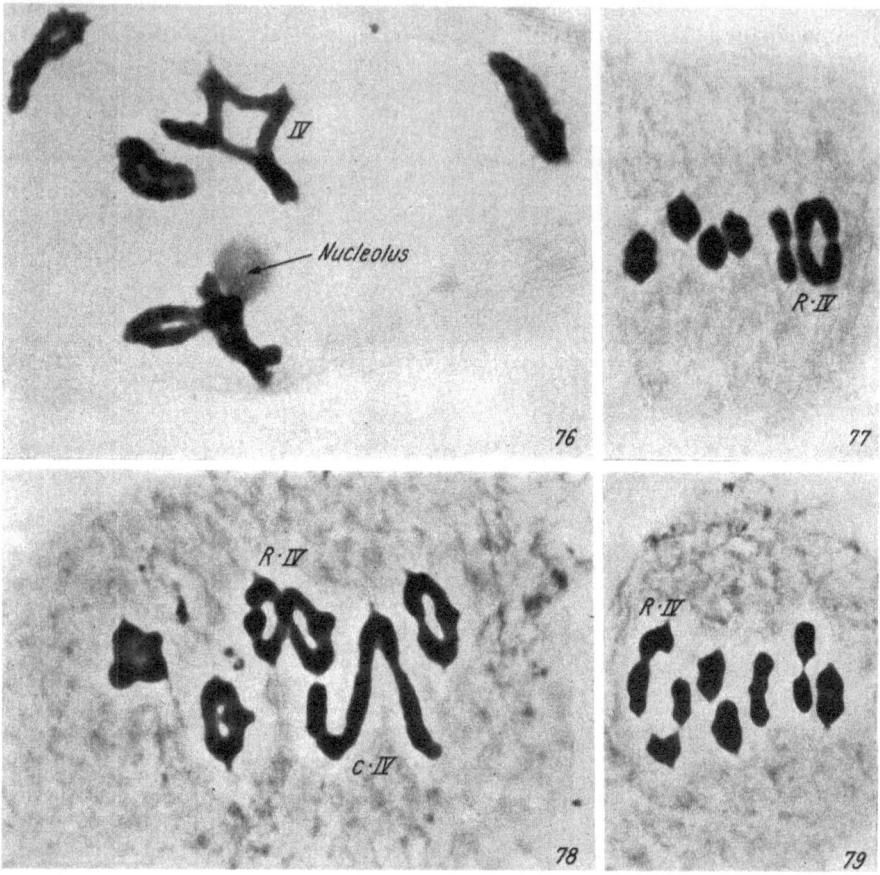

Figs. 76–79. Interchange associations in inbred rye (all carmine preparations, ca. × 2000); Figs. 78 and 79 taken with phase contrast. Fig. 76. Diplotene with a single interchange cross. Fig. 77. Metaphase-I with a ring-of-four showing adjacent orientation. Fig. 78. Metaphase-I, a chain-of-four and a ring-of-four in alternate orientation. Fig. 79. Metaphase-I with a discordant ring-of-four.

interstitial segments does not improve fertility but it does transfer the onus onto the second meiotic division for the products of first anaphase are identical and disjunction depends on the random orientation of the chromosomes at second anaphase. There is one further difference (Table 20): all the balanced gametes now carry one cross-over and one non-crossover chromatid following adjacent non-homologous orientation while with alternate orientation half of them carry two cross-over chromatids and the other half two non-crossover chromatids. Of course, as with inversion heterozygotes, reciprocal chiasmata within the interstitial segment will cancel one another.

When chromosomes, or their homologues are involved in two or more interchanges, multiples higher than four result. Fig. 80 illustrates how two or three serial interchanges can lead to the production of pairing associations of six and of eight respectively (Figs. 81–84). This system is of course capable of extension so that finally every member of the chromosome complement can be involved in one giant concatenated system as in fact has occurred in *Oenothera* and in *Rhoeo* (Fig. 85). In such large multiples one can again distinguish pairing and interstitial segments. But, in addition to these, there is a third type of segment

Table 20. *The Influence of Ring and Chain-of-Four Multiples on Meiosis.*

Chiasma formation	%-balanced Gametes			Fertility Dependent on Orientation at		Interstitial Segment Crossover Chromatids in Balanced Gametes	
	Adjacent Homologous	Adjacent non-homologous	Alternate	Div. I	Div. II	Adj. Hom.	Alternate
1. In pairing segments only	0	0	100	+	−	Not applicable	Not applicable
2. In pairing and interstitial segments (a) Single chiasma in one interstitial segment	0	50	50	−	−	1 per gamete	None
(b) Single chiasma in both interstitial segments	0	50	50	−	+	1 per gamete	½ with 0 ½ with 2

This is the differential segment which is produced only when a chromosome pair is involved in two interchanges for it is the segment between the two points of interchange. If the interchanges occur one in each arm the differential segment will of course include the centromere but if they both involve the same arm it will not.

It is true that the differential segment can be defined in terms of the interstitial segments but they are distinguished from these because chiasma formation within them leads to the production of chromatids which are new with respect to their ends. In fact a chiasma here is equivalent to a reverse interchange because it leads to the breakdown of the multiple (Table 21). Thus in the ring-of-six illustrated the balanced non-crossover gametes are of two kinds, AB . CE . FD and BC . EF . DA. If only balanced gametes survive, the viable gametes after crossing-over in the differential segment will be AD . *BF* . CE and *CD* . AB . EF. If either of these unite with those normally produced by the heterozygote in question then the products will be ring-of-four heterozygotes.

In complex interchange configurations the variety of possible orientations is greatly increased, at least in theory, so that it should be more difficult to achieve disjunctional separation. There is some evidence, however, which suggests that

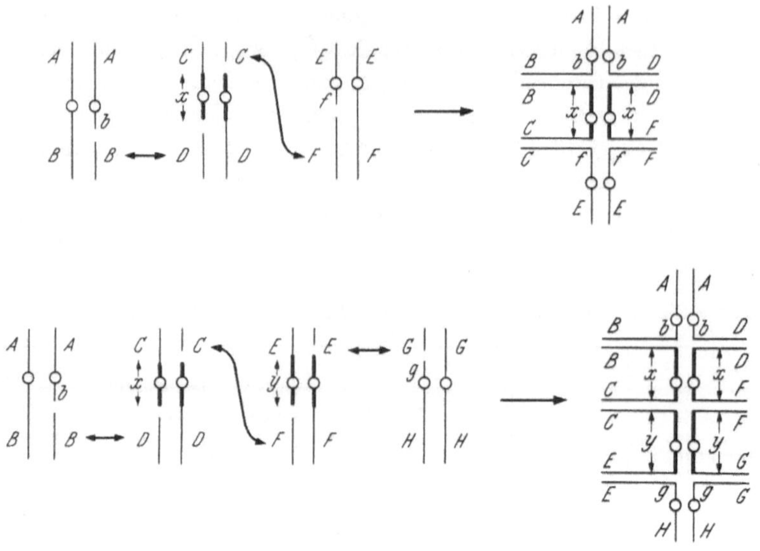

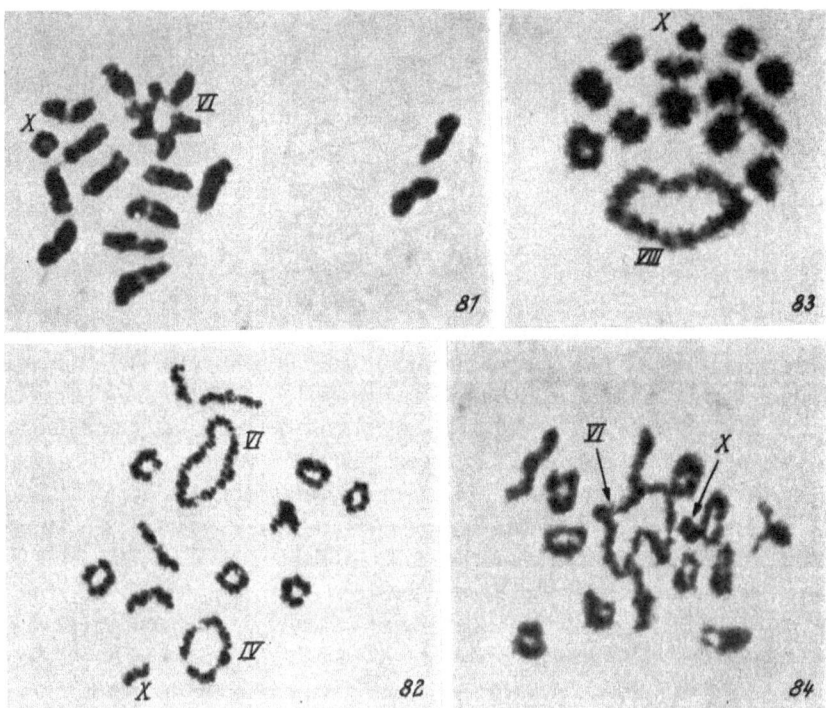

Fig. 80. The origin of interchange associations of six and eight following two and three serial interchanges respectively. Note b, f and g represent interstitial segments while x and y represent differential segments.

Figs. 81–84. Interchange hybrids in *Periplaneta americana* (all orcein, ca. × 1500). Figs. 81. Pachytene cross of six (compare Fig. 80). Fig. 82. Ring-of-six and a ring-of-four at c-diakinesis. Fig. 83. Ring-of-eight at c-diakinesis. Fig. 84. Oriented ring-of-six at pre-metaphase.

this is not always so (John and Lewis 1958). Three main factors have been held to affect the type of orientation shown by interchange multiples. First, the structural properties of the chromosomes themselves. Thus, multiples composed of equal sized, isobrachial chromosomes are expected to show a high frequency of disjunction, for these conditions facilitate the equal interaction between a given centromere and both its neighbours—the interaction on which co-orientation depends. Second, since they too will affect the equality of this interaction and the flexibility of the multiple, the frequency and distribution of chiasmata; a high frequency of chiasmata and especially of interstitial chiasmata will tend to reduce the frequency of disjunctional separation. These mechanical properties of chiasma formation, unlike the structural properties, are genotypically controlled aspects of multiple morphology and, thus, behaviour. Third, genotypically controlled properties other than those affecting the morphology of the multiple. These are not easily specified but they are likely to include such properties as the time available for orientation (and re-orientation) and the occurrence and extent of pre-metaphase stretching (see pg. 172). Thus in 11 out of 13 plants of *Secale kupri-*

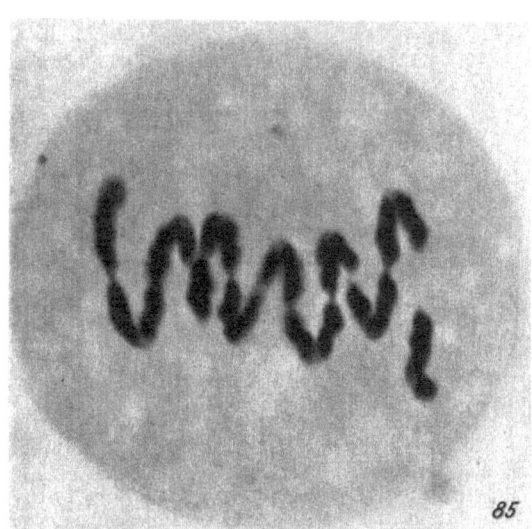

Fig. 85. Disjunctional chain-of-twelve in *Rhoeo discolor*, a mono-typic genus with five serial interchanges in the heterozygous condition. Photograph kindly supplied by Dr. Keith Jones.

janovii heterozygous for the same interchange (Hrishi and Müntzing 1960) the percentage of PMC's with zig-zag chains at first metaphase ranged from 56.5 to 92.8 percent (Table 22). That factors other than these may also be important follows from Snell's work on radiation induced interchanges in mice (Snell 1946). Here 5 of the 6 translocations tested showed a higher percentage fertility in the male sex (Table 23), but the basis for this difference was never established. It is worth noting, however, that chiasma frequency is higher in female mice (see Table 6, p. 26).

A particularly interesting situation is exemplified by the recent study of fertility in interchange hybrids of the screw-worm fly, *Cochliomyia hominivorax* (La Chance, Riemann and Hopkins 1964). In back-crosses with wild-type flies, male interchange heterozygotes were found to have a fertility of ca. 66% while equivalent females were only 50% fertile. It was concluded that adjacent homologous orientation did not occur in the males and that alternate orientation was twice as common as the adjacent non-homologous arrangement. The absence of adjacent homologous separation would appear to favour the view (which we earlier denied—see page 78) that homologous centromeres segregate in preference to non-homologous ones. It must be remembered, however, that the situation in the male of this dipteran is peculiar in that crossing-over does not occur. In its absence, the maintenance of the multiple at metaphase depends on prolonged parallel pairing and this means greater communication and interaction between homologous centromeres. But it is not homology *per se* that matters but the morphology of the association.

Table 21. *The Consequences of Crossing-over in the Differential Segment of a Ring-of-six Multiple.*

Cross-over Chromatids Italicised. Compare with Fig. 80.

AB CE FD \ / \ / \ BC EF DA	Non-cross-over Gametes	
	AB · CE · FD	BC · EF · DA
Viable differential segment cross-over Gametes AD · *BF* · CE	AB FD CE \ / \ ‖ BF DA, CE	BC EF AD \ / \ ‖ CE FB, AD
CD · AB · EF	DC EF AB \ / \ ‖ CE FD, AB	BC DA EF \ / \ ‖ CD AB, EF

Table 22. *Disjunctional Behaviour of Eleven Individuals of Secale kuprijanovii Heterozygous for a Single Interchange.*

(Data of HRISHI and MÜNTZING 1960.)

Configuration Type	Plant Number										
	11	14	22	1	8	23	15	10	16	13	25
Non zig-zag	43.5	41.7	40.0	33.3	31.3	18.8	18.2	16.7	16.7	9.1	7.2
Zig-zag	56.5	58.3	60.0	66.7	68.7	81.2	81.8	83.3	83.3	90.9	92.8
%-good pollen	80.0	75.0	68.0	72.0	78.0	87.0	82.0	75.0	82.0	89.0	92.0

Note: The correlation coefficient between pollen fertility and the frequency of zig-zag orientation is $+ 0.72$ ($P = 0.02 - 0.01$) indicating that pollen fertility is positively correlated with the frequency of zig-zag orientation at first metaphase.

The peculiar morphology of the multiple in the male is important in another direction. LA CHANCE et al. attribute the bias towards alternate orientation to the increased flexibility of the multiple in the male and this to the absence of chiasmata. But owing to the morphology indicated above, the situation can hardly be compared with that in chiasmate systems where the chromosomes open-out at diplotene-diakinesis. Indeed, where pairing is prolonged, the twist which alternate orientation requires has more in common with the relational twist that can form between the two centromeres of a dicentric chromosome at mitosis. Actually, where multiples of this type are concerned, a correlation between the frequency of alternate orientation and the combined lengths of the interstitial segments would not be surprising. In the case studied, the interchange points were close to the centromeres and in respect of both departures from random orientation behaviour, the degree of pairing near the centre of the cross is expected to be of importance.

A dearth of adjacent homologous orientation in the male is not unexpected but La Chance et al. concluded that a similar situation obtained in the females as well. And, on the assumption of an equal frequency of adjacent non-homologous and alternate orientation, a good fit with the observed results was obtained. Even so we do not think that the results warrant the conclusion which is not

Table 23. *Differences in Fertility Between the Sexes in Six Radiation Induced Interchanges in Mice.*
(Data of Snell 1946.)

Translocation	%-fertility of Semi-sterile Females	Difference from 50%	P	%-fertility of Semi-sterile Males	Difference from 50%	P	Difference Between Males and Females	P
T − F$_1$ 194	38.4 ± 2.9	− 11.6	0.0001	41.3 ± 1.8	− 8.7	< 0.0001	2.9	0.39
T − F$_1$ 111	42.9 ± 3.3	− 7.1	0.032	45.0 ± 1.9	− 5.0	0.0085	2.1	> 0.5
T (2:?) d	49.1 ± 2.1	− 0.9	> 0.5	47.3 ± 2.5	− 2.7	0.31	− 1.8	> 0.5
T (1:?) c								
cross 1	44.7 ± 2.6	− 5.3	0.042	52.7 ± 2.4	+ 2.7	0.26	8.0	0.023
cross 2	45.4 ± 2.2	− 4.6	0.037	53.0 ± 2.6	+ 3.0	0.25	7.6	0.025
T (5:8) a								
cross 1	50.4 ± 1.9	+ 0.4	> 0.5	61.8 ± 1.7	+ 11.8	< 0.0001	11.4	< 0.0001
cross 2	46.2 ± 3.7	− 3.8	0.31	59.8 ± 6.5	+ 9.8	0.13	13.6	0.07
cross 3	50.5 ± 2.7	+ 0.5	> 0.5	59.5 ± 3.8	+ 9.5	0.013	9.0	0.06
T − R$_1$ 8	62.2 ± 31.1	+ 12.2	0.0002	68.9 ± 1.9	+ 18.9	< 0.0001	6.7	0.07

a likely one on more general grounds. Thus, on crossing interchange hybrids in this species one expects—duplication/deficiency zygotes, interchange homozygotes, basic homozygotes and interchange heterozygotes with a frequency of 14 : 2 : 2 : 6 on the basis of the hypothesis. But on the basis of 1 Adjacent Homologous : 2 Alternate : 1 Adjacent Non-homologous orientation on the female side (and again, 66% alternate and 33% adjacent non-homologous on the male) the expectations are not greatly different, namely, 15 : 2 : 2 : 5. The results actually obtained were:—

(i) Failure to hatch (dup./def.) = 61%   (58.4 versus 62.5%)
(ii) Pupal heterozygotes   = 52%   (60.0 versus 55.5%)
(iii) Adult homozygotes   = 26.6% (25.0 versus 28.6%)

In brackets we have put the expectations on the basis of La Chance et al.'s hypothesis and ours in that order. Ours gives a better fit except in the third case. However, the expectations given here are those on the basis of total inviability of interchange homozygotes as adults and the frequency of adult homozygotes was that determined on the basis of the exophenotype (the interchange is linked with a dominant mutant gene). But, in a sample of 74 adults with the mutant character, four were found to be interchange homozygotes.

And when one takes this into account, our suggestion gives a better fit in respect of the third observation also.

One of the most thorough analyses so far made of the behaviour of interchange multiples has recently been carried out by SHONG SUN (1963) using inbred lines and species hybrids of rye. Two distinguishable interchanges, A and B, are known to occur in inbred lines of rye (THOMPSON 1956), one of them (A) involving the nucleolar organising chromosome. By crossing two inbred lines (line 1 × line 3) SUN obtained three F 1—families. From these, heterozygous

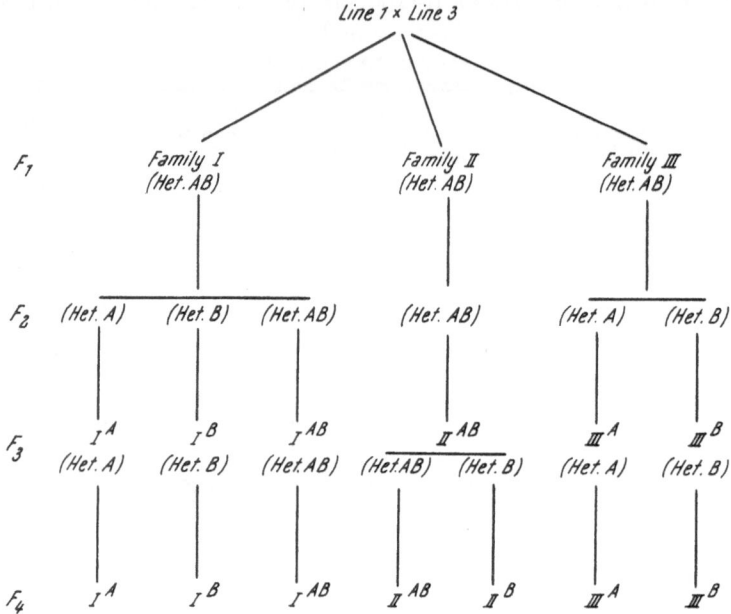

Fig. 86. The breeding programme practised by Dr. S. SUN for interchange hybrids of rye.

progeny were followed through F 2 — F 4 generations (Fig. 86). A comparative study of these heterozygotes showed that in rye the disjunction frequency of interchange heterozygotes decreases with increasing chiasma frequency in individual PMC's (Fig. 87), in individual plants and in individual generations. And this applies to all three kinds of heterozygote (Het. A, Het. B and Het. AB). There is, however, no correlation between disjunction frequency and the chiasma frequency of the bivalents of the same cells, showing that it is the chiasma frequency of the interchange associations themselves which has a direct influence on their disjunction (Fig. 88). In this property, however, rye differs from *Chorthippus* (LEWIS and JOHN 1963b).

By establishing high and low selection lines for A, B and AB heterozygotes, SUN practised direct selection for disjunction frequency in the F 2 — F 4 series. Selection was carried out in conjunction with forced inbreeding by self-pollination. Rather surprisingly he found that in all three cases the disjunction frequency increased both in the high and the low selection lines (Fig. 89). Now in rye, as in many other species which normally outbreed, one of the characteristic consequences of forced inbreeding is a reduction in chiasma frequency with increasing homozygosity. And in both high and low selection lines the chiasma frequency of the multiples showed a marked reduction from F_2 to F_4. Since chiasma frequency and disjunction frequency are known to be negatively cor-

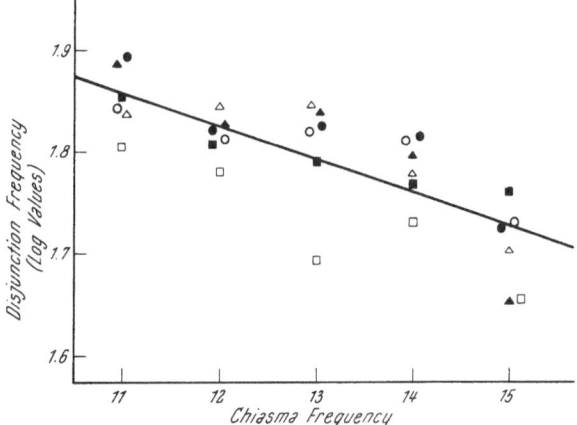

Fig. 87. Relationship between disjunction frequency (in logs. of angular values) and the chiasma frequency of individual P. M. C.'s in interchange hybrids of rye. Note:

$\triangle$ = F_2—Family I : $\bigcirc$ = F_2—Family II : $\square$ = F_2—Family III
$\blacktriangle$ = F_3—Family I : $\bullet$ = F_3—Family II : $\blacksquare$ = F_3—Family III

(compare with Fig. 86). Unpublished data kindly supplied by Dr. SHONG SUN (see SUN 1963).

related in rye, the increased disjunction in the high selection lines would appear to be due, in a measure at least, to the fall in chiasma frequency which results from inbreeding. And this suggests that the effectiveness of selection must

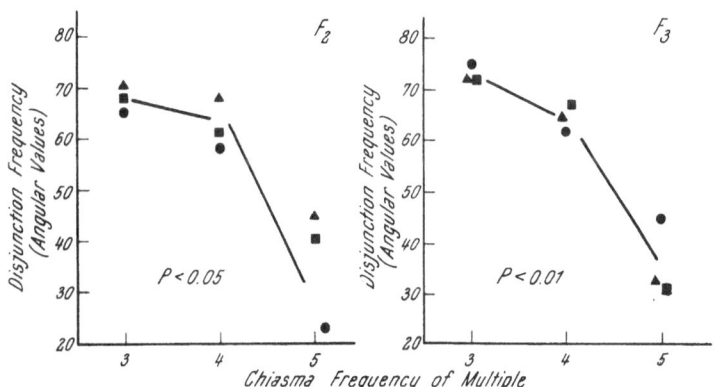

Fig. 88. Relationship between disjunction frequency (in angular values) and the chiasma frequency of interchange multiples in the F_2 and F_3 generations of SUN's breeding programme. (See REES and SUN 1965). Note:

$\blacktriangle$ = Heterozygous AB
$\blacksquare$ = Heterozygous B
$\bullet$ = Heterozygous A

depend, to a large degree, on the adjustments in chiasma frequency, which in turn influences the mechanics of chromosome behaviour.

SUN also performed a diallel cross between *Secale anatolicum* and six other rye species which differed from it by at least two interchanges involving three chromosomes. From a study of such hybrids he showed (Fig. 90) that:

(1) The disjunction frequency varied according to the size of the interchange configuration. The smaller the multiple the higher the frequency of disjunction.

(2) In all multiples, however, the disjunction frequency was considerably higher than that predicted on a random basis. Clearly rye chromosomes are pre-adapted to give regular disjunction.

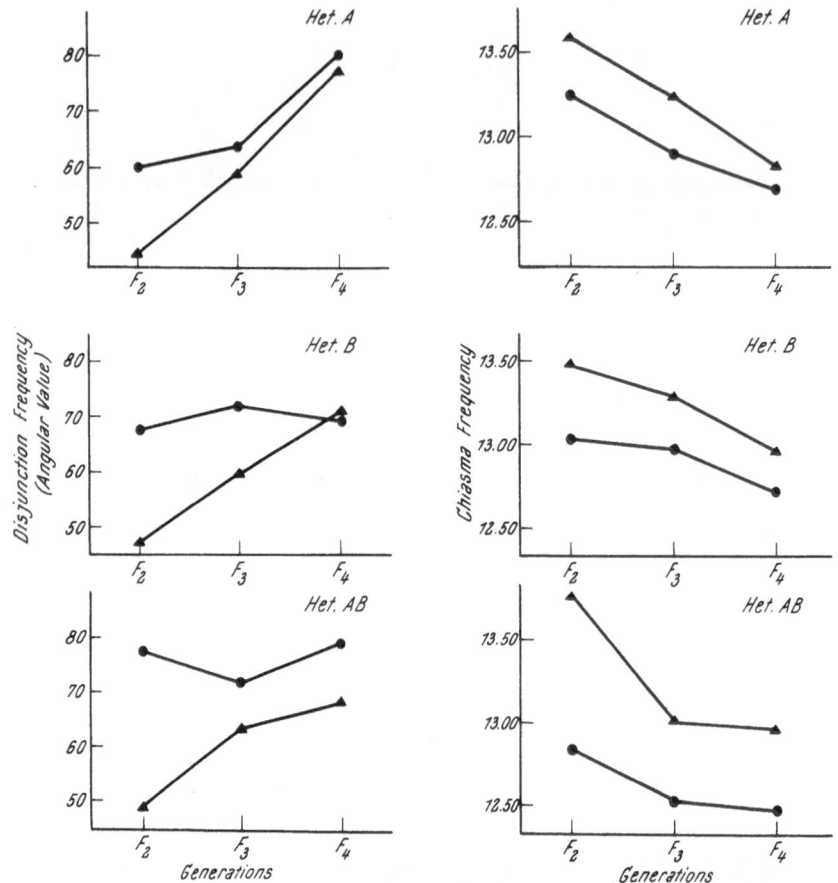

Fig. 89. The consequences of selection over three generations for disjunctional behaviour in rye interchange hetero-zygotes. Selection was practised in conjunction with forced inbreeding by self-pollination and the effect of this on chiasma frequency is shown alongside the effect on disjunction.   Note:
△ = Low selection line
◯ = High selection line
Unpublished data kindly supplied by Dr. SHONG SUN (see SUN 1963).

However, while rings involving equal-sized, isobrachial chromosomes and terminal chiasmata etc., are expected to show higher frequencies of alternate orientation, these do not appear to be sufficient for ensuring them. Further, associations which, on morphological grounds, are not expected to orientate dis-junctionally, frequently do so (BURNHAM 1956). Thus, LAWRENCE (1962) studied the behaviour of interchanges induced by pre-meiotic X-irradiation of four inbred lines of rye and their six first generation hybrids. The majority of the 882 inter-changes scored must, therefore, have been of independent origin. He found that:
(1) The disjunction frequency was high, varying between 68 and 82%,
(2) That of rings and chains did not differ significantly, and

(3) The frequency was not correlated with the proportion of paired arms having terminal chiasmata in cells without interchanges in the same line.

In all ten types, inbreds and hybrids, the chiasmata were close to the ends, the majority of the interchanges were equal and no obvious differences could be observed between families with regard to preferred points of exchange. But though the ten types did not differ in respect of the structurally and genotypically determined morphology of the multiples induced in them, the frequency of alternate orientation did vary significantly.

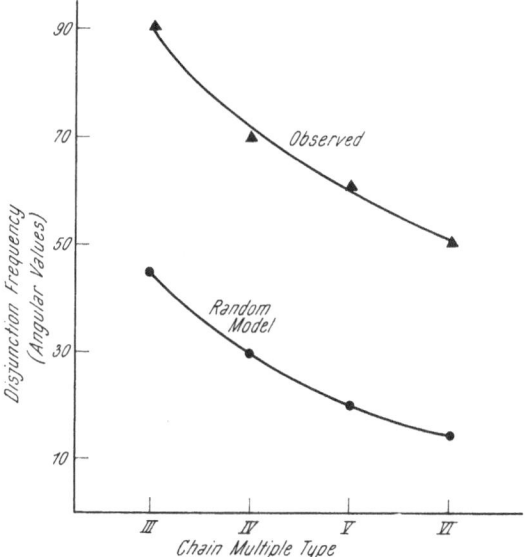

Fig. 90. Relationship between the average disjunction frequency and the number of chromosomes involved in interchange multiples recovered from species hybrids between *Secale anatolicum* and six other rye species. Unpublished data kindly supplied by Dr. Shong Sun (see Sun 1963).

Garber and Dhillon (1961), on the other hand, found that for interchanges induced by X-irradiation in *Collinsia heterophylla* ring multiples of four showed 70% zig-zag orientation compared with 90% for chains, while for associations of six the frequency of alternate orientation was 74% for rings and 96% for chains. The high frequency of alternate orientation shown by these irradiation-induced multiples is in marked contrast with those of 17 colchicine-induced interchanges in the same species (Soriano 1957). Further, while the interchange multiples of some inter-specific hybrids showed a high frequency of zig-zag orientation those of others did not (see Bell and Garber 1961).

Now, in both these comparisons, a lower frequency of zig-zag orientation was associated with a higher frequency of "bivalent" formation on the part of the chromosomes involved in interchange. It was suggested, therefore, that the variation in pairing and disjunction were due to differences in the points of exchange relative to the distal segments to which chiasmata are confined. Interestingly, the chromosomes in a multiple association had a higher chiasma frequency per chromosome than those not affected by interchange.

There is one final point worth mentioning about interchanges. On selfing or crossing a single interchange heterozygote, the expected products and their ratios are as shown in Table 24. And these expectations are certainly borne out in some interchanges. Thus Ahloowahlia (1926) has described an interchange between chromosomes 5 and 6 in rye where the segregating progeny of the heterozygote were in the ratio:

14 basic homozygotes : 32 interchange heterozygotes : 12 interchange homozygotes.

When compared with an expectation of $1:2:1$ the $\chi^2 = 0.77$ shows no significant departure between observed and expected ($P = 0.5$).

But these expectations are not always fulfilled. As DARLINGTON and LaCOUR (1950) first showed, when heterozygotes of the same interchange type in *Campanula* are crossed, the proportion of structural homozygotes obtained in the

Table 24.   *Expectations on Selfing or Crossing a Single Interchange Heterozygote.*

AB CD BC DA		♀ Gametes from Alternate Orientation	
		AB · CD	BC · DA
♂ Gametes from Alternate Orientation	AB · CD	AB CD · · AB CD basic homozygote	AB CD BC DA interchange heterozygote
	BC · DA	AB CD BC DA interchange heterozygote	BC DA · · BC DA interchange homozygote
Total expected progeny		1 basic homozygote: 2 interchange heterozygotes: I interchange homozygote	

progeny is less than the mendelian expectation since interchange homozygotes are totally eliminated and basic homozygotes are less frequent than expected. On selfing, all homozygotes are eliminated and only heterozygotes survive

Table 25.   *Progeny of Single and Double Interchange Rings-of-four in Crossed and Selfed Campanula persicifolia.*
(Data of DARLINGTON and LaCOUR 1950.)

Progeny	Parental Type	One Ring-of-four			Two Rings-of four		
		Basic Homozygote	Interchange Heterozygote	Interchange Homozygote	Basic Homozygote	Interchange Heterozygote	Interchange Homozygote
Expected		1	2	1	1	12	3
Observed	Crossing	32	97	0	3	18	0
	Selfing	0	35	0	0	15	0

(Table 25). The elimination of homozygotes in these experiments is due to the action of homozygous genes linked with the locus of interchange. The same principle is seen also in the interchange system of *Oenothera*. Indeed races of *Oenothera* breed true not because they are pure lines but rather because they are permanent heterozygotes.

d) Centric Fusion and Fission

When the two chromosomes involved in interchange are acrocentric and both breakage points occur close to the centromere, though in opposite arms, symmetrical union gives rise to a large and a small metacentric. This event is described as a centric fusion. With the loss of the small metacentric the homologies of the remaining chromosomes in the fusion heterozygote will lead to the production of a chain of three multiple. Indeed the production of multiples of this sort provides evidence for the occurrence of centric fusion and they have

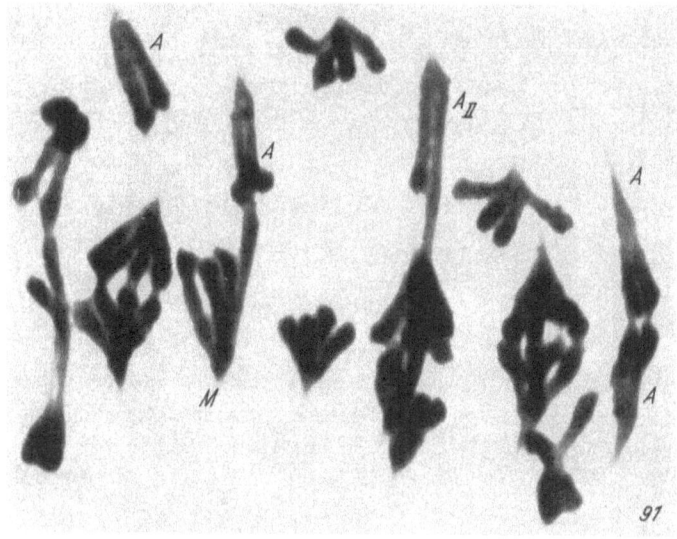

Fig. 91. Meta-anaphase-1 in *Northoscordum fragrans*. There are six metacentric pairs (two of which have separated) and two acrocentric pairs. In addition one metacentric element (M) is attached to a acrocentric while a second acrocentric is free. These three chromosomes form a multiple of three (carmine).

been observed in the whelk *Purpura lapillus* (STAIGER 1954) and the shrew *Sorex araneus* (FORD, HAMERTON and SHARMAN 1957).

Centric fusion thus leaves the total number of major chromosome arms within a complement unaffected and until recently it has been customary to assume that cases where arms show ROBERTSONIAN relationships of this sort have involved centric fusion. WHITE (1957), however, suggested that two of the acrocentric pairs in the 17-chromosome race of *Moraba scurra* are a product of the dissociation or fission of the large AB metacentric members of the 15-chromosome race. Mis-division, of course, is one type of fission process in which one centromere splits into two. WHITE's dissociation process is not, however, of this kind and he postulates a donor chromosome which supplies one new centromere and two new telomeres. Dissociation is thus a special type of translocation and though it increases the chromosome number in one sense it must be accompanied by a compensatory reduction in another sense and, as yet, really critical evidence for this does not exist. It is of course conceivable that B-chromosomes (see page 196) could serve as donors, for in this event no reduction in the standard complement would be involved.

An interesting example of the sort of problem created where ROBERTSONIAN relationships occur is found in *Northoscordum fragrans*. The chromosome complement of this species was first described by KOERPERICH (1930) who found eight pairs of large metacentric elements (2n = 16). Five years later, however, LEVAN reported a diploid complement of 2n = 18 for the same species including

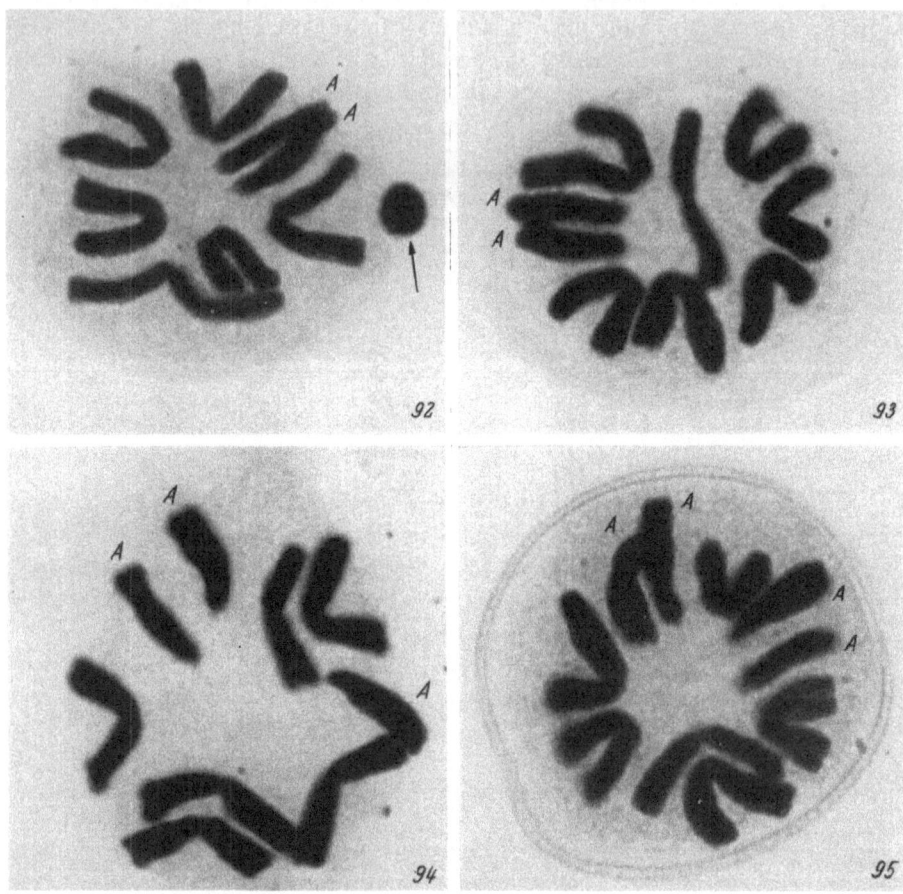

Figs. 92–95. First pollen grain mitoses in *Northoscordum fragrans*. Fig. 92—6M + 2A + micronucleus (arrow). Figs. 93—7M + 2A. Fig. 94—6M + 3A. Fig. 95—6M + 4A (carmine).

seven metacentric (L) elements and four acrocentric (S) ones, each S-chromosome being about the length of one of the arms of the metacentrics. Numerous authors (LEVAN and EMSWELLER 1938, SATO 1942, D'AMATO 1949, SATO and ASANO 1951, KURITA 1953 and TANDON and KAPOOR 1963) have subsequently and invariably observed 2n = 19 with thirteen metacentric and six acrocentric chromosomes (Fig. 91) and it would appear that over the last twenty-five years the complement has become stabilised as a structural heterozygote.

Whilst the history of observations on this type suggests an origin of acrocentric chromosomes from those with median centromeres (i. e. fission versus fusion) a direct origin by mis-division would seem to be excluded by the presence of short arms in the acrocentrics, the long arms of which pair with the meta-centric member. A donor would seem to be required here, as in *Moraba*, and while B-chromosomes have not been reported in *Northoscordum*, a member of the normal complement in the polysomic condition, could have served as a donor. Polysomic pollen grains are, in fact, common in this species (Figs. 92–95) and the extra-chromosomes are not confined to those involved in the multiple association of three. In fact an additional association of four is sometimes observed giving 1 IV + 1 III + 6 II.

Until recently it has been assumed that fusions could take place only by interchange. This view has, however, been challenged by observations of two kinds. First WOLF (1960) found that in male embryos of the ephemeropteran *Cloeon dipterum* the two telocentric sex chromosomes undergo a temporary and reversible fusion during the cleavage mitoses, a fusion which WOLF attributed to a simple association of centric heterochromatin. More recently MATTHEY (1963) has claimed a comparable, though irreversible, fusion in the autosomes of female hybrids obtained from a cross of *Acomys cahirinus* ♀ × *A. minous* ♂. Since this claim is based on hybrid material we shall leave a detailed discussion of it until section VII.

2. The Meiotic Behaviour of Abnormal Chromosomes

a) Dicentric Chromosomes

We have seen how dicentric chromosomes may arise at meiosis as a consequence of crossing-over in inversion hybrids or in duplications. They may also arise by breakage and reunion at meiosis, in the mitotic divisions of the pollen grain or the primary divisions of seedling roots. Finally, they can be induced both by chemicals and X-rays (Table 26).

At mitosis, dicentrics invariably orientate so that both centromeres lie on the spindle equator and, according to the length of the intercentric segment, may give parallel, criss-cross or interlocked separation (Fig. 96). It has often been assumed that, since criss-cross and interlocked separation lead to breakage, they will be fatal to the cells concerned. But, as McCLINTOCK (1941) first showed, the fractured ends can undergo repeated cycles of reunion so producing a breakage-fusion-bridge cycle leading to the persistence of the dicentric, though as an iso-dicentric (Fig. 97). Theoretically, of course, the dicentric can also persist as it arose given predominantly parallel separation.

Most of the dicentrics so far studied have been in plants and the majority of these have been confined to mitotic tissues for induced dicentrics rarely survive the hazards of the long vegetative period between germination and meiosis. Indeed they are usually eliminated in a few cell generations. In only three cases has the behaviour of the dicentrics at meiosis been described, viz., in the F_2-progeny of *Triticum aestivum* × *T. dicoccum* (MORRISON 1954b), in *Agropyron* (HAIR 1954) and *Triticum aestivum* (SEARS and CAMARA 1952). We shall deal with the two latter cases in some detail, for between them they illustrate most of the important aspects of dicentric behaviour.

(1) The *Agropyron* dicentric ($2n = 6x - 1 = 41$).—Here the dicentric (*LC*) occurred in the nucleolar organising chromosome and probably arose by an

inversion cross-over in an embryo-sac mother cell. It persisted by a breakage-fusion-bridge cycle giving variants with little or no intercentric segment, *SC* and *TC* respectively (Fig. 98). Loss of the greater part of the intercentric segment had no serious results for the dicentric or for the plant. Indeed the most successful product of breakdown was the *SC*-derivative which spread through an entire part of the clone.

Table 26. *The Occurrence of Dicentric Chromosomes.*
(After DARLINGTON and WYLIE 1953.)

Type		A. Induced	B. Spontaneous
1. Temporary	X-Rays	*Crocus* (2x) — MATHER and STONE (1933) *Pisum* (2x) — KOLLER (1934) *Allium* (2x) — SAX (1941) *Tradescantia* (2x, 3x) — SAX (1941) *Drosophila* (2x) — PONTECORVO and MULLER (1941)	Pollen grain divisions in many plants, e.g. *Tulipa* (2x) — UPCOTT (1937) Primary divisions in seedling roots, e.g., *Crepis* (2x) — NAWASCHIN (1933) *Allium* (2x) — NICHOLS (1941) *Pisum* (2x) — D'AMATO (1951) *Elymus farctus* — HEENIN (1963) Cleavage divisions of flatworm eggs — MELANDER (1963a) Ehrlich mouse ascites tumor — MELANDER (1963b)
2. Persistent		*Tulipa* (2x) — MATHER and STONE (1933) Maize (2x)[1] — McCLINTOCK (1942) Axolotl (2x) — DALTON and HALL (1950)	*Godetia* (3x)[1] — HÅKANSSON (1950) *T. aestivum* × *T. dicoccum* (F₂) — MORRISON (1954) *Narcissus* (3x) — DARLINGTON and WYLIE (1953)
3. Permanent	Chemicals	Rat tumour — KOLLER (1953)	*Scilla* (2x)[1,2] — BATTAGLIA (1949) *Triticum aestivum* (6x)[3] — SEARS and CAMARA (1952) *Agropyron* (6x)[1] — HAIR (1953)

[1] Iso-dicentric, i. e. both arms homologous.
[2] Effectively monocentric, i. e. two centromeres placed next to one another.
[3] Heterocentric and effectively monocentric, i. e. centromeres of unequal strength.

At meiosis both the *LC* and the *SC* were univalent and did not pair with the other chromosomes. However they are iso-chromosomes, one half being a mirror-image of the other, each half containing a centromere. Thus, they had the capacity for internal pairing. In 15 out of 60 cells examined (25%), the *LC*-univalent had a chiasma in the distal segments (i. e. outside the inter-centric region). This was followed by the co-orientation of the centromeres of the dicentric. It thus simulated a bivalent though, in a sense, it was a ring-univalent albeit a dicentric one.

At first anaphase the *LC* formed a 2-stranded bridge in all but one of 61 cells examined and subsequently the bridges broke at different places. The broken ends had the opportunity of joining after telophase but did not always do so because, in Hair's view, of the reduced period between the time of breakage and the beginning of the second meiotic division in this species. In those cases

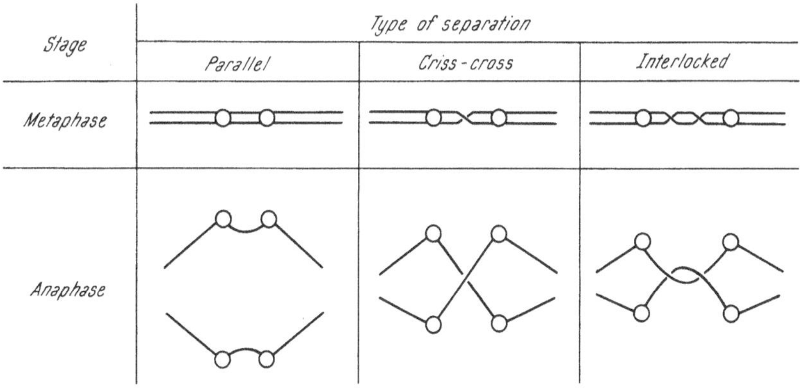

Stage	Type of separation		
	Parallel	Criss-cross	Interlocked
Metaphase			
Anaphase			

Fig. 96.   The behaviour of dicentric chromosomes at mitosis.

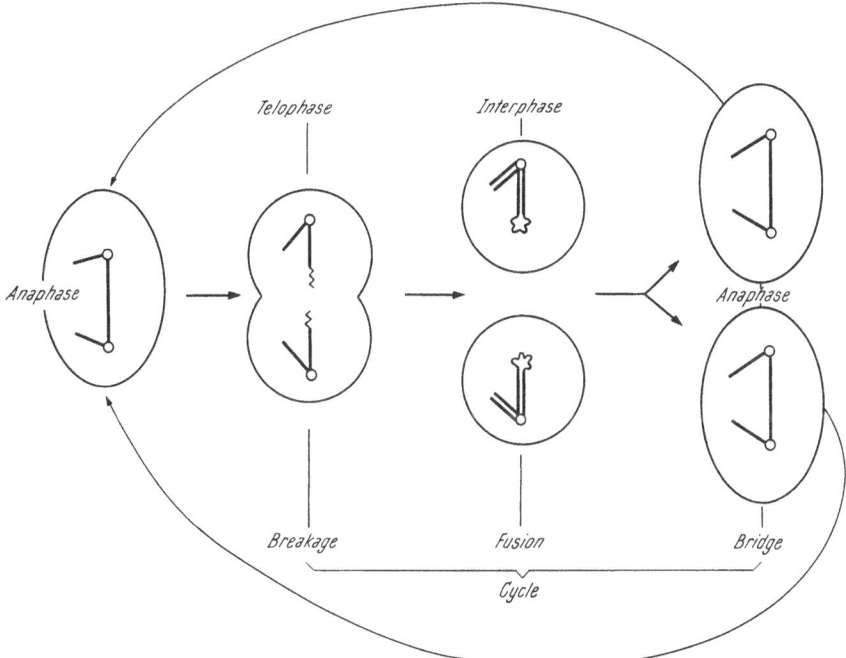

Fig. 97.   The breakage-fusion-bridge cycle.

where union did occur it would, of course, be followed by breakage at second anaphase.

The short dicentric almost always behaved as a univalent for, even when internal pairing took place, co-orientation failed, simply because the two centro meres were too close to give efficient co-orientation. And here, as in those cells where no inter-arm pairing occurred, the dicentric lay off the plate. Thus, the SC usually lagged and divided at first anaphase. Sometimes it passed intact to one pole or rarely was lost. In both *LC* and *SC* plants, therefore, the pollen was almost completely defective.

(2) The *Triticum* dicentric.—Though derived from an iso-chromosome for the short arm of chromosome VII, this wheat dicentric had two very unequal free arms and the centromeres also were of very different potencies. It was therefore a hetero-dicentric. But the influence of the submedian centromere usually overrode that of the subterminal one (Fig. 99).

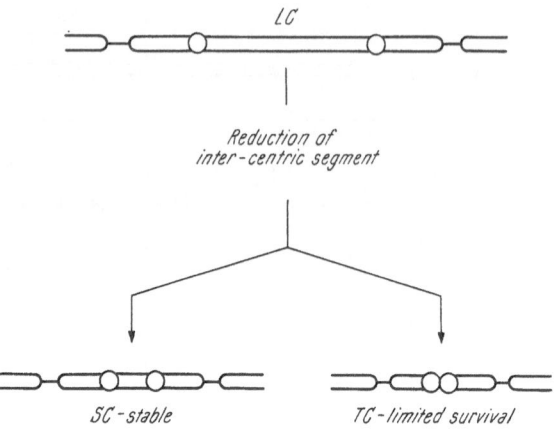

(i) Unpaired dicentrics at meiosis.—The two centromeres of the unpaired dicentric almost always opposed one another and so produced a co-orientation. But the dicentric ordinarily lay off the plate and towards that pole to which

Fig. 98. The new dicentric types produced by breakage and fusion in the LC-dicentric chromosome of *Agropyron* (HAIR 1954).

the median centromere was directed. At first anaphase the intercentric region became attenuated as the stronger, median centromere moved poleward but the secondary centromere made little if any polar progress and usually the entire dicentric was included in one telophase group. In about 1% of the first meiotic cells the dicentric moved onto the plate late and divided. When it failed to divide at first division it did so at the second and this division was usuall ynormal with sister half-centromeres being directed to opposite poles.

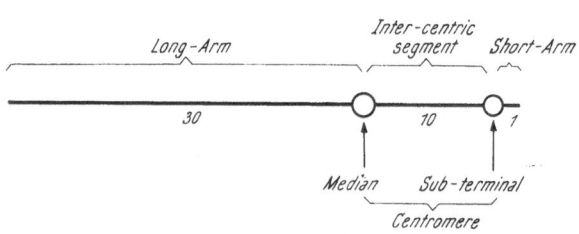

Fig. 99. The structure of the *Triticum* dicentric (SEARS and STEINITZ-SEARS 1953).

(ii) Pairing behaviour of the dicentric.—Several plants had two dicentrics and in 93% of the 200 PMC's classified the two were paired, most usually with a chiasma between the two long arms. When two unpaired dicentrics were present, both centromeres were active but where they were paired, only the median centromere of each chromosome showed activity.

The dicentric was also present in some other plants with a normal VIIth chromosome. Pairing occurred between them in only 59.5% of the 168 PMC's studied, the long arm of the dicentric pairing with the short arm of chromosome VII. In these bivalents again only the median centromere of the dicentric was active. But derived chromosomes, with only the secondary centromere, behave normally at second division. The weak centromere thus shows its weakness only when opposed by the stronger one.

In both these cases we have considered, indeed in all such cases, the dicentrics are present only as occassional anomalies in otherwise monocentric forms. No case has yet been established where the dicentric condition characterises a species though, as we shall see, such a claim has been made for *Luzula purpurea*, scorpions and the *Hemiptera-Homoptera* (see section VI). It has been made also for the Y-chromosome of *Forficula auricularia* (Callan 1941, see section VIII-2).

b) Ring-chromosomes

The behaviour of ring chromosomes at meiosis has been relatively well studied in only three organisms. The study in maize deals with an X-ray-induced ring (Schwartz 1953, 1955) whereas the ring in *Antirrhinum* was spontaneous in origin (Michaelis 1958). Finally in *Drosophila* a variety of rings have been synthesised experimentally (Morgan 1933, Novitski 1955).

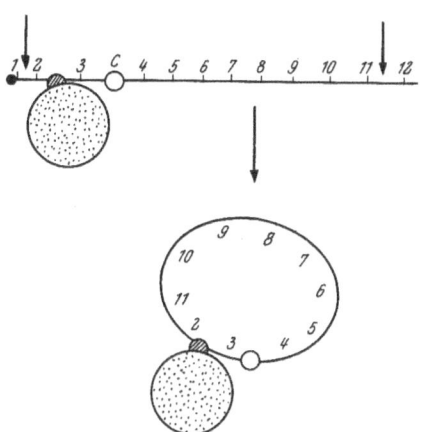

Fig. 100. The origin of Schwartz's X-ray induced ring chromosome-6 in *Zea mays*. The nucleolus is stippled and its organiser cross-hatched. The centromere (c) is a clear circle.

Most of the spontaneous rings found in maize have been quite small. Consequently they survive only in trisomic plants which carry two normal representatives of the chromosome from which the ring originated. In such circumstances the rings behave as univalents at meiosis and show little or no association with the homologous rods. By selection of the progeny of X-irradiated plants, however, Schwartz (1953) was able to isolate a ring involving almost the whole of chromosome 6 (Fig. 100), so that gametes carrying the ring in addition to the other nine maize chromosomes were viable. Pachytene configurations of such ring-rod chromosome 6 heterozygotes show close pairing between these two chromosomes throughout most of their length.

A variety of possible anaphase configurations can be produced by different patterns of crossing-over in such ring-rod associations. Some of these lead to loss of the ring, others to the production of single or double bridges at first or second anaphase. In Schwartz's case chromosome 6 is acrocentric and, since crossing-over occurs very infrequently in the short arm, the expectations following crossing-over in the long arm are reasonably simple (Fig. 101). Thus:

(a) Single cross-overs produce single anaphase-I bridges only.

(b) Two strand doubles do not produce bridges at either anaphase.

(c) Three strand doubles involving two non-ring and one ring chromatid produce single bridges at first anaphase.

(d) Three strand doubles involving two ring and one non-ring chromatid produce single bridges at first and second anaphase, and

(e) Four strand doubles produce double bridges at anaphase-I.

The four kinds of double exchange should be equally frequent if no chromatid interference occurs and, in normal maize at least, the available evidence indicates that no chromatid interference does occur.

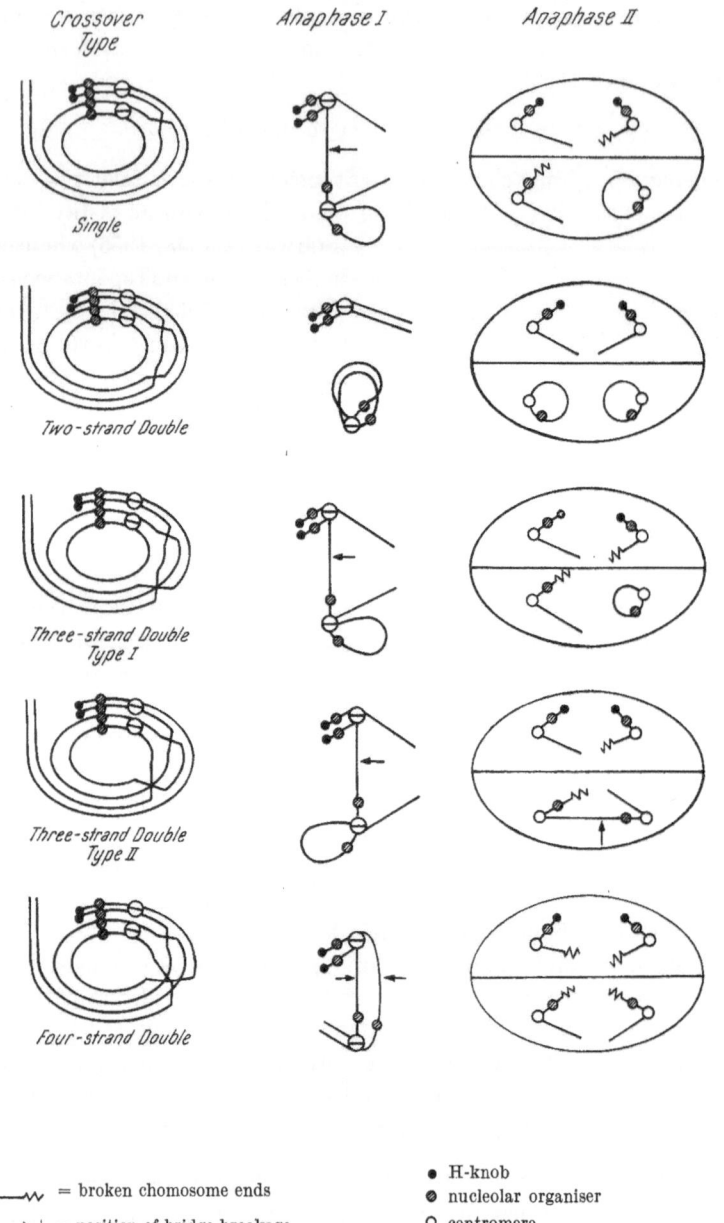

Crossover Type — *Anaphase I* — *Anaphase II*

Single

Two-strand Double

Three-strand Double Type I

Three-strand Double Type II

Four-strand Double

_____�_ = broken chomosome ends

→| = position of bridge breakage.

● H-knob

◉ nucleolar organiser

○ centromere

Fig. 101. Anaphase configurations resulting from crossing-over between the ring chromosome-6 and its normal homologue in *Zea mays*. (SCHWARTZ 1953.)

Summarising, on the basis of single and double cross-overs only, one would expect the same frequency of single bridges at anaphase-II (d) to double bridges

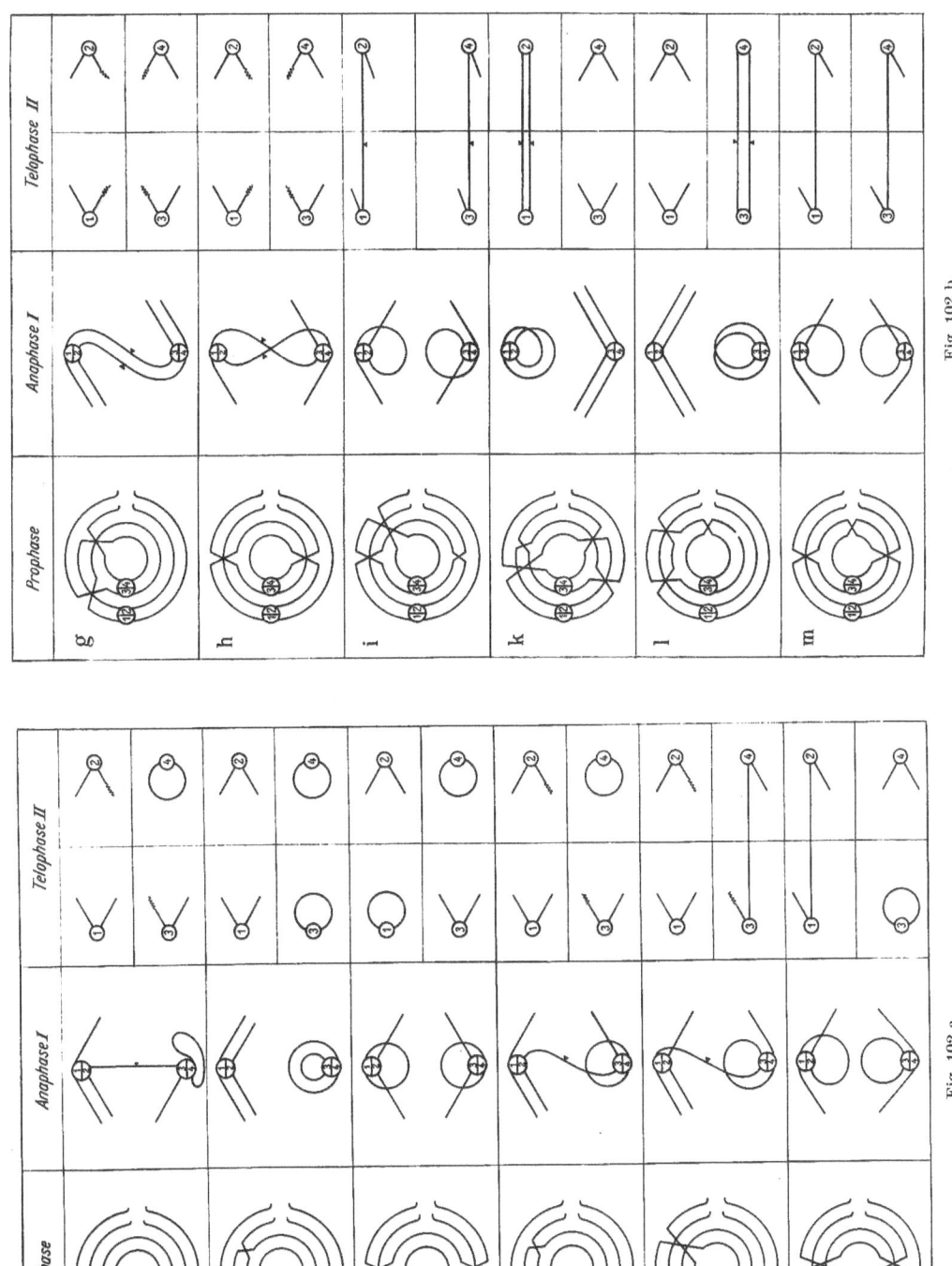

Fig. 102 b.

Fig. 102 a.

at anaphase-I (e), while single anaphase-I bridges (a + b + d) should be three times as frequent as double bridges at anaphase-I (e). But these expectations were not realised. In the first place, single anaphase-II bridges were much more common than double bridges at anaphase-I. Secondly, double bridges at anaphase-II are not expected from either single or double cross-overs and from only 1/16th of triple cross-overs. Yet they were found in 10% of the second division tetrads. Finally single anaphase-I bridges were more common than expected by comparison with double bridges at first anaphase (Table 27).

Table 27. *Meiotic Anaphase Configurations Observed in Ring-rod Heterozygotes.*

Data		Anaphase-I				A-II (daughter cell pairs)				
		Single Bridge	Double Bridge	No Bridge	Total	Single Bridge	Two Single Bridges	Double Bridge	No Bridge	Total
1. *Zea mays* (SCHWARTZ 1953)										
Totals from 4 plants	No.	368	81	171	620	166	0	47	262	475
	%	59	13	28	100	35	0	10	55	100
2. *Antirrhinum majus* (MICHAELIS 1958)										
Plant no. H 1		1567	39	1346	2952	927	219	16	1059	2221
H 3		78	5	53	136	258	29	8	273	571
H 5		121	5	108	234	194	34	6	173	407
H 8		237	12	173	422	189	36	8	218	451
Totals from 4 plants	No.	2003	61	1680	3744	1568	318	38	1726	3650
	%	53.5	1.6	44.9	100.0	43.0	8.7	1.0	47.3	100

In *Antirrhinum* the situation is rather more complex since the chromosome involved in the formation of the ring is a metacentric. The expectations are thus more numerous since cross-overs may occur in either or both arms. Thus there is a completely new class involving the production of two single bridges at anaphase-II (Fig. 102). But, like SCHWARTZ, MICHAELIS claimed clear departures from expectation (compare Tables 28 and 27), thus:

(1) Only about 1/10th as many double anaphase-II bridges occur as are expected.

Fig. 102. The meiotic consequences of crossing-over in a metacentric ring-rod heterozygote of the type found by MICHAELIS (1959) in *Antirrhinum majus*. Note: a—Single cross-over in one arm only; b—2-strand double in one arm only; c—Single cross-overs in each of the two arms involving the same strands in both cases; d—3-strand double in one arm only (type I); e—3-strand double in one arm only (type II); f—Single cross-overs in each of the two arms involving one common and one distinct chromatid; g—4-strand double in one arm only; h—Single cross-overs in one of each of the two arms involving distinct strands; i—One of the types involving a 3-strand double in one arm combined with a single cross-over in the other arm, leading to the production of two single bridges at second anaphase; k—One of the types involving double cross-overs in both chromosome arms which leads to the production of double anaphase-II bridges; l—Production of a double anaphase-II bridge following a 2-strand double cross-over in one arm combined with a sister-strand cross-over between the two ring chromatids; m—Production of two single anaphase-II bridges following a single cross-over in each of the two chromosome arms combined with a single distal sister-strand cross-over between the two ring chromatids; ▼ = position of bridge breakage.

Table 28. *Expected % Anaphase Configurations Following Various Cross-over Patterns in Metacentric-Derived Ring Heterozygotes in Antirrhinum majus.*
(After Michaelis 1958.)

No. of Xta			A I			A II			
In One Arm	In the Other Arm	Sister-strand Cross-overs	Single Bridge	Double Bridge	No Bridge	Single Bridge	Two Single Bridges	Double Bridge	No Bridge
1	—		100	...	...	...	...	...	100
2	—		50	25	25	25	...	...	75
3	—	Absent	75	12.5	12.5	25	...	6.25	68.75
1	1		...	50	50	25	...	...	75
1	2		62.5	18.75	18.75	25	6.25	...	68.75
2	2		50	25	25	31.25	3.125	3.125	62.5
1	—	1 s/s cross-over in ring	50	...	50	50	...	...	50
2	—	1 s/s cross-over distal in ring	25	50	25	25	...	25	50
2	—	1 s/s cross-over proximal in ring	25	50	25	25	...	25	50
2	—	1 s/s cross-over between the two non-sisters	50	25	25	25	...	25	50
1	1	1 s/s cross-over distal in ring	...	50	50	25	25	...	50
1	1	1 s/s cross-over proximal in ring	...	50	50	25	25	...	50
1	2	1 s/s cross-over distal in ring	50	25	25	50	6.25	...	43.75
1	2	1 s/s cross-over proximal in ring	56.25	25	18.75	43.75	6.25	...	50
1	2	1 s/s cross-over between the non-sisters	50	25	25	50	6.25	...	43.75
2	2	1 s/s cross-over distal in ring	50	25	25	31.25	3.125	9.375	56.25

Table 29. *Theoretical Ratio of Anaphase Configurations Resulting from Double (2-, 3- and 4-strand) Non-sister Strand Cross-overs Associated with Sister-strand Crossing-over.*
It is assumed that no chromatid interference occurs and in making the calculations all possible combinations of the four non-sister double cross-overs have been taken into account.
(After Schwartz 1953.)

No. of Sister Ring Chromatid Cross-overs	Single Bridge at A I only	Single Bridge at A I and A II	Double Bridge at A I	No Bridge at A I		Total
				No Bridge at A II	Double Bridge at A II	
0	1	1	1	1	0	1
1	1	1	1	0.5	0.5	1
2	1	1	1	0.5	0.5	1
3	1	1	1	0.5	0.5	1
Many	1	1	1	0.5	0.5	1

(2) The frequency of single anaphase-II bridges is about 1½ times the expected value, and

(3) The proportion of cells without bridges is higher than expected.

Now the precise ratios of anaphase configurations produced in ring-rod heterozygotes will depend on the precise combination of cross-over types which

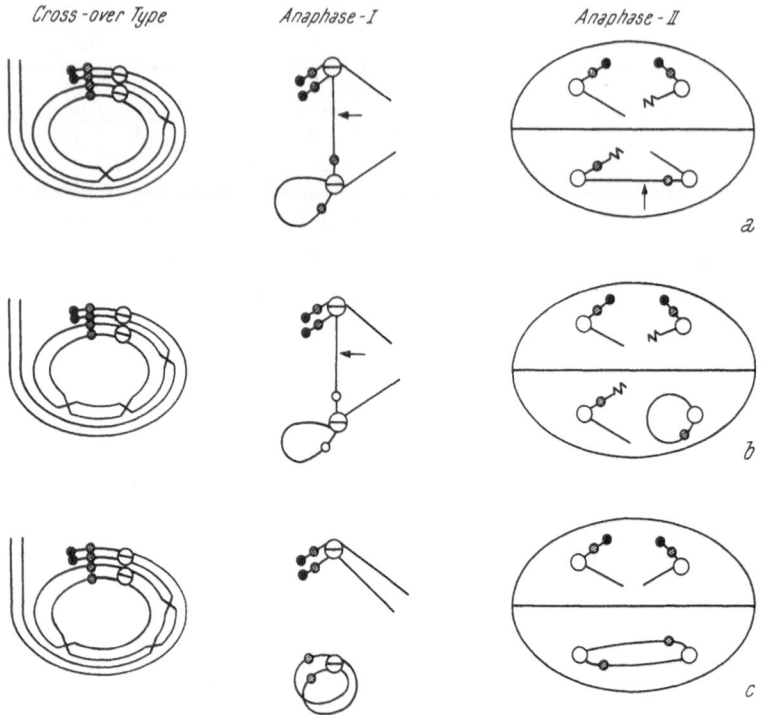

Fig. 103. Anaphase configurations resulting from a combination of non-sister and sister cross-overs in the rod-ring-6 heterozygote of *Zea mays*. Note: *a*—Production of single bridges at anaphase-I and anaphase-II following a single non-sister cross-over associated with a sister-strand cross-over between the ring chromatids; *b*—production of a single anaphase-I bridge following a single non-sister cross-over associated with two sister-strand cross-overs between the ring-chromatids; *c*—production of a double anaphase-II bridge following a 2-strand double non-sister cross-over associated with a single sister-strand cross-over between the ring-chromatids; ——⌇ = broken chromosome ends →| = position of bridge breakage.

occur. And to determine these, diplotene observations are required. In *Antirrhinum* this is not possible since the diplotene chromosomes pass into a diffuse stage (see pg. 24) and in maize no such observations were attempted. Instead both SCHWARTZ and MICHAELIS argued that their discrepancies could be accounted for by assuming that crossing-over could occur between sister, as well as between non-sister, chromatids (see section II–1 pt. II). Thus in the maize case (Fig. 103):

(1) A single sister-strand exchange between ring chromatids in conjunction with a single non-sister cross-over produces a single bridge both at anaphase-I and at anaphase-II.

(2) Two sister-strand exchanges, involving ring-chromatids, accompanying a single non-sister cross-over leads to the production of a single first-anaphase bridge, and

(3) A single sister-strand exchange between ring-chromatids together with a 2-strand double non-sister exchange gives a double bridge at second anaphase.

The full theoretical ratios expected from sister-strand crossing-over associated with double non-sister exchanges are summarised in Table 29. Clearly sister-strand exchange in the ring can lead to an increase in single anaphase-II bridges, can explain the production of double anaphase-II bridges and the excess of single bridges at first anaphase over double anaphase-I bridges.

In a like manner the occurrence of sister-strand exchanges in *Antirrhinum*, in conjunction with non-sister exchanges, also go some way towards bringing the observed frequency of anaphase bridges into approximate correspondence with the theoretical expectation of the author. But it does not explain the low value of the double anaphase bridges (Table 27).

By contrast with this work in maize and *Antirrhinum*, studies on ring X-chromosomes of *Drosophila* show no suggestion of sister-strand exchange or rather, it would be more accurate to say that the *Drosophila* studies do not require sister-strand exchange.

V. Meiosis in Numerical Mutants

1. Haploids

In animals, spontaneously occurring haploid individuals usually die in early organogeny. In flowering plants, on the other hand, haploids occur extensively

Table 30. *Chromosome Behaviour in*

Species		Configuration Type					
	All I's	1. II	2. II	3. II	4. II	1. III
1. *Hordeum spontaneum* (n = 7)	322	49	1	...	...	1
Tsuchiya (1962)	86.1	13.1	0.26	...	...	0.26
2. *Oenothera blandina* (n = 7)	938	219	17	...	...	11
Catcheside (1932)	79	18.4	1.4	...	...	0.9
3. *Oryza glaberrima* (n = 12)	265	94	21	3	...	12
Hu (1960)	65.8	23.3	5.2	0.7	...	2.9
4. *Antirrhinum majus* (n = 8)	3247	3053	915	164	21	159
Rieger (1957)	42.34	39.82	11.94	2.14	0.27	2.07

in pure lines, field cultures and among hybrid populations from both intra- and inter-specific crosses. Thus they are known among species of at least seven different families, viz., *Solanaceae, Gramineae, Compositae, Malvaceae, Cruciferae, Portulaceae* and *Onagraceae*. Crossing distantly related species seems especially to promote the occurrence of haploids. It would appear, therefore, that while disharmony between the gametes prohibits or impedes fertilisation it never-theless stimulates the unfertilised egg cell to develop. Alternatively the presence of two embryo sacs in one ovule, and perhaps too the presence of two egg cells in one embryo sac, may lead to polyembryony because the fertilisation of one apparently creates conditions favourable for the development of the other without the necessity of fertilisation (IVANOV 1938).

(a) Monohaploids.—In all the haploid plants of diploid species studied to date, meiosis in the PMC's is irregular because many or indeed all the chromosomes remain univalent, having no partner with which to pair. Complete lack of pairing is known, for example, in haploids of *Datura stramonium, Lycopersicum esculentum, Crepis capillaris, Nicotiana glutinosum* and *Zea mays*. In other cases, however, chromosome pairing does occur, though its extent and frequency varies considerably (Table 30). The most marked pairing found to date is in a haploid of *Antirrhinum majus* ($2n = x = 8$) which is characterised by 57.6% pairing between chromosomes, with associations of up to V and, in the extreme case, 4 bivalents (RIEGER 1957).

Pairing of this sort is understandable where haploids arise from auto- or allo-polyploids, i. e., in polyhaploids, since they are then likely to have homologous or at least homoeologous members in the haploid complement (see section V–3). Thus haploids of *Solanum nigrum* consistently form $12 \, \text{II} + 12 \, \text{I}$ at meiosis. But when haploid plants of diploid origin show pairing there are only two possible explanations. Either interchromosomal duplications exist within the com-

Some Haploid Plants of Diploid Origin.

and Frequency										Total Cells Analysed
1. III + 1. II	1. III + 2. II	2. III	2. III + 1. II	1. IV	1. IV + 1. II	1. IV + 2. II	1. V	1. V + 1. II		
...	...	...	1	...	...	...	...	...	No.	374
...	...	...	0.26	...	...	...	...	...	%	
1	...	...	...	1	...	...	...	...	No.	1187
0.1	...	...	...	0.1	...	...	...	...	%	
4	2	1	1	...	...	...	...	...	No.	403
1.0	0.5	0.3	0.3	...	...	...	...	...	%	
71	8	8	...	15	4	1	1	1	No.	7668
0.93	0.10	0.10	...	0.20	0.05	0.01	0.01	0.01	%	

plement or pairing is due to secondary association of a non-homologous kind (see section VII). A number of authors have come out in favour of the former interpretation. Thus CATCHESIDE (1932) claimed that the pairs he found in haploid *Oenothera blandina* were genuine bivalents. Likewise RIEGER (1957) claimed that about 40% of the associations found by him in *Antirrhinum majus* were

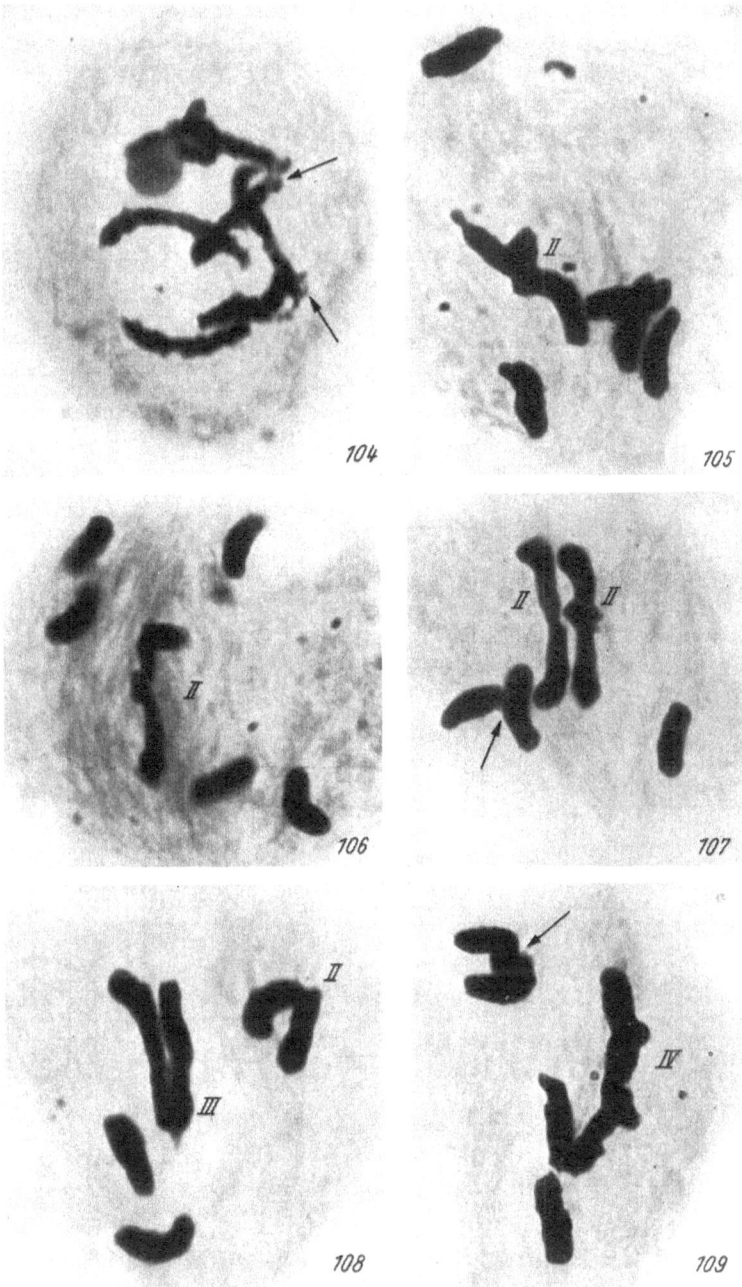

Figs. 104–109. Meiosis in haploid rye (carmine, ca. × 1500). Fig. 104—Prophase showing seven univalent chromosomes and the nucleolus. Note association of chromosome ends (arrows). Figs. 105–109—"Metaphase-I" showing various univalent associations. In 105 and 106 there is a single association of two univalents. 107 shows three such associations, one of which (arrow) involves a centromere -end (c-e) relationship. In 108 there is an association of II and one of III while in 109 there is a (c-e) relationship (arrow) and a compound arrangement of four univalents. Compare Fig. 105 with Fig. 66c and 107 with 66b and c. Photographs from a slide kindly loaned by Dr. Hubert Rees.

the result of chiasma formation between inter-chromosome duplications. The remainder were held to be due either to the same cause or else to an exchange between non-homologous segments though RIEGER did not exclude the possibility of association by stickiness.

Table 31. *A Comparison of the Numbers of Unreduced Meiotic Products Expected on the Basis of Random Anaphase Movement of Univalents in Haploids with those Actually Observed.*
(After IVANOV 1938.)

Species	n	Expected $\frac{1 : 2n - 2}{2}$	Observed	Reference
1. *Triticum monococcum*	7	1 : 63	6 : 500 ≡ 1 : 83.3	KATAYAMA (1935)
2. *Oenothera franciscana*	7	1 : 63	14 : 244 ≡ 1 : 17.4	EMERSON (1929)
3. *Oenothera blandina*	7	1 : 63	2 : 110 ≡ 1 : 55	CATCHESIDE (1932)
4. *Nicotiana glutinosa*	12	1 : 2,047	115 : 3291 ≡ 1 : 28.6	WEBBER (1933)
5. *Triticum aestivum*	21	1 : 1,048,575	2 : 300 ≡ 1 : 150	YAMAMATO (1936)
6. *Nicotiana rustica*	24	1 : 8,388,607	24 : 596 ≡ 1 : 24.8	IVANOV (1938)

In rye, as in *Antirrhinum*, extensive pachytene pairing has been held to occur in the haploid state (LEVAN 1940). But LEVAN regards most of this as non-homologous in character. Thus at zygotene the chromosomes are usually paired over only 50–80% of their length and foldback pairing may occur (see

Table 32. *Distribution of Univalent Chromosomes in Haploid Datura* (n = 12; data of BELLING and BLAKESLEE 1927).
For the Purpose of the χ^2-analysis the First Two Classes (11—1, 10—2) Have Been Pooled.

Item	Anaphase-I Distribution						Total Cells
	11—1	10—2	9—3	8—4	7—5	6—6	
Observed	1	3	12	19	38	27	100
Random expected	1	3	11	24	38	23	100

$$\chi^2_{[4]} = 1.828, \ P = 0.8 - 0.7.$$

section VII). A small number of the chromosomes paired at diakinesis were, however, claimed to contain chiasmata some of which were held to be asymmetrical. Through the generosity of Dr. HUBERT REES of the Department of Agricultural Botany, University College of Wales, Aberystwyth, we have had the opportunity of examining a preparation of haploid rye (Figs. 104–109). We have seen associations of II, III, IV and occasionally V but in no case have we found any discernible chiasmata. We have already seen (section III-3) that univalents produced by asynapsis may frequently give rise to quasi-bivalents which represent sticky non-chiasmate associations. And we have also seen that such quasi-bivalents co-orient just like ordinary bivalents. Indeed the appearance of the paired associations we have found in haploid rye are identical to the quasi-bivalents figured by Östergren and Vigfusson in asynaptic rye (compare Figs. 105–109 with Fig. 66). All the pairing associations we have found are near terminal and, significantly, rye chromosomes have heterochromatic end segments which frequently come together at zygotene. Thus in haploid rye we have seen quite unmistakable non-specific associations of end segments at pachytene (Fig. 104). Centromeres and ends are also known to show a non-specific and mutual attraction (Ribbands 1941, see also section VII) and some of the associations in haploid rye are unquestionably of this sort (Fig. 107 left).

Table 33. *Pairing and Association in*
(Data of Riley

Type	Origin	No. of Cells Examined
1. Haploid *Aegilops longissima* (2n = x = 7)	A parthenogenetic product from the cross *A. longissima* × *T. durum*	100
2. Polyhaploid *Triticum timopheevi* (2n = 2x = 14)	One of twin embryos produced in the cross *T. timophevi* × *T. monoccocum*	60
3. Polyhaploid *Triticum aestivum* (2n = 3x = 21)	One of the progeny from a rye chromosome III addition line	104

It is true that some associations look more regular and superficially give the appearance of interstitial chiasmata (e. g., Fig. 107 right). But critical examination of these shows that they are not in fact chiasmate. Anyone familiar with rye cytology will surely look with caution on Levan's so-called chiasmata (his Figs. 3*p* and *r*). Chiasmata of this quality cannot be demonstrated in paired bivalents with modern techniques. We are thus unconvinced by any of the claims for genuine pairing and chiasma formation not only in rye but in all other haploids of diploid origin, i. e., monohaploids, many of which are far less suitable for critical cytological study than rye. We are equally unconvinced by claims for chiasma formation between non-homologous segments which some have held possible (Alexander 1956). Critical evidence is lacking in support of both these claims.

The persistence beyond prophase of non-homologous associations in haploids may seem surprising at first sight since most forms of non-homologous pairing lapse before metaphase-I. It should be remembered, however, that the prophase of meiosis in haploids is not easily described in terms of the stages generally employed for diploids. It is even difficult to distinguish first metaphase of meiosis from first anaphase for no clear plate is formed. Moreover, the lack of internal balance, which is presumably one of the reasons why haploids are not common, is likely also to lead to disturbances of meiosis. And stickiness is a common meiotic abnormality.

As far as the fate of the univalent chromosomes of haploids is concerned, in some cases they divide at first anaphase. Thus in *Oenothera* and *Matthiola incana* such division appears to be quite common. In other cases, however, it appears to be exceptional, e. g., *Nicotiana tabacum* and *Crepis capillaris*. Those univalents which do not divide at this time are generally believed to pass at random to the poles. On this basis the probability of obtaining a haploid gamete through chance distribution is expected to be $1 : \dfrac{2^n - 2}{2}$, n being the haploid number.

Related Haploids and Polyhaploids.
and CHAPMAN 1957.)

Freq. Bivs./cell					Mean Prop^n. of the Complement as			Mean Association of Univalents	
0	1	2	3	4	I's	II's	III's	s — s	e — e
0.89	0.09	0.02	...	...	0.96	0.04	0.00	0.21 ± 0.13	0.15 ± 0.11
0.77	0.21	0.02	...	...	0.96	0.04	0.00	0.45 ± 0.08	0.32 ± 0.07
0.15	0.45	0.28	0.10	0.02	0.86	0.13	0.01	1.32 ± 0.12	0.75 ± 0.07

Where adequate investigations have been made, however, there appears to be a considerable discrepancy between expectation and observation (Table 31). These discrepancies can be explained by the fact that the first meiotic division is frequently suppressed. Thus in haploid *Datura* there can be no doubt that the distribution of chromosomes at first anaphase is at random (Table 32). Nevertheless up to 12% of cells undergo non-reduction leading to the production of two cells each of which have the normal gametic number of chromosomes.

Few studies have been made on meiosis in the megaspore mother cell in haploids but all the indications are that the first meiotic division proceeds in essentially the same manner as in the PMC's and consequently, only very few of the megaspores are able to develop into embryo sacs.

(b) Polyhaploids.—One of the most revealing studies on the meiosis of polyhaploids was carried out by RILEY and CHAPMAN (1957). They studied "haploids" of diploid, tetraploid and hexaploid types in the closely related genera *Aegilops* and *Triticum* (Table 33). A number of points of interest emerge from this work. First, when the amount of pairing is considered as a proportion of the chromosome number, there are more than three times as many bivalents in the polyhaploid *Triticum aestivum* (2 n = 3 x = 21) as in the haploid *Aegilops longissima* (2 n = x = 7) and the polyhaploid *Triticum timopheevi* (2 n = 2 x = 14). But there is no difference between the two latter. Likewise trivalents were found only where inter- and intra-genome homologies permitted.

Secondly, univalents were usually associated in pairs and such associations were predominantly of two types:

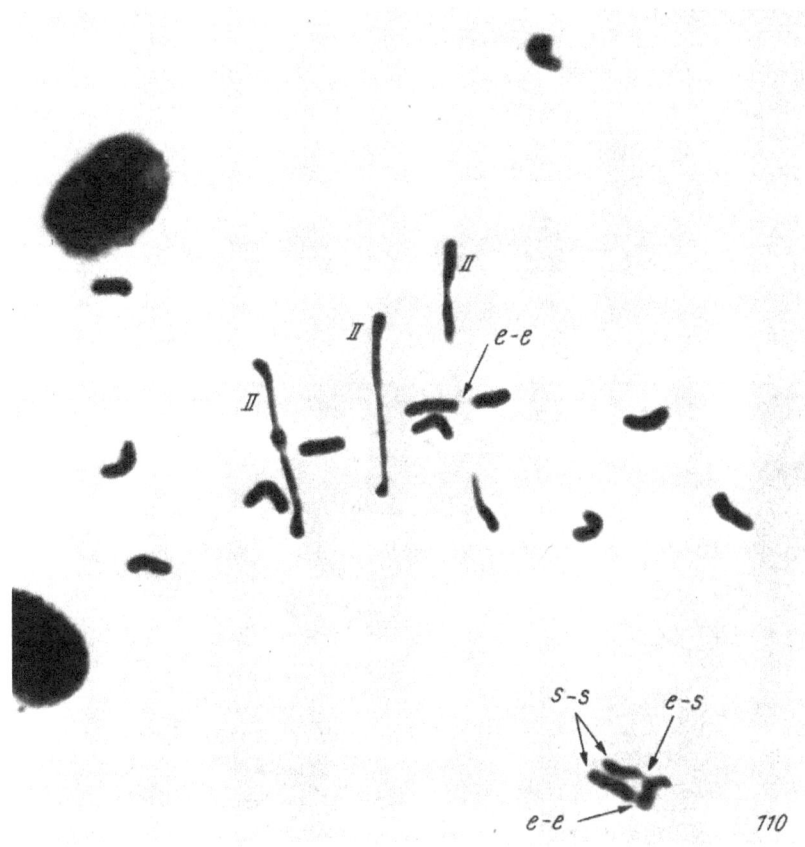

Fig. 110. A cell of polyhaploid *Triticum aestivum* (2n = 3x = 21). Three bivalents are present, one of which has disjoined. Two univalents at the equator show an end-to-end (e-e) association with a third univalent in close proximity. There is also a group of three univalents (bottom right) showing a combination of e-e, side-to-side (s-s) and side-end (s-e) associations. Photograph kindly supplied by Dr. RALPH RILEY (see RILEY and CHAPMAN 1957).

(1) Side to side (s-s), where the two univalents lay alongside and parallel to one another or else similarly flexed.

(2) End to end (e-e), where the two univalents lay in tandem, sometimes connected by a single chromatic thread.

In the *T. aestivum* polyhaploid (2 n = 3 x = 21) there were occasional multi end-to-end associations, a few end-to-end pairs and mixed multiple associations combining various arrangements of s-s, e-e and e-s types (see Fig. 110). There were also significant negative correlations between the numbers of chromosomes in s-s and chiasmate associations and between the numbers in s-s and e-e associations. Similar, though less marked correlations existed in the two other types

(Table 34). RILEY and CHAPMAN interpreted this to mean that chromosomes which frequently participate in chiasmate associations may also, on different occasions, be found in frequent s-s associations. But, though chromosomes which are frequently found in e-e arrangements may sometimes participate in s-s associations, they rarely form chiasmate associations. This interpretation is supported by the fact that an aneupolyhaploid ($2n = 3x + 1 = 22$), having a rye chromosome-III in addition to the standard haploid $T.$ *aestivum* complement, showed a significant increase in the number of e-e associations but the mean number of s-s associations was not significantly higher than in the polyhaploid $T.$ *aestivum*. Since the number of e-e associations may be increased by the pres-

Table 34. *Relationship Between Association Types in Related Haploid and Poly-haploid Types.*
(Data of RILEY and CHAPMAN 1957.)

Type	d.f.	X ta − s/s		X ta − e/e		s/s − e/e	
		n	P	n	P	n	P
1. Haploid, *A. longissima* ($2n = x = 7$)	98	− 0.0474	0.7 − 0.6	− 0.0703	0.5 − 0.4	− 0.1478	0.2 − 0.1
2. Polyhaploid, *T. timopheevi* ($2n = 2x = 14$)	58	− 0.1451	0.3 − 0.2	− 0.0851	0.6 − 0.5	− 0.0279	0.9 − 0.8
3. Polyhaploid, *T. aestivum* ($2n = 3x = 21$)	102	− 0.2810	0.02 − 0.01	− 0.1184	0.2 − 0.1	− 0.2202	0.05 − 0.02
4. Aneupolyhaploid, *T. aestivum* ($2n = 3x + 1 = 22 = 21 +$ 1 rye chromosome III)	38	− 0.0293	0.9 − 0.8	− 0.1384	0.4 − 0.3	− 0.5332	< 0.001

ence of an additional non-homologous chromosome, this type of pairing cannot depend on structural similarity (But see pg. 153 *et seq.*).

Thus participation in s-s associations tends to be between segmentally homologous partners. By contrast e-e arrangements are formed predominantly between non-homologous chromosomes, as was first claimed by PERSON (1955). Indeed RILEY and CHAPMAN suggest that e-e associations in $T.$ *aestivum* are due to the fusion of heterochromatic regions since both terminal and intercalary heterochromatic segments can be demonstrated in this species by cold-treatment.

The observations of RICHARDSON (1935), RIBBANDS (1937) and ÖSTERGREN and VIGUFUSSON (1953) showed that where chiasma formation fails after successful zygotene pairing the two chromosomes have a positional advantage in any subsequent secondary association which ensues (see section III-3). In these terms, fusions in e-e and e-s pairs would take place between segments of chromosomes which lie adjacent by chance, whereas s-s would follow chiasma failure following homologous pairing.

The mechanism which distributes chromosomes during division is also subject to errors. These are of two principal kinds and so two major kinds of numerical variants are defined. First, individual chromosomes may misbehave in the company of their well-behaved brethren on an apparently normal spindle

(Fig. 111), so leading to states of aneuploidy where different chromosomes of the basic set are represented by different numbers. Thus in 137 PMC's of normal diploid *Datura*, Belling and Blakeslee (1924) found 8 cases of 11–13 disjunction so producing about 0.4% of (x + 1) pollen grains. Second, the spindle itself may fail to form and function in which case the whole complement will be subject to numerical modification producing a polyploid condition. Polyploidy may also result from meiosis being replaced by an essentially mitotic mechanism in which a normal spindle is developed.

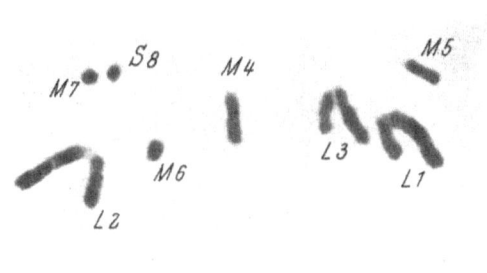

These terms, aneuploid and polyploid, may be applied to cells, tissues, individuals, races or species whose nuclei show such variation.

2. Aneuploids

Two main classes of aneuploids are known. In the first (hypodiploid) one or more chromosomes are missing from the standard complement. Where one of a pair of homologues is absent (monosomic condition) univalents are formed at meiosis and their behaviour is identical with that of other spontaneously arising univalents (see section III-3). In the second class (hyperdiploid or polysomic) one or more extra chromosomes are present and meiosis may then be complicated by the formation of multivalent associations which may develop between the

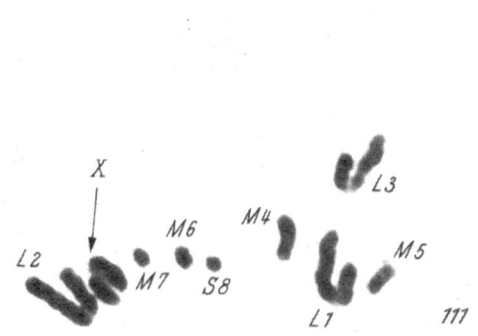

Fig. 111. Non-disjunction of the X-chromosome at second anaphase in *Chorthippus brunneus*. Both X-chromatids have moved together to the lower pole, other chromosomes normal (orcein, ca. ×1800). Photograph kindly supplied by Dr. S. A. Henderson.

extra chromosomes and their homologous counterparts (Figs. 112 and 113). The types of multivalent formed will depend on the precise number and homology of the extra chromosomes and the number and positions of the chiasmata which form (Table 35). The commonest conditions are the trisomic (2 n = 2 x + 1) and the tetrasomic (2 n = 2 x + 2).

At meiosis in trisomics the extra chromosomes tend to be lost and, in plants at least, (2 x + 1) seeds often do not germinate as well as (2 x) seeds. Einset (1943), who studied eight of the ten possible primary trisomics of maize found a positive correlation between the length of a chromosome and the frequency with which it was transmitted through (x + 1) eggs in (2 x + 1) × (2 x) crosses (Table 36). In Einset's study, failure of the extra chromosomes to be transmitted through the egg to 50% of the progeny was due to the elimination of the extra chromosome as a univalent in the meiotic divisions. This, in turn, was correlated with the desynapsis of one member of the trivalent, an event which occurred more frequently in plants trisomic for the shorter chromosomes. Univalents produced in this way lagged at first anaphase and were not incorporated in the daughter nuclei produced at this division.

In *Datura*, too, particular extra chromosomes have specific effects on the viability of spores, gametes, embryos and seeds. Differentials are also manifest in their rate of elimination from trisomics.

The aneuploids so far considered were not merely hyper- or hypo-diploid, they also contained more or less chromosome material compared with their diploid progenitors. However, two opposite, indeed inverse, changes—namely, centric "fusion" and centric fission can effect a numerical change without affecting the quantitative balance (see page 90).

3. Polyploids

a) Meiosis in Autopolyploids

(i) Autotriploids (2 n = 3 x)— At meiosis in an autotriploid each group of three homologues may form a trivalent, a bivalent and a univalent or else three univalents (Fig. 114). And different combinations of these events take place in different cells. In maize (2 n = 3 x = 30) the longer chromosomes form trivalents more frequently than the

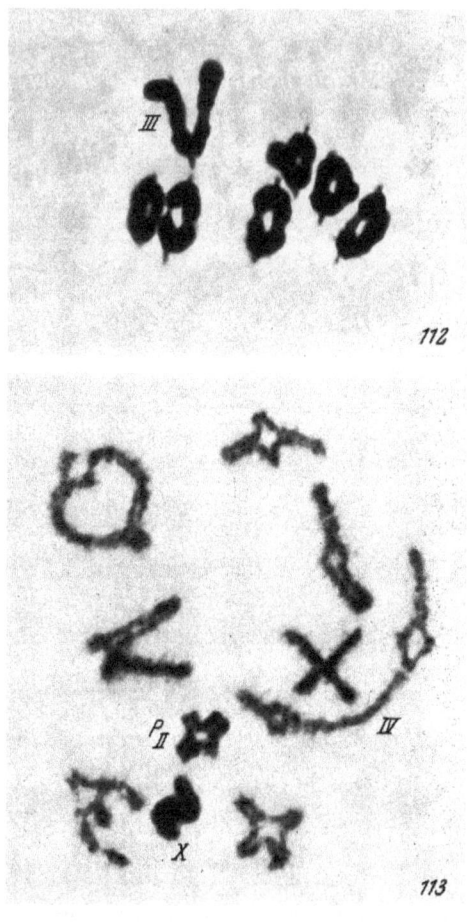

Figs. 112 and 113. Polysomic cell types. Fig. 112. A rye trisomic (2n = 2x + 1 = 15) in which the three homologous members have formed a trivalent to give 6 II + 1 III (carmine, ca. × 1500; kindly supplied by Dr. HUBERT REES). Fig. 113. A tetrasomic cell (2n = 2x + 2 = 21) in *Pyrgomorpha kraussi* with 1 IV + 8 II + x (orcein, ca. × 1500).

shorter ones but no cell was found with more than four univalents (McCLINTOCK 1929) and the most common arrangement is 9 III + II + I. And in species differing in chromosome length an equivalent difference in trivalent frequency is found (Table 37).

The most complete study of chromosome disjunction in an autotriploid is that on triploid *Datura stramonium* (Tables 38–40). These data reveal a number of interesting facts:

(1) There is a strong tendency for the chromosomes of the "extra" set to pass together to one pole rather than separate at random (Table 38).

Table 35. *Behaviour of Primary and Secondary Trisomics in Datura.*
(Data of BELLING and BLAKESLEE 1924.)
Note the non-trivalent associations* represent separate samples.

Trisomic Type	Trivalent Type and Frequency						Total	II + I Type*	
	U	ᠺ	Y	(	◯	◖		Ring II + Rod I	Ring II + Ring I
1. Primary (extra chromosome showing complete homology)	48	33	17	9	1	1	109	9	—
2. Secondary (extra chromosome is an iso-chromosome, homologous with one arm of a normal member of the complement)	26	13	1	5	53	0	98	0	20

Table 36. *Chromosome Length and Frequency of Transmission through the Egg from Trisomic Plants in* $(2x + 1) \times 2x$ *Crosses Together with the Pairing of the Extra Chromosomes in P. M. C.'s and their Transmission to Pollen by* $(2x + 1)$ *plants in Zea mays.*
(Data of EINSET 1943.)

Relative Chromosome Length at Pachytene (Chromosome 1 ≡ 100)	Trisomic for Chromosome No.	%-trisomics in Progenies of Trisomic Plants	No. of Observations	% P.M.C's with Univalents at First Metaphase	No. of Observations	% P.G's with (x + 1) Chromosomes	No. of Observations
Long							
80	2	47	320	20	602	48	454
74	3	45	91	28	342	41	167
73	5	52	89	14	209	50	198
Average for long chromosomes		48	...	21	...	46	...
Medium							
60	6	38	155	28	109	34	109
56	7	41	80	26	246	50	193
57	8	31	191	40	267	36	245
Average for medium chromosomes		37	...	31	...	40	...
Short							
52	9	22	113	44	214	23	218
45	10	28	198	43	672	34	299
Average for short chromosomes		25	...	44	...	29	...

(2) Lagging chromosomes are particularly frequent on the female side where over 50% of the nuclei possess laggards (Table 39). This results in greater loss of chromosomes on the female side emphasising once more the difference in chromosome behaviour between the "sexes" (see section II-3).

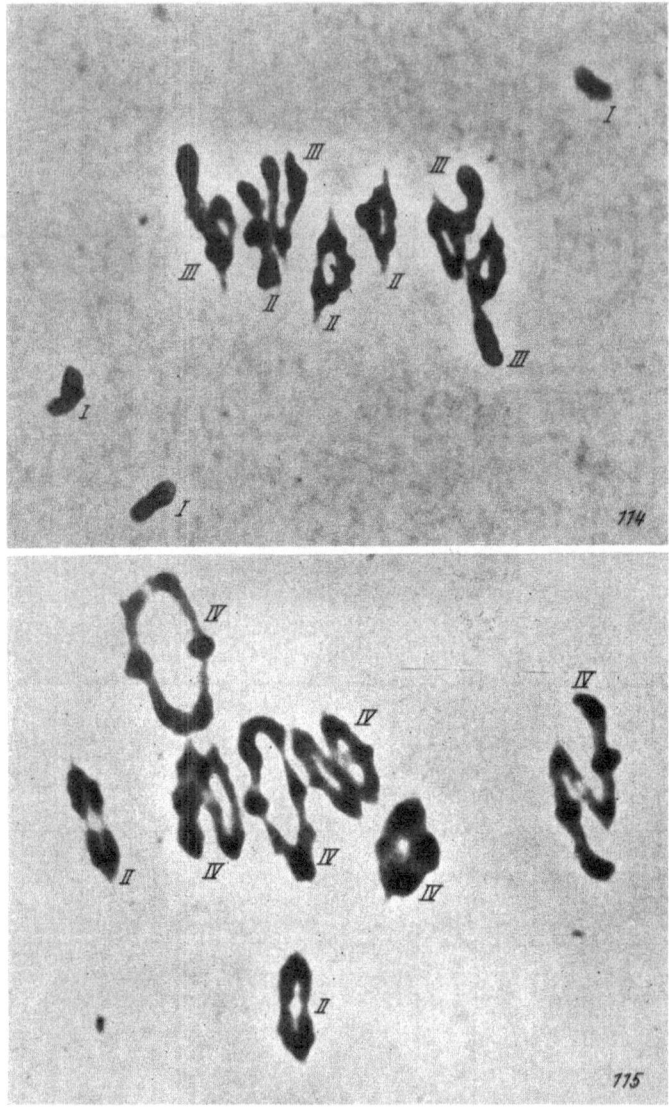

Figs. 114 and 115. Autopolyploid cells in *Secale cercale* (carmine, phase contrast, ca. ×1500). Photographs kindly supplied by Dr. HUBERT REES. Fig. 114—A triploid cell with 4. III + 3. II + 3. I (2n = 3x = 21). Fig. 115—A tetraploid cell with 6. IV + 2. II (2n = 4x = 28).

(3) In *Datura* there is only one mitosis in the pollen grain itself and this occurs after the separation of the four meiotic products. Counts from 500 grains (Table 40) showed that though division occurs in all grains, regardless of their chromosome content, yet the frequency of grains with counts lower than 17 ex-

ceeds calculated values based on the assumption of random behaviour. Indeed, the number of haploid (x) grains recorded (2.6%), is over one hundred times the number expected from random assortment (0.025%). After the nucleus of the pollen grain has divided between 40 and 50% of the pollen aborts, presumably

Table 37. *Comparative Behaviour of Triploids.*

Species	No. of III's per cell											Total Cells	Mean no. of III's per Cell	Chromosome Type	Reference
	2	3	4	5	6	7	8	9	10	11	12				
1. Lycopersicon esculentum (n = 12)	...	5	13	17	10	5	...	...	...	...	...	50	4.9	Relatively short	Upcott (1935)
2. Lilium tigrinum (n = 12)	...	...	...	...	...	...	5	10	25	27	29	2	98	9.7	Long	Chandler et al. (1937)

Table 38. *Disjunction Frequency in* 3x (= 36) *Datura. Cells with Lagging Univalents
(see Table 39) Have not Been included.*
(Data of Satina and Blakeslee 1937a and b.)

Sex	Division	%-cells with Distributions of							Total Cells	Totals of First 4 Classes
		(+ 12) 12 + 24 (+ 1)	(+ 11) 13 + 23 (+ 2)	(+ 10) 14 + 22 (+ 3)	(+ 9) 15 + 21 (+ 4)	(+ 8) 16 + 20 (+ 5)	(+ 7) 17 + 19 (+ 6)	(+ 6) 18 + 18		
♂	I	0.8	4.5	8.5	14.5	22.9	30.8	18.0	1000	28.3
	II	3.0	5.5	11.0	16.5	20.0	30.5	13.5	200	36.0
♀	I	0	3.5	9.0	14.0	21.5	34.5	17.5	200	26.5
		0.05	0.6	3.2	10.7	24.2	38.7	22.6	100%	14.55
Random expectation (½)¹²		Higher than expected from random distribution				Lower than expected from random distribution			—	

as a result of chromosome unbalance. This further increases the number of haploid grains; so too does the failure of certain unbalanced pollen to germinate and the abnormal growth of the pollen tube in other cases. The net result is that the proportion of haploid male gametes which are effective in fertilisation is some 3000 times as large as that expected had random behaviour prevailed throughout the development of the male gametophyte.

The *Datura* situation, however, does not apply to all triploids. Thus WILSON (1959) analysed the behaviour of triploids in *Endymion nonscriptus* and the closely related *E. hispanicus*, both with 2 n = 3 x = 24. The pairing behaviour of these was fairly comparable (Table 41), but whereas there is a significant

Table 39. *Frequency of Laggards in* 3x *Datura.*
(Data of SATINA and BLAKESLEE 1937a and b.)

Division	I		II	
	♂	♀	♂	♀
%-cells with laggards	4.5	52.5	2.5	52
No.-cells with laggards	$89/2{,}000$	$105/200$	$20/800$	$52/100$
No. Chromosomes lost owing to lagging	109	218	27	128
Loss per nucleus	0.05	1.1	0.03	1.3

departure from random behaviour at first anaphase in *E. hispanicus* there is not in *E. nonscriptus* (Table 42).

Again, while chromosome counts lower than 1.5 (x) exceed random expectation in pollen grains of *Hyacinthus* (DARLINGTON 1926), *Allium* (LEVAN 1933) and *Narcissus* (NAGAO 1935) this is not so in *Endymion* (WILSON 1958). Here, as in

Table 40. *First Pollen Grain Counts at Metaphase in* 3x *Datura.*
(SATINA and BLAKESLEE 1937a.)

Chromosome no. per grain		12	13	14	15	16	17	18	19	20	21	22	23	24	Totals
Total grains	No.	13	20	36	55	77	80	56	54	46	25	19	13	6	500
	%	2.6	4.0	7.2	11.0	15.4	16.0	11.2	10.8	9.2	5.0	3.8	2.6	1.2	100

most other triploids, there were no sub-haploid pollen grains undergoing mitosis but the frequency distribution at first pollen grain mitosis was towards an increase in grains with more than 1.5 (x) chromosomes and not, as in the other genera, towards an increase in those with less than this value.

(ii) Autotetraploids (2 n = 4 x)—The analysis of chromosome behaviour in autotetraploids can be carried out in two ways. First, by studying meiosis in autotetraploid individuals and second by considering autotetraploid cells in an otherwise diploid individual. The former is used commonly for plants, while the latter is used predominantly in animals, where polyploid individuals are far less

8*

common. Tetraploid cells may arise in two ways. On the one hand spindle failure may lead to the production of tetraploid resting nuclei and in these, or their mitotic products, multivalents form at meiosis (Fig. 115). Alternatively failure of cytoplasmic division following a normal mitosis may lead to the produc-

Table 41. *Pairing in Endymion Triploids* $(2n = 3x = 24)$.
Data *of* WILSON (1959).

Species	%-frequency of Associations			
	8 III	7 III + II + I	6 III + 2 II + 2 I	5 III + 3 II + 3 I
E. nonscriptus	56.25	32.29	9.38	2.08
E. hispanicus	60.51	27.84	9.66	1.90

tion of multinucleate cells. Multivalent formation cannot occur in these for, at the time of pairing and chiasma formation, each diploid group is still invested by its own nuclear membrane (Figs. 116 and 117). However, the nuclei can fuse after pairing is completed, usually on the meiotic spindle and so produce a true polyploid condition as opposed to a multinucleate one.

Table. 42. *Chromosome Disjunction at Anaphase-I in Endymion Triploids.*
(Data of WILSON 1959.)

Distribution Pattern	(3x) nonscriptus		(3x) hispanicus	
	Observed	Expected (Random behaviour)	Observed	Expected (Random behaviour)
8 + 16	2 } 11	0.79 } 6.93	5 } 23	1.91 } 16.73
9 + 15	9	6.14	18	14.82
10 + 14	29	21.68	63	52.34
11 + 13	37	43.26	79	104.44
12 + 12	22	27.03	74	65.25
Totals	99	98.90	239	238.76
$\chi^2_{[3]}$	7.02, P > 5%		11.89, P = 0.01	

Where multivalent formation is possible within an autotetraploid complement, the number and kinds of quadrivalent formed per cell appears to be governed by two main factors—one structural and the other known to be under genotypic control. First, the size of the chromosomes (DARLINGTON 1932, WHITE 1954, JOHN and HENDERSON 1963 and see Table 43). Second, chiasma frequency and position (ROSEWEIR and REES 1962).

Of course, the frequency of bivalents and multivalents varies from cell to cell since chiasma frequency varies also in this way. Theoretically there are five possible types of association—quadrivalent, trivalent plus univalent, two bivalents, one bivalent plus two univalents and four univalents (Figs. 118–120). However in no organism yet studied are these associations formed at random. Quadrivalents and bivalents are common whereas trivalents and univalents are much less so (Tables 43 and 44). In general, autotetraploids do not breed true for chromosome number. Thus in maize (DOYLE 1963) only 60.7% of a total of 557 plants obtained as progeny of autotetraploids had 40 chromosomes—the rest were hypo- or hyperploid.

There is one further distinction worth making in discussing autotetraploid behaviour. Established autotetraploids, as compared with newly produced ones, having been subject to natural or artificial selection, may very well have changed their pattern of behaviour since the time of their origin. Thus ROSEWEIR (1964) has recently compared the chiasma and quadrivalent frequency in experimentally produced autotetraploids of rye with that of an autotetraploid subject to selection for fertility over a three year period. His results, summarised in Fig. 121, indicate quite clearly that quadrivalent frequency has increased in the selected line. This result stands in marked contrast to the fact that selection for fertility did not affect the frequency of quadrivalents significantly in barley and maize (McCOLLUM 1958, MORRISON and RAJATHY 1960, Table 44)

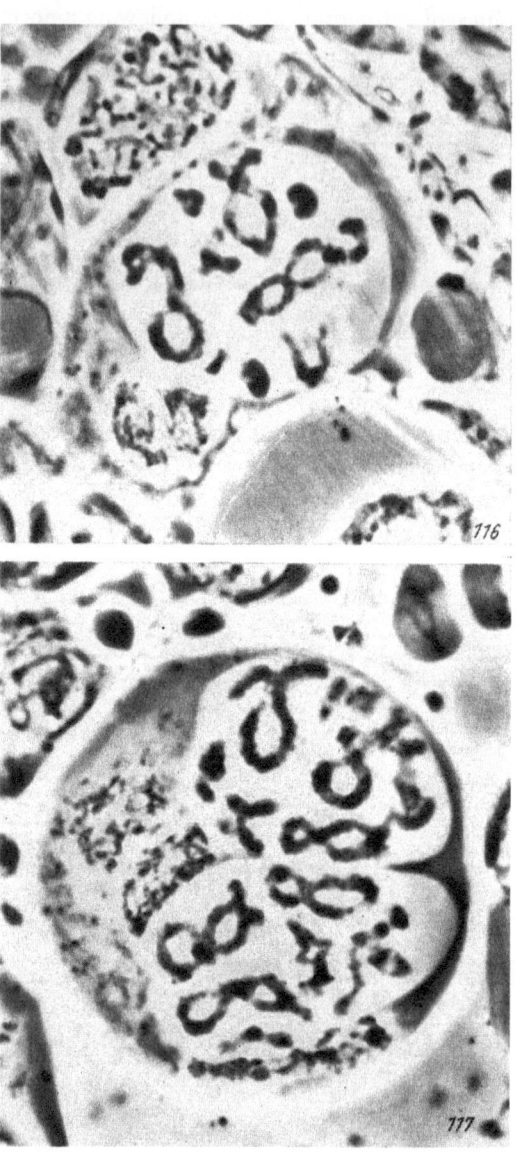

Figs. 116 and 117. Diploid and binucleate sperm mother cells at diplotene in *Chorthippus brunneus* as seen unfixed and unstained with phase contrast (ca. ×1500). The binucleate cell (Fig. 117) will appear as a tetraploid cell without multivalents at metaphase-1.

or indeed in other experiments on rye (MÜNTZING 1951, MORRISON 1956). However the different results of such experiments may depend, in part at least, on the precise properties of the individuals used in the experiments. Thus the mean

Table 43. *Frequency of Chromosome Association in Newly Arisen Autotetraploid Cells of Diploid Male Orthopterans.* (Data of John and Henderson 1962 and Callan 1949.)

Species and diploid Karyotype	IV				III + I				II				I				Total Cells Analysed	Mean X ta/cell	
	L	M	S	Total	L	M	S	Total	L	M	S	Total	L	M	S	Total		2 x	4 x
1. *Schistocerca paranensis* 3 L II + 5 M II + 3 S II	70	92	13	175	...	1	4	5	8	64	115	187	...	...	3	3	25	19.33	44.44
2. *Pyrgomorpha kraussi* 2 L II + 5 M II + 2 S II	10	9	...	19	...	...	...	...	20	80	40	140	...	2	...	2	10	9.8	22.3
3. *Forficula auricularia* 11 S II	—	—	—	37	—	—	—	12	—	—	—	206	—	—	—	4	14	11.1	24.71

number of quadrivalents per cell in autotetraploid rye ranges from 1.1 (Chin 1943) to 3.9 (Muntzing 1951, Morrison 1956). Similarly the equivalent values for natural autotetraploids of *Dactylis glomerata* range from 2.83 to 3.99 quadrivalents per cell (McCollum 1958).

The most effective demonstration of the influence of chiasma position on quadrivalent frequency in autotetraploids comes from organisms with localised chiasmata. In diploid species of *Allium* with such chiasmata, pachytene pairing is, as we have seen (section II-1), complete. In 1940 Levan described an autotetraploid form, *Allium porrum* $(2 n = 4 x = 32)$, with proximally localised chiasmata. Here pairing associations of four were regularly formed at pachytene but by diakinesis bivalents predominated and quadrivalents were rare. Thus in 380 cells there were only 10 quadrivalents present whereas in autotetraploids of *Allium* species with distributed chiasma formation there are on average 1–5 quadrivalents per cell. Moreover the quadrivalents in *A. porrum* differ from ordinary quadrivalents in the same ways as bivalents with localised chiasmata differ from other bivalents (Fig. 122). Most of the pollen grains in *A. porrum* contain 16 chromosomes so that the few quadrivalents which do form in this species produce little disturbance of meiosis. Similarly, in *Paris quadrifolia* $(2 n = 4 x$, Darlington 1941) and *Paris japonica* $(2 n = 8 x$, Haga 1937), both of which show proximal chiasma localisation, only bivalents form at meiosis and both species are sexually fertile.

There are, of course, some cases in animals where meiosis is chiasmate (see section II-2) and in one of these, *Callimantis antillarum*, Hughes-Schrader (1943) has made a study of autotetraploid cells during meiosis in the diploid male germ line. The advantage of studying polyploid cells in otherwise diploid tissues is that variation owing to the genotype, or even the environment, can be largely ruled out. In tetraploid spermatocytes of *C. antillarum* (2 n = 4 x = = 34 = 32 + X Y) a high proportion of bivalents form though there may be from 2–6 quadrivalents per cell (Table 45, Fig. 122). Trivalent formation is again rare and the associations of four homologues which do form are maintained to first metaphase without any chiasmata. At first metaphase 38% of the rings form alternate and 57% adjacent orientations.

It is commonly held that the infrequent occurrence of autopolyploids depends, at least in part, on the reduced fertility which may follow multivalent formation. This is understandable in the case of trivalents and perhaps, to some extent, in the case of chain quadrivalents where linear and indifferent orientation lead to nondisjunction. But in the case of ring quadrivalents both alternate and adjacent orientation should give numerically equal separation for only when rings are discordant should univalents be left at the equator (see also page 78).

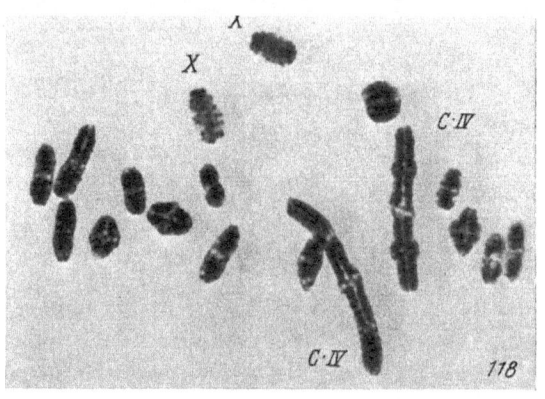

b) Meiosis in Allopolyploids

Interspecific hybridisation followed by chromosome doubling leads to the production

Figs. 118. A tetraploid cell in *Pyrgomorpha kraussi* at metaphase-I with two chain quadrivalents in linear orientation (orcein, ca. ×850).

of an allopolyploid. Since some degree of relationship is necessary for the cross itself to have succeeded, it is to be expected that at least partial homology exists between the chromosomes belonging to the different parental species. Chromosomes that are only partially homologous are called homoeologous, a term first introduced by HUSKINS. This partial homology means that, in addition to the pairing which will occur between homologues, homoeologues may also associate. The extent of homoeologous pairing will vary according to the relationship of the species involved in the cross. At one extreme one might expect to find those allopolyploids which show little or no homoeology and, consequently, form only bivalents. At the other extreme, there are those forms in which the chromosomes are almost completely homologous and these should be virtually equivalent in their behaviour to autopolyploids *sensu stricto*. And between these two extremes lie what STEBBINS has called the segmental allopolyploids.

These expectations, however, are disturbed by the fact that some allopolyploids, whose diploid ancestors show clear genetical and cytological relationships, form only bivalents at meiosis. The best understood instance of this is in common breadwheat, *Triticum aestivum*. This is an allohexaploid (2 n = 6 x = 42 = = AABBDD, see Table 46) combining the complements of three diploid pro genitors which in all probability are represented in present-day forms by *Triticum monococcum* (A-genome), *Aegilops speltoides* (B-genome) and *Aegilops squarrosa* (D-genome). Although the chromosomes of these three species are unquestionably homoeologous, this homoeology is not expressed as pairing in the allohexaploid. And the amphidiploidising role is known to be the function of a gene system located in the long arm of the metacentric chromosome V which

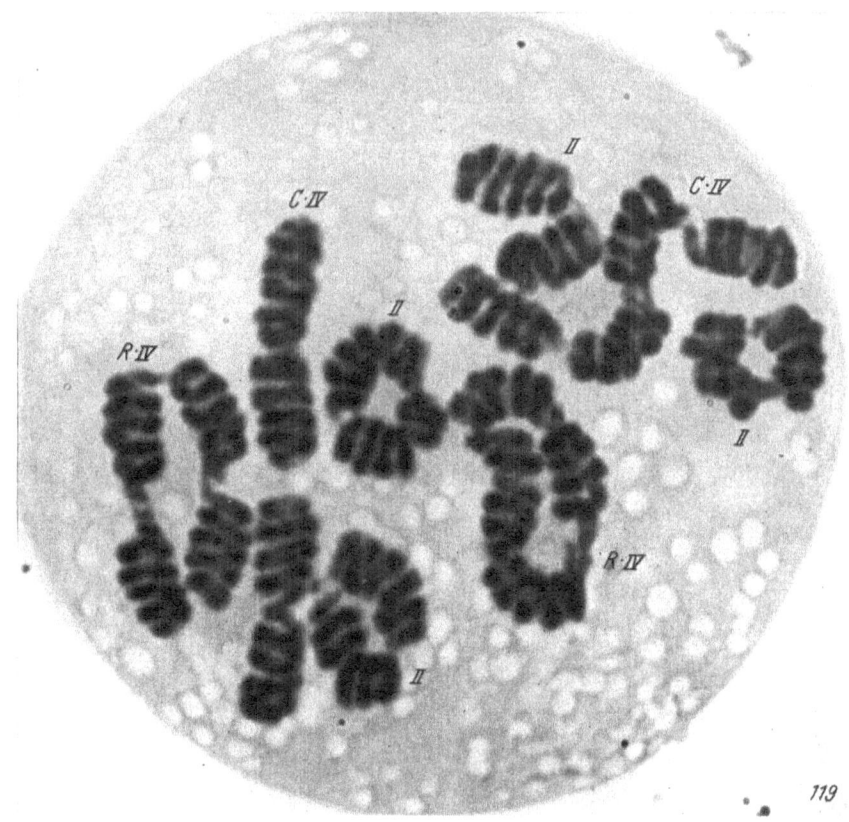

Fig. 119. P. M. C. of tetraploid *Tradescantia* (2n = 4x = 24) with 2 R. IV + 2 C. IV + 4 II. Photograph kindly supplied by Mr. C. G. Vosa (Darlington and Vosa 1963).

has been introduced into the hybrid by the supposed B-genome donor, *Ae. speltoides* (Table 46). Riley and his co-workers, following the clue first provided by Sears, showed that when V is completely missing from the allohexaploid or its polyhaploid derivatives, homoeologous pairing occurs (Figs. 123–124). But a single representative of V, even a long telocentric V, is sufficient to maintain the control so that the mechanism is effective when hemizygous (Riley 1960).

Riley, Kimber and Chapman (1961) are of the opinion that this system, which discriminates between homoeologous and homologous affinity and thus brings about an essentially disomic system of inheritance, represents a single mendelian mutation selected for its obvious role in improving fertility and genetic stability. Since the presence or absence of V has no influence on pairing in *T. aestivum* × *Ae. speltoides*, it would appear that the genetic activity of the *speltoides* genome *suppresses* the activity of chromosome V or, to put the matter another way, that the amphidiploidising gene system on V must be recessive to its allele in *speltoides*. This, in turn, suggests that the mechanism arose by muta-

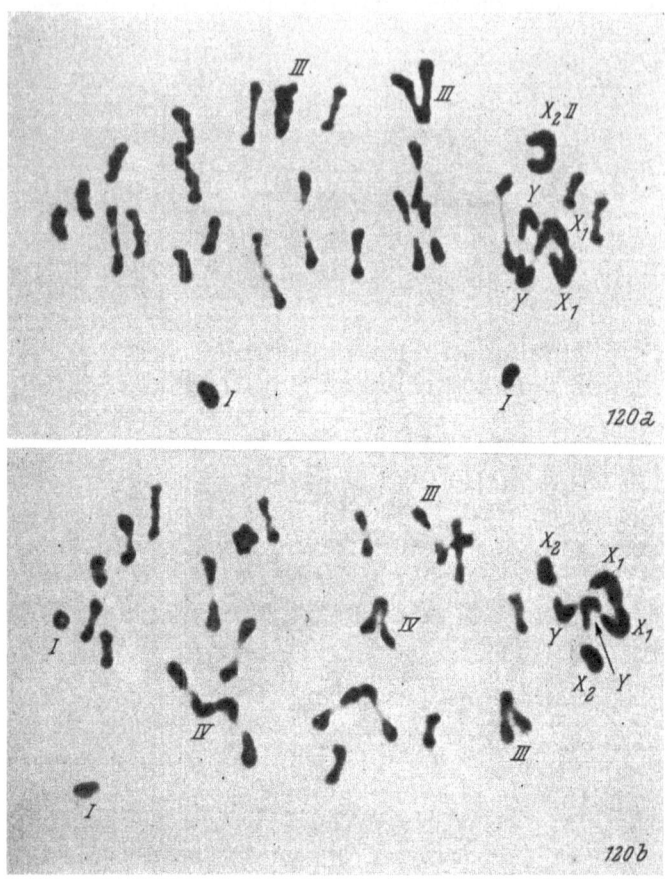

Figs. 120*a* and *b*. Tetraploid primary spermatocytes of *Sphodromantis gastrica*. This species has an $X_1 X_2 Y$ sex-chromosome mechanism ($2n = 2x = 12 + X_1 X_2 Y$ see Fig. 158) and in tetraploid cells a sex-sexivalent may form as in 120*b*. Fig. 120*a*—The autosomes form 2 III + 20 II + 2 I while the sex chromosmes have formed an X_2-bivalent and an $X_1 X_1 Y Y$ association of four. Fig. 120*b*—The autosomes have formed 2 IV + 2 III + + 16 II + 2 I while the sex chromosomes combine as an association of six (orcein). Photographs kindly supplied by Dr. S. A. HENDERSON (see HENDERSON 1965).

tion after the incorporation of the B-genome into tetraploid wheat, though the importance of atypical individuals in speciation cannot be ignored.

The homoeologous equivalents of chromosome V of the B-genome are IX (A-genome) and XVIII (D-genome). Nulli-V, tetra-XVIII and nulli-V, tetra-IX combinations show that extra doses of IX and XVIII do not compensate for the absence of V since in both states there is considerable pairing of homoeologues (KEMPANNA 1962). Thus, although tetrasomy for IX or XVIII removes the sterility otherwise caused by a deficiency of V (pollen deficient for V rarely functions) there is no influence on the pairing of homoeologues which occurs in the absence of V.

Riley (1960) is of the opinion that an equivalent amphidiploidisation has been involved in other allopolyploids for many existing polyploids have been successful in becoming functionally diploid. Thus, although there are forms in

Table 44. *Chromosome Associations in Six Autotetraploid Cereal and Grass Species all with* 2n = 4x = 28.
(Summarised from Morrison and Rajhathy 1960.)

Species	Associations per Cell				Total Cells Analysed
	IV	III + I	II	2 I	
1. *Avena strigosa*	4.4	< 0.1	5.0	0.1	125
2. *Secale cereale*	3.7	0.2	6.1	0.5	50
3. *Hordeum vulgare*	3.9	0.1	5.7	0.3	125
4. *Hordeum bulbosum*	4.0	0.2	5.6	0.1	200
5. *Arrhenatherum elatus*	4.8	...	4.3	...	150
6. *Triticum monococcum*	5.1	0.1	3.5	0.2	260

the *Rosa canina* complex with 28 and 42 chromosomes, pentaploids with 2 n = 5 x = 35 (AABCD) are by far the most common. All three forms produce only 7 (AA) bivalents at meiosis and 14, 21 (BCD) or 28 chromosomes are left as univalents. Functional pollen grains invariably receive 7 chromosomes all derived from the 7 bivalents and none of the univalents are incorporated. By contrast megaspores receive all the univalents together with 7 chromosomes derived from the bivalents. Egg cells thus have 28 chromosomes (ABCD) and these when fertilised by 7-chromosome (A) pollen restore the pentaploid level (AABCD).

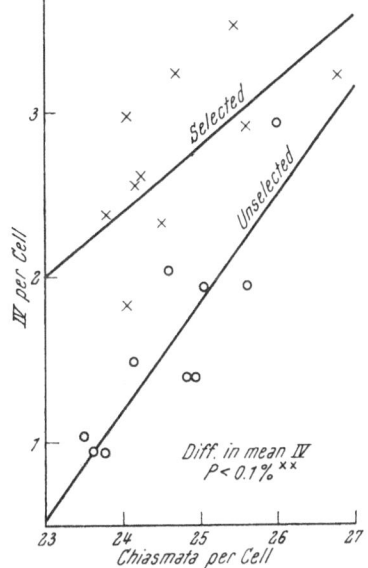

Fig. 121. Chiasma frequency and quadrivalent formation in a selected and unselected line of autotetraploid rye. Each point is a mean value for twenty cells. Unpublished data kindly supplied by Mr. John Roseweir.

However, when pentaploid *R. canina* is crossed with the distantly related diploid *R. rugosa* (2 n = = 2 x = 14) the resulting hybrids have 11–14 bivalents. Since 7 of the *R. rugosa* chromosomes can pair with those from *canina*, it is reasonable to suppose that 7 of the *canina* chromosomes have probably been contributed by *R. rugosa*. By analogy with wheat this can be interpreted to mean that recessive genetic factors restricting pairing may reside in that part of the *R. canina* genome corresponding to *R. rugosa*. Thus the *R. rugosa* set has a dominant or epistatic effect on *R. canina* equivalent to that which *Ae. speltoides* shows on *T. aestivum*. Selection for a mutation in the *R. rugosa* genome, after incorporation into the *R. canina* complex might therefore have led to the evolution of a system preventing intergenomic pairing in the that complex.

KIMBER (1961) has suggested that the situation in tetraploid cotton favours a similar explanation. The cultivated cottons (*Gossypium hirsutum, G. barba-dense* and *G. tomentosum*) are natural allotetraploids ($2n = 4x = 52$) composed of two subgenomes, designated A and D, derived from two geographically

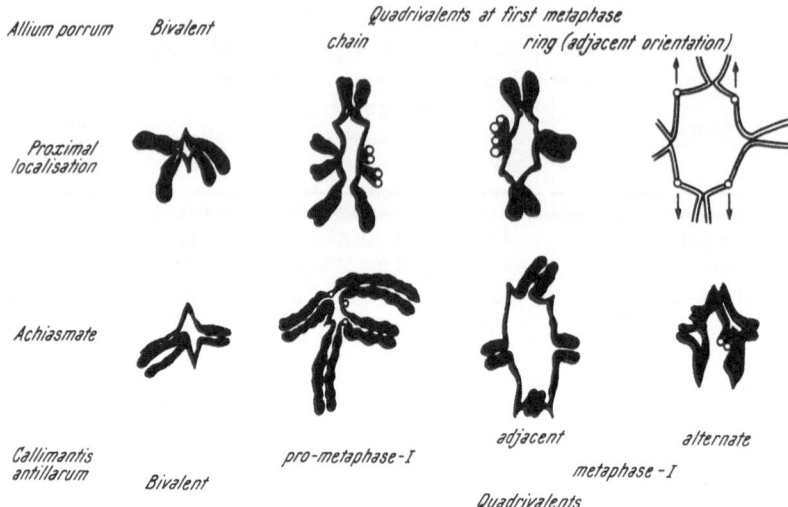

Fig. 122. The morphology of bivalents and quadrivalents in *Allium porrum*, with proximally localised chiasmata, and *Callimantis antillarum*, with a non-chiasmate meiosis. The illustrations for *A. porrum* are taken from LEVAN (1940) and those for *C. antillarum* from HUGHES-SCHRADER (1943).

isolated diploid species. A species from Asia has donated the A-genome; a second from America has supplied the D-genome. The chromosomes of these diploid progenitors differ in length, those of the A-genome being longer than those of the D. This size difference is maintained both in the natural polyploids and in

Table 45. *Quadrivalent Frequency per Cell in 21 Tetraploid Spermatocytes of Calli-mantis antillarum.*
(Data of HUGHES-SCHRADER 1943.)

Quadrivalents per Cell					Total Quadrivalents
2	3	4	5	6	87
2	2	9	7	1	21
Number of Cells					Total Cells

all hybrid combinations of the diploids (ENDRIZZI and PHILLIPS 1960). It can thus be used to categorise meiotic configurations in interspecific hybrids and to distinguish the univalent chromosomes in polyhaploids derived from the tetraploid.

KIMBER found that in cotton haploids ($2n = 2x = 26 = AD$) the 26 uni-valents often associate in pairs as though synapsis had occurred but no chiasmata had formed. And these univalent associations occurred in three types—LL(AA),

LS(AD) and SS(DD). Since there are equal numbers of L- and S-chromosomes the ratio of LL : LS : SS should be 1 : 2 : 1 if the chromosomes associate at random. In fact there was a significant departure from random expectation (Table 47); LS(AD) associations are in excess of prediction. This behaviour in the polyhaploid, together with the fact that duplicate loci have been frequently demonstrated in the allotetraploid, suggests that A and D genomes are homoeologous. But in the tetraploid the chromosomes form only bivalents and this led KIMBER to the suggestion that a genetic amphidiploidisation system operates here. Unfortunately this suggestion cannot be tested in the way it has been in wheat, because a complete range of aneuploids is not available in cotton.

Table 46. *The Homoeology of the A-, B- and D-genomes of Triticum aestivum.*
(Data of SEARS 1953 and OKAMOTO 1962.)

Group	Homoeologous Members		
	A-genome	B-genome	D-genome
1	XIV	I	XVII
2	XIII	II	XX
3	XII	III	XVI
4	IV	VIII	XV
5	IX	V	XVIII
6	VI	X	XIX
7	XI	VII	XXI

However in A × AD triploid hybrids the chromosomes of the A-genome pair with the long chromosomes of the AD-allotetraploids leaving 13 short univalents, while in D × AD triploid hybrids there are 13 L univalents (Table 48). Moreover the cytological equivalence between the two diploids is shown by the high mean synapsis observed in the AD hybrid *G. arboreum × G. thurberi*.

ENDRIZZI (1962), on the other hand, has proposed a quite different explanation. At pachytene in "haploids" of *G. hirsutum* there is an average of 7–9 closely paired chromosomes, each of which though equal in length differ in the amount of heterochromatin they contain. But the two members of each bivalent subsequently show differential contraction. Presumably these bivalents represent associations between A and D homoeologues. Similarly in hybrids between species with large and small chromosomes, pachytene partners are equal in length despite the size difference which become evident in associated homoeologues at first metaphase.

The chromosomes of the different genomes thus appear to be of the same length in the dispersed state but different in the condensed, a situation quite different from that in the three genomes of wheat. And ENDRIZZI has in fact suggested that it is this difference in the condensation cycle of the A and D genomes after pairing which determines the diploid-like pairing behaviour of the tetraploid cottons. This assumes that exchange will be infrequent between pairing partners which coil asynchronously. However, RILEY and KIMBER (1964) claim that chiasma formation takes place before the differential contraction in which

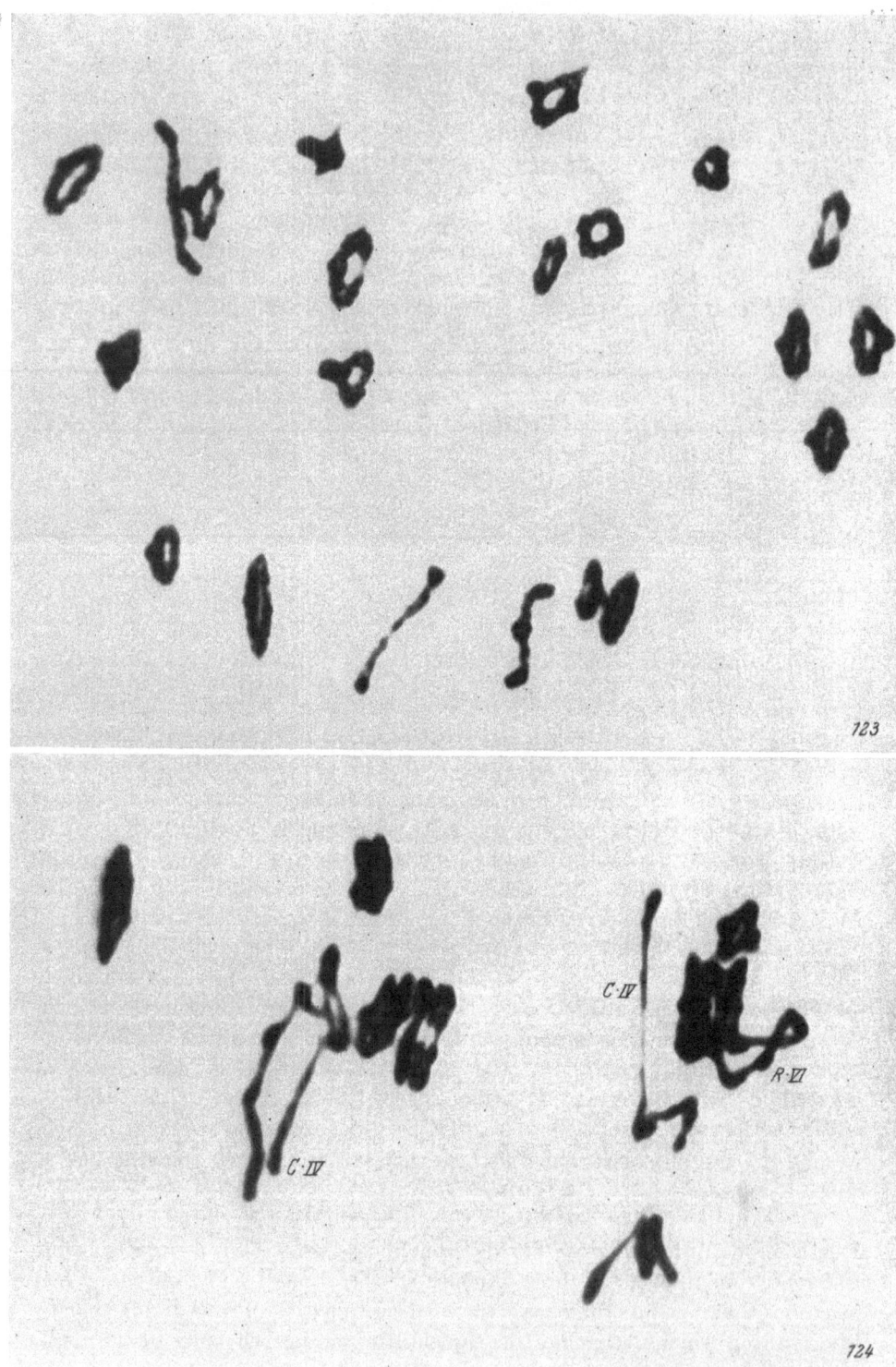

Figs. 123 and 124. The influence of chromosome V on homoeologous pairing in bread wheat, *Triticum aestivum*. In the normal allohexaploid (2n = 6x = 42) only bivalents form (Fig. 123). But in nullisomic-V (2n = 6x — 2 = 40) homoeologous pairing, in the absence of chromosome V, leads to the formation of multivalent configurations and in the cell shown (Fig. 124) there is 1 R VI + 2 C IV + 2 C IV + 13 II. Photographs (ca. ×2000) kindly supplied by Dr. RALPH RILEY.

event it can hardly be considered as the basis of the low frequency of multivalent formation.

Diploidisation through gene control of the kind found in *T. aestivum* would be expected to proceed rapidly once initiated and stands in marked contrast to the opinion, often expressed in the past, that the amphidiploidisation of a polyploid depends on a gradual loss of homology through the accumulation of small

Table 47. *Types and Frequencies of Univalent Associations in Haploid Gossypium* (2n = 2x = 26 = AD).
Data of KIMBER (1961).

Association	LL (AA)	LS (AD)	SS (DD)	Total
1 Observed	65	202	62	329
2. Expected (1 : 2 : 1)	82.25	164.5	82.25	329

$$\chi^2_{[N=2]} = 17.15, \ P < 0.001$$

changes such as minor gene mutations or the production of cryptic structural alterations.

There is also some evidence which suggests that large chromosome rearrangements may be involved in the stabilisation of meiosis in some allopolyploids. For example, SHAVER (1963) has recently studied the influence of a long paracentric inversion in the long arm of chromosome 3 (*In . 3a*), on the amphidiploidisation

Table 48. *Meiotic Behaviour of the A and D Genomes in Gossypium.*
Data of KIMBER (1961)

Type	Genome Constitution	Chromosome Pairing	Reference
3 x	A × AD	13 I, 9 II and 2 IV	GERSTEL (1953)
3 x	D × AD	13 I, 12.15 II, 0.51 III and 0.04 IV	ENDRIZZI (1957)
2 x	A × D	9.6 I, 7.8 II, 0.2 III and 0.05 IV	SKOVSTED (1937)
Polyhaploid 2 x	AD	26 I	KIMBER (1961)

of the allotetraploid hybrid (4 x) maize × (4 x) teosinte and has found that this inversion exerts a profound effect on gene segregation. His results are interpretable in terms of preferential pairing between homologous as opposed to homoeologous chromosomes. This leads to a reduction in the frequency with which chromosome 3 forms quadrivalents because there is no effective pairing between the relatively inverted segments. The net result of this production of preferential bivalents is that chromosome 3 is more faithfully transmitted to the progeny of this allotetraploid structural hybrid than it is in the normal allotetraploid. Moreover, although a known inversion was introduced into this cross, no inversion bridges could be detected in the allotetraploid inversion heterozygote which indicates that structural hybridity can exist in an allopolyploid and yet be cytologically cryptic. Preferential pairing of homologues was also demonstrated by DOYLE (1963), both genetically and cytologically, in his recent study of the effect of *In. 3a*, on trisomic, triploid and autotetraploid plants of maize.

So far we have considered two possible avenues open to the allopolyploid by means of which its meiosis might function efficiently:

(1) Restriction of pairing to complete homologues only, as a consequence of gene mutation

(2) Restriction of pairing by gradual or abrupt differentiation of the chromosomes so that homoeologues cease to show any significant pairing potential. Indeed, it is conceivable that subsequent change could amphidiploidise autopolyploids and some apparent allopolyploids may have originated in this way. Inbreeding would be implicated in this sequence. Subsequent change may also contribute to polymorphism at the polyploid level (but see also below).

JONES (1964) has suggested a third possibility, namely the retention, or even the increase, of multivalent association, together with the ability to undergo regular orientation. This suggestion considers the quadrivalent to constitute an advantage rather than a threat to fertility. Thus, in a number of polyploid forms the frequency of alternate orientation of the quadrivalents is very high (70–97%). This applies to *Anthoxanthum odoratum*, *Sorghum australiense* and *plumosum*, *Agrostis canina*, *Dactylis glomerata*, *Agropyron cristatum*, *Hordeum bulbosum* and others. It has been claimed (McCOLLUM 1958) that this behaviour has been selected to ensure correct numerical separation. But, as we have already mentioned, numerically-even separation is expected from adjacent orientation (see pg. 119). JONES therefore suggests that the real role of such a high frequency of alternate separation is to control the segregation of particular chromosomes so giving a non-random segregation of parental differences.

The absence of strict homology may obtain from the start if the type is of hybrid origin or the state may be acquired or intensified by subsequent mutation. But zig-zag orientation can be effective only if the chromosomes occur in a fixed order in the "multivalent". And if they do the "multivalent" is, in fact, comparable with an interchange multiple. This would require all four of the chromosomes to be different for if chromosomes of only two types exist then preferred bivalents would be as effective as zig-zag rings given that there are two chromosomes of each kind.

Thus, in a segmental polyploid of the type (AB AB BC BC), unprefered bivalent formation (ie. AB—BA with BC—CB—autosyndesis, or AB—BC with AB—BC—allosyndesis, in equal numbers) will give 75% ABBC gametes (100% from the former and 50% from the latter) which, on fusion, would give true breeding. But even this value (75%) is not exceeded by consistent multiple formation unless zig-zag orientation exceeds 50%. This is because only one type of adjacent orientation ("adjacent non-homologous") gives the ABBC gamete which zig-zag orientation produces exclusively. This assumes, of course, that no preference is expressed between the two kinds of adjacent orientation. From the point of view of true breeding therefore, multiples are preferable to bivalents only if the bias towards zig-zag orientation in the first case exceeds that towards autosyndetic bivalents in the second. Thus, the frequency of production of ABBC gametes is $x + 1/2$ where x is the frequency of zig-zag orientation, in the case of consistent multiple formation, and where it represents the frequency of autosyndetic bivalents where only bivalents are formed. What this presumably means is that the initial degree of homology will canalize subsequent change. Of course, long-term advantage may accrue from the recombination between incompletely homologous chromosomes on which multiple formation depends. This too can be a source of chromosomal polymorphism (JONES 1964).

In 1954 GARBER suggested that "if a species does not display directional orientation of the chromosomes in an interchange complex at metaphase-I, the quadrivalents in an autotetraploid of this species will likewise not display

Table 49. *Orientation of Interchange Complexes at First Metaphase in Gossypium hirsutum.*
(Data of ENDRIZZI 1958.)   See also Fig. 125.

Translocation No.	Rings		Chains		% Alternate Orientation			% Rings
	Alt.	Adj.	Alt.	Adj.	Rings	Chains	Total	
Z 872	51	37	88	60	57.95	59.46	58.90	37.29
Z 1036	17	23	35	29	42.50	54.69	50.00	38.46
Z 1040	26	22	28	28	54.17	50.00	51.92	46.15
Z 1589	8	12	16	7	40.00	69.57	55.81	46.51
Z 1587	10	18	14	11	35.71	56.00	45.28	52.83
Z 850	41	44	35	18	48.24	66.04	55.07	61.59
Z 10586	32	11	17	9	74.42	65.38	71.01	62.32
Z 1358	37	22	19	14	60.66	57.58	59.57	64.13
Z 1039	47	21	20	18	69.12	52.63	63.21	64.15
Z 1052	13	3	4	1	81.25	80.00	80.95	76.19
Z 1043	54	74	17	22	42.19	43.59	42.51	76.65
7-2 B M 8	95	78	11	13	54.91	43.83	53.81	87.82

a directed orientation. ... the two types of configuration occurring with equal frequencies." At first sight this might appear to be a reasonable suggestion and one which implies, as LAWRENCE (1963) does, that there is one gene system controlling the orientation of complex associations.

We have already pointed out that it is arguable whether a 50% frequency of disjunction is the one expected on the basis of random behaviour for "expectation" is likely to vary with the structurally and genotypically controlled aspects of multiple morphology. Indeed, in *Collinsia*, multiples of independent origin differ in respect of disjunction frequency (see page 88), a difference attributed to structural causes. Comparable observations were made on the interchange heterozygotes of *Gossypium* examined by ENDRIZZI (1958) to test GARBER's hypothesis. In this case the interchange heterozygotes were isolated from species crosses or from irradiated material of *G. hirsutum*. For the twelve interchanges he studied (Table 49), ENDRIZZI found that:

(1) The frequency of ring versus chain formation varied from 37.3% to 87.8% for different interchanges

(2) The disjunction frequency of rings varied from 35.7% to 81.3% and that for chains from 43.6% to 80.0%. The over-all variation was from 42.5% to 81.0%.

An examination of ENDRIZZI's data shows that where a given interchange is concerned, its disjunction behaviour when it forms a ring is not correlated with the disjunction frequency it shows as a chain (Fig. 125).

Thus, one can expect GARBER's contention to hold, in the first place, only if multiples and multivalents of similar structure are concerned. And secondly, in view of the genetic variation within the species, the comparison would have to be made between groups considerably smaller than a species. In the *Gossypium*

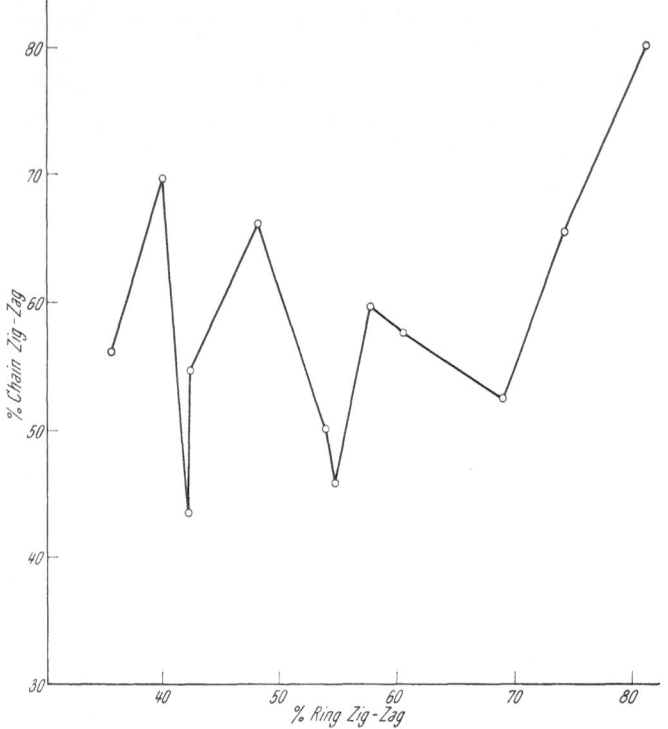

Fig. 125. Relationship between the disjunction behaviour of ring and chain-of-four multiples in twelve interchange heterozygotes of *Gossypium hirsutum* (data of ENDRIZZI 1958 from Table 49).

case there was doubtless both structural variation between the different interchanges and genotypic differences between the individuals in which they occurred. Results consistent with the hypothesis have, however, been obtained in *Sorghum* (GARBER 1948, 1954, 1956; ENDRIZZI 1958).

c) Meiosis in Polyploid Interchange Heterozygotes

Polyploid interchange heterozygotes combine the multiple configurations made possible by their hybridity with the variable multivalent-forming capacity characteristic of numerical variants carrying additional sets of homologues (Table 50). One of the most complete studies on interchange polyploids has been made by JONES (1962) in *Dactylis*. This genus includes diploid (2 n = = 2 x = 14), tetraploid (2 n = 4 x = 28) and hexaploid (2 n = 6 x = 42) forms and interchange heterozygotes have been found in all of them. In the tetraploid subspecies—*glomerata* and *hispanica*—12.1% of the plants examined had chromo-

Table 50. Pairing in Interchange Polyploids.

Species	%-configurations												Total Chromosomes	No. Inter-changes	Reference
	I	II	III	IV	V	VI	VII	VIII	IX	X	XI	XII			
Anthoxanthum odoratum (x = 5)	5.8	59.9	2.4	20.4	0.7	4.8	0.0	3.8	0.0	1.4	0.0	1.4	120	2	KATTERMANN 1931
Rosa relicta (x = 7)	1	62	3	23	2	2	2	5	0	0	0	0	380	1	ERLANSON 1931
Aucuba japonica (x = 8)	0.3	34.9	0.8	52.7	0.1	5.2	0.1	4.8	0.1	0.8	0	0	2496	2	MEURMAN 1929

Dactylis subspecies	Mean metaphase-I pairing												Mean Xta	No. Interchanges	Reference
	I	II	III	IV	V	VI	VII	VIII	IX	X	XI	XII			
Diploid *Dactylis*	...	ALL	...	...	...	...	...	...	...	...	...	...	9.6 – 12.57	0	JONES 1962
Tetraploid *Dactylis* subsp. *glomerata*															
Austria	0.23	7.86	0.15	2.70	...	0.32	0.04	0.08	...	...	...	0.04	23.13	2	
Silesia	0.08	7.50	0.05	3.16	...	...	...	0.04	...	...	...	...	22.69	1	
Finland	0.18	8.62	0.06	2.58	...	0.08	0.04	0.04	...	...	...	...	23.06	1	
subsp. *hispanica*															
Portugal (a)	0.63	7.74	0.11	2.85	0.04	...	...	...	...	...	...	...	} 23.09	1	
(b)								0.04						1	
Israel (a)	0.18	7.90	0.09	2.83	...	...	0.04	0.08	...	...	...	...	} 23.21	1	
(b)					0.04	0.08	...	0.04						1	
Turkey	0.09	8.23	0.08	2.79	...	0.12	...	0.04	...	...	...	...	23.70	1	
Hexaploid *Dactylis*															
plant 1	0.32	9.31	0.96	1.76	0.36	1.0	0.20	...	...	0.04	...	...	34.28	1	
plant 2	...	17.30	...	1.00	...	0.40	...	...	...	...	...	...	34.05	1	

some associations higher than quadrivalents. These ranged from associations of V to associations of XII but the most common were sexivalents and octavalents. Similarly in the hexaploid, associations of VII and X were found in addition to sexivalents. Indeed interchanges were more common in polyploid than in diploid *Dactylis*. This can be related to the fact that in polyploids they are not eliminated so rapidly because their effects on fertility are less marked.

Theoretically, the chromosomes of an interchange polyploid may show two types of pairing (STEBBINS 1950):

(1) Preferential—when the rearranged chromosomes associate mostly or exclusively as bivalents

(2) Non-preferential—when mainly multivalents are formed.

AHLOOWAHLIA (1963) has recently made a study of this problem in an interchange tetraploid of rye. A spontaneous interchange between chromosomes V (5,5') and VI (6,6') led to the development of two distinctive chromosomes— 5,6' which became the shortest, and 6,5' which became the longest, member of the complement. Tetraploids of this interchange heterozygote were synthesised by colchicine treatment of seeds from the heterozygote. At first metaphase in this interchange autotetraploid the 20 normal chromosomes formed II's, III's and IV's while the 8 chromosomes involved in the interchange formed associations of up to VIII. The interchange associations could be distinguished from the others by the nature of the chromosomes involved.

A comparison of the percentage of chromosomes involved in multivalent associations showed that 71.8–89.3% of the interchange chromosomes formed multivalents but only 28.6–45.6% of the non-interchange ones did so. Clearly, in this case, there is a strong tendency for the rearranged chromosomes to pair non-preferentially.

These interchange tetraploids of rye were more or less sterile. This contrasts both with normal tetraploid rye and with the progenitor interchange heterozygote which showed high fertility. Most of the associations involving the interchange chromosomes in the interchange polyploid were in fact non-zig-zag. In this instance, the interchange has a much better chance of surviving in the diploid as compared with the tetraploid state.

d) Secondary Pairing in Allopolyploids

At diploid mitosis in many dipterans, homologous chromosomes lie parallel and in pairs, a phenomenon referred to as somatic pairing. Somatic pairing is known also between the chromosomes of polyploid plants at mitosis. It is conspicuous but spurious, in endo-mitotically derived polyploid cells whether plant or animal, natural or induced (Fig. 126). And in these cases, as in polyploid somatic cells of dipterans, more than two chromosomes may lie in homologous association.

A similar juxtaposition of different bivalent chromosomes has long been known at meiosis in allopolyploid plants and the term secondary association was introduced by DARLINGTON and MOFFETT (1930) to describe this close association of bivalents into pairs or groups. This secondary pairing is frequently described as a post-synaptic phenomenon since it is first clearly apparent at pro-metaphase-I (but see page 133). It involves chromosomes of similar size and shape. It may be continued between pairs of daughter half-bivalents at first anaphase and is often most marked at metaphase-II, though this is probably a reflection of the first division association. Like somatic pairing, secondary pairing is variable in its occurrence, is most pronounced in species with small chromosomes and the associated chromosomes are not materially connected.

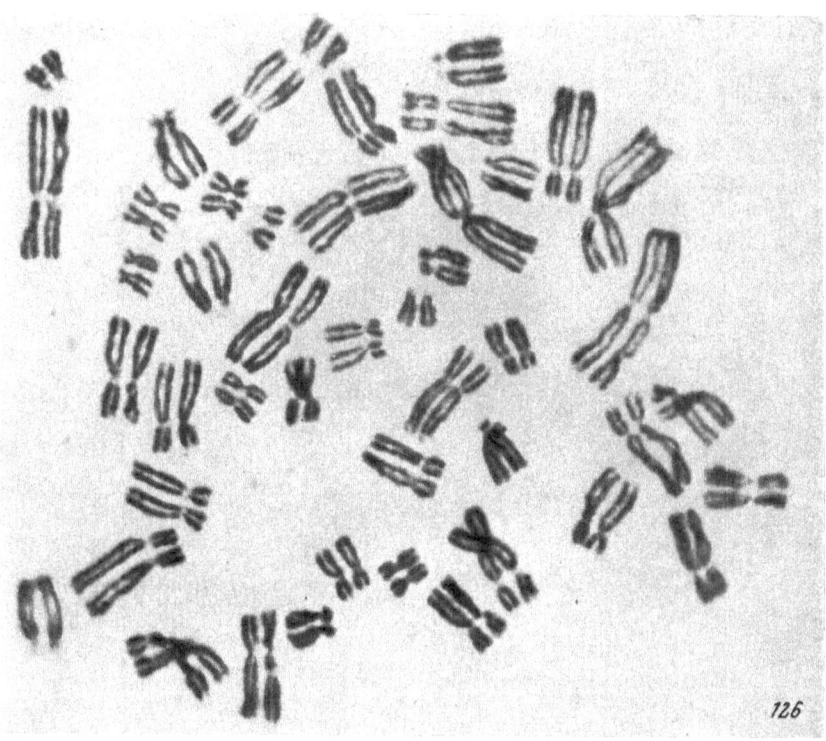

Fig. 126. A tetraploid cell produced by endo-mitosis and observed at C-mitosis from a human blood culture. Note spurious somatic pairing actually derived from the splitting of diplochromosomes. Taken from a preparation made by Dr. M. D. HAYWARD.

Most of the earlier workers, following the lead set by DARLINGTON (1928) and by LAWRENCE (1931), assumed this secondary pairing to be a residual attraction between distantly related chromosomes but this was never really established for most forms showing it had small chromosomes which did not lend themselves to critical study. HEILBORN (1936), on the other hand, maintained that secondary association depended on equality of size irrespective of homology.

One further complication arose from the failure to draw a distinction between secondary pairing due to homology and secondary pairing due to heterochromatic behaviour. Thus, as THOMAS and REVELL (1946) showed in *Cicer arietinum*, there is a marked precocity of primary pairing in heterochromatic regions which is followed, at pachytene, by secondary fusion of these regions. And these pachytene fusions persist through a diffuse diplotene and diakinesis to produce a secondary association at first metaphase. Here then fusions between heterochromatic regions at pachytene appear to be directly responsible for secondary pairing. But this fusion itself is not always at random. In autotetraploids induced by colchicine there is a marked preference for secondary association between homologous bivalents, whereas in the diploid state the heterochromatic fusions at pachytene are indiscriminate.

Recently RILEY (1960) has carried out an experimental analysis of secondary pairing in *Triticum aestivum*. Chromosomes III, XII and XVI are homoeologous and belong to group-3 (see Table 46). Three lines were produced carrying each of these homoeologues, in turn, as a pair of telocentrics. The three lines were then crossed in pairs, so producing F_1's heteromorphic for each of the two telocentrics introduced into the cross. Thus:

$$\text{Telo}_{III} \times \text{Telo}_{XII} \rightarrow F_1 \text{ Heteromorphic}_{III} \text{ Heteromorphic}_{XII}$$

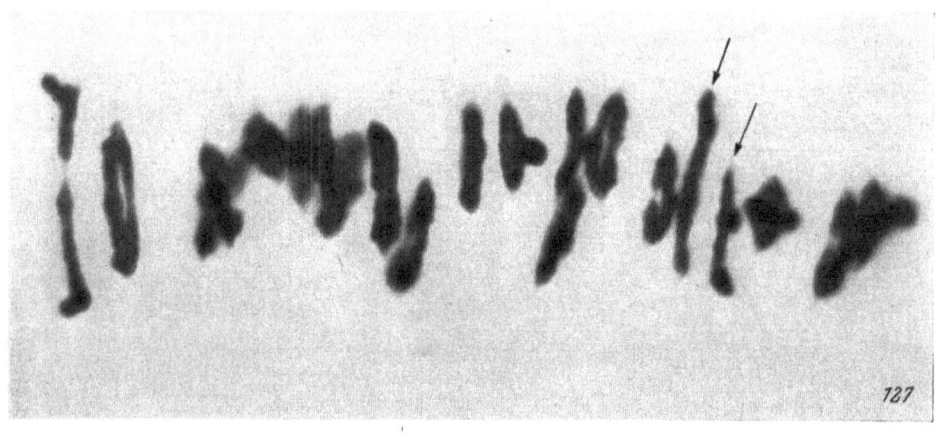

Fig. 127. Secondary association in *Triticum aestivum* heterozygous for two homoeologous bivalents (arrows) Photograph (ca. ×2000) kindly supplied by Dr. RALPH RILEY.

The relative positions of the two marked heteromorphic bivalents were then recorded in first metaphase cells with strictly linear alignments of bivalents by recording the number of unmarked bivalents that intervened between the marked heteromorphic homoeologues (Table 51). The considerable excess of cells with adjacent marked homoeologues may be taken as reasonable evidence that secondary pairing does in fact bring together homoelogous pairs—at least in *T. aestivum* (Fig. 127). An equivalent demonstration has also been made for two further sets of homoeologues, groups 1 and 7 (KEMPANNA and RILEY 1964).

Presumably zygotene pairing commonly brings homoeologues into proximity, though the action of chromosome-V prevents them from actually engaging in pairing (see page 120). This initial approximation of homoeologues would, however, result in their lying near to one another during post-zygotene and this, in turn, might be a pre-requisite for the eventual expression of secondary association. In terms of this explanation secondary pairing is a residual expression of prophase attraction, but this would be an attraction without actual contact. The only other possibility is that secondary association results from distinct forces of attraction that become manifest during the congression of the bivalents.

Table 51. *Relative Positions of Marked Homoeologues at First Metaphase in Triticum aestivum.*
(Data of Riley 1960.)

No. of intervening bivalents	Cells per intervening class					χ^2
	Telo's III + XII	Telo's III + XVII	Telo's XII + XVI	Total	Random expectation[1]	
0	21	40	23	84	39.22	62.55
1	9	21	15	45	37.29	1.59
2	8	15	12	35	25.30	1.12
3	9	10	8	27	33.33	1.20
4	6	10	6	22	31.39	2.81
5	6	11	8	25	29.42	0.66
6	6	9	6	21	27.48	1.52
7	6	8	5	19	25.50	1.66
8	9	3	5	18	23.57	1.32
9	9	11	5	25	21.59	0.54
10	5	4	1	10	19.61	4.71
11	6	4	5	15	17.67	0.40
12	4	3	4	11	15.69	1.40
13	5	3	3	11	13.72	0.54
14	3	6	1	10	11.78	0.27
15	4	4	3	11	9.81	0.14
16	3	3	6	12	7.87	2.17
17	1	2	2	5 ⎫	5.89 ⎫	
18	3	4	1	8 ⎬ 19	3.96 ⎬ 11.83	4.35
19	0	3	3	6 ⎭	1.98 ⎭	
Total	123	174	122	412	412	88.95

[1] Calculated from the formula $\dfrac{2(n-1-r)}{n(n-1)}$ where, n = total no. bivs.
r = no. intervening bivs.

$$\sum \chi^2_{[17]} = 88.95, \ P < 0.001$$

VI. Meiosis in Organisms with Non-localised Centric Systems

So far we have dealt with chromosomes whose centric activity is strictly localised. Most of these are monocentric, a few dicentric, so that at the most only two centric units are functional at any one time and these occupy constant positions. There is, however, a number of species which posses a non-localised centric system. The precise structural differentiation underlying this system is not well understood but the term "non-localised" is applied to cases where:

(1) The chromosomes do not show any primary constriction or any regions with a special cycle of division.

(2) The spindle fibres are—or at least appear to be—attached along the length of the chromosome so that the chromatids separate simultaneously along their entire length and then tend to move parallely to the poles, and

(3) Chromosome fragments, whether spontaneous or induced, tend to move normally at mitosis and meiosis and are not lost.

Two variants of the non-localised condition are known:

(1) The multiple or polycentric (polykinetic) condition, in which each chromosome is equipped with many centromeres individually separated by small non-

Table 52. *The Occurrence of Holokinetic Systems.*

Type	Species	Reference

1. Plants	(1) *Algae — Conjugales — Zygnemataceae*
	    Spirogyra crassa
	    Desmidaceae
		GODWARD 1961
		KING 1960

	(2) Angiosperms — Monocotyledons
	    Juncales — Juncaceae (probably all)
	      Luzula purpurea
	
	      Luzula Campestris
	    Cyperales — Cyperaceae (probably all)
	      Eleocharis palustris
		CASTRO, CAMARA and
		MALHEIROS 1949
		NORDENSKIÖLD 1961, 1962
		NORDENSKIÖLD 1961
		
		HÅKANSSON 1954, 1958

2. Animals	*Arthropoda* (1) *Scorpionidea — Tityus*	PIZA 1943 — see SCHRADER and HUGHES- SCHRADER 1961

	(2) *Acarina — Pediculopsis graminum*
	      Pediculoides ventricosus
	
	      Eylais setosa
		COOPER 1939
		HUGHES-SCHRADER and
		RIS 1941
		KEYL 1957

	(3) *Chilopoda — Thereuonema hilgendorfi*	OGAWA 1962[1]

	(4) *Insecta*
	  (a) *Hemiptera — Homoptera*
	               — Heteroptera
		see HUGHES-SCHRADER 1948
		SCHRADER and HUGHES-
		SCHRADER 1961

	(b) *Anoplura — Pediculus corporis*
	      Haematopinus suis
	  (c) *Mallophaga — Gyropus ovalis*
	  (d) *Odonata*
	  (e) *Lepidoptera*[2]
		BAYREUTHER 1955
		SCHOLL 1955
		see OKSALA 1943[1]
		see SUOMALAINEN 1953[1]

[1] Confirmation is required in these cases.

[2] BAUER (personal communication) found a higher frequency of viable than inviable translocations following the X-irrad. of eggs in *Pieris brassicae*. Clearly has this is the reverse of the situation expected for monocentric chromosomes.

centric segments and the chromosome is, in reality, a multiple or compound structure. The only certain instance of this is the germ line chromosomes of some ascarid nematodes.

(2) The diffuse or holocentric (holokinetic) condition where every point along the whole length of the chromosome seems to possess the property of active mobility. This system is much more widespread (Table 52).

Clearly, the distinction between these two systems is far from clear cut. At the least they can differ only in the spacing of the organising particles. In a truly polycentric system one might expect to produce minute fragments lacking centric ability which should therefore fail to maintain themselves. At the most, however, the two systems may differ, not merely in the spacing of their organising particles, but in their division cycles.

1. The Polycentric Compound Chromosomes of *Parascaris*

In the horse threadworm *Parascaris equorum* the gametic chromosomes, unlike those of the somatic cells, are compound polycentric units. During the early segmentation mitoses the collective germ-line chromosomes break up into a large number of small somatic chromosomes. The collective or compound chromosomes are, however, retained in thegerm-line. The numerous somatic chromosomes so produced each have their own functional centromere and are perfectly stable units.

The heteropycnotic end segments, however, are acentric and they are lost in the soma. Meiosis has not been described in detail. But if each centric unit was associated with regions having a special cycle of division (similar to those which obtain in the case of localised centromeres) difficulties would arise in connection with the terminalisation and resolution of chiasmata. What is more, chiasma formation is not expected to occur in highly condensed heteropycnotic segments. And, in the female at least, the end segments exist in this state at meiotic prophase. Indeed from some of the published figures (eg. Figs. 33 and 34, LIN 1954) it would appear that female meiosis is achiasmate in *Parascaris equorum*.

2. The Holocentric Chromosome

In most holocentric forms, pairing and chiasma formation are regular (but see section II–2) though in many species only one chiasma forms per bivalent and most commonly this is terminal. The details of post-diakinetic behaviour varies, however, in different forms. Two types have been particularly well studied and for this reason we shall concern ourselves predominantly with these.

a) *The Luzula system.*—This type is best known in *Luzula purpurea* (CASTRO, CAMARA and MALHEIROS 1949, NORDENSKIÖLD 1962) where only three bivalents are present at meiosis. These have either one or two chiasmata per bivalent (Fig. 128 for *L. maxima*) and interstitial chiasmata can persist to first metaphase. At first anaphase these chiasmata remain unresolved and thus persist as relict or half-chiasmata. As chromatid pairs move poleward these relict connections begin to lapse and they are eventually replaced by an active pairing of homologous chromatids during the interkinetic period.

These observations can be accomodated only on the assumption that the two chromatids of each of the chromosomes comprising a bivalent orient to opposite poles. That is, that the paired chromatids auto-orient in a manner reminiscent of mitotic chromosomes. This assumption is supported by observations of four kinds:

(1) The separation of rings of four in X-ray induced interchange heterozygotes, where chromatid rings of four pass to each pole at first anaphase (CASTRO, NORONHA-WAGNER and CAMARA 1954, LACOUR 1952).

(2) The separation of X-ray induced heteromorphic bivalents (LACOUR 1952).

(3) The behaviour of heteromorphic associations between a normal chromosome and its fragmented homologue in a 2n = 7 aneuploid produced in the progeny of X-rayed material (NORDENSKIÖLD 1962, see Fig. 129).

(4) The study of chromosome pairing and distribution in hybrids between *Luzula campestris* and related forms. NORDENSKIÖLD (1961) studied the course of meiosis in hybrids between *L. campestris* and one aneuploid and two polyploid derivatives in which the numerical differences were due to breakage rather than an actual quantitative increase. The normal diploid *campestris* has 12 chromosomes referred to as AL in size. In the aneu

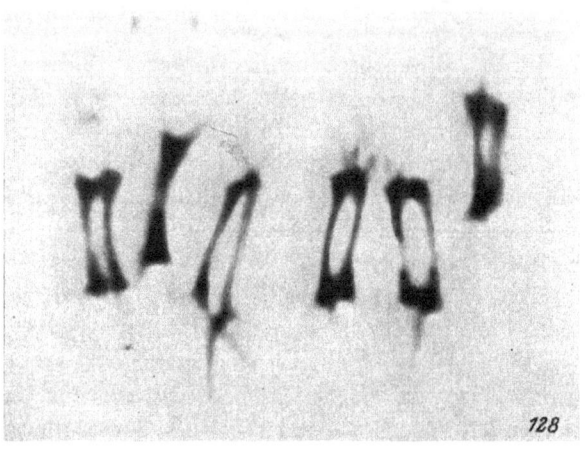

Fig. 128. First meta-anaphase of meiosis in *Luzula maxima* (2n = 12; carmine, ca. ×1000).

ploid (2n = 16) four of these are replaced by eight half-sized BL chromosomes. In the polyploids, on the other hand, there are 24 BL type chromosomes in the tetraploid *L. australasica* and 48 quarter sized CL-chromosomes in the octoploid *L. sudetica*. In her hybrids NORDENSKIÖLD was able to show that 2 BL chromosomes or 4 CL chromosomes correspond to and pair with one AL type (Table 53) which suggests that the most probable origin of the polyploids is through the fragmentation of the large AL chromosomes of the basic diploid. Moreover, since all four meiotic products remain together in the *Juncaceae*, it is possible to carry out a cytological tetrad analysis on the products of a single meiosis. And from the distribution of chromosome types between the four cells of a tetrad, NORDENSKIÖLD was able to confirm that the chromosomes must auto-orient at the first division so dispelling the earlier claim of BROWN (1954) that first division auto-orientation does not take place in *L. campestris*.

b) *The coccid-aphid system.*—This is seen in its simplest form in the coccid *Puto* (HUGHES-SCHRADER 1948). Here the single chiasma which forms in each bivalent is completely terminalised by first metaphase when each chromosome of the bivalent auto-orients as in a somatic metaphase. During first anaphase the two chromatids which move to the same pole retain a terminal connection. At interkinesis, however, the two chromatids of each half-bivalent disjoin completely and then re-associate in a side-by-side secondary pairing.

The main difference between the *Puto* sequence and that in *Luzula* is the fact that in the woodrush the two homologous chromatids of each half-bivalent re-pair without losing their terminal connection. One further difference is that in the male of *Puto* there is an unpaired sex (X) chromosome which, as one might expect from its auto-orientation, divides at first anaphase (see section VIII-1).

In *Puto*, as in most other coccids so far studied, the chromosomes are too small and too condensed to distinguish readily between their axes and hence their mode of orientation. The sequence of orientation can, however, be analysed with certainty where heteromorphic bivalents and/or trivalents are present at meiosis. The indirect evidence certainly points to an equational first division as a basic feature of the aphid-coccid type of meiosis. But the most cogent

Diplotene   *Methaphase-I*   *Anaphase-I*

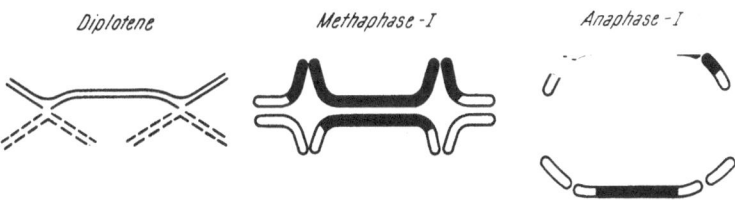

Fig. 129. The behaviour of an association of three produced by X-ray induced breakage in one homologue of *Luzula purpurea* (2n = 7) by NORDENSKIÖLD (1962).

evidence pointing to this conclusion comes from a study of irradiation-induced gynogenetic triploid females (2n = 3x = 15) in the mealy bug *Planococcus* (= *Pseudococcus*) *citri* (CHANDRA 1962). These originate from the activation of the triploid polar nucleus formed by the fusion of the three polar bodies. Under normal circumstances this participates in the formation of a highly specialised structure, the mycetome, which contains intracellular symbionts necessary for sexual differentiation.

In such females a varying number of trivalents, bivalents and univalents were seen in the triploid cells and all the trivalents formed had only two chiasmata. At first anaphase segregation was always 15 : 15. This is possible only if all the chromosomes, irrespective of their valency, auto-orientate at first metaphase.

In other coccids many deviations are known from the *Puto* pattern. These include the subdivision of the prophase nucleus into vesicles, asynapsis of the autosomes, complete diassociation of chromatids at first metaphase and asynchrony of first anaphase separation. Thus in *Llaveia* the principal difference lies in the vesiculation of the nucleus at meiotic prophase. This begins before the chromosomes can be resolved. But at diakinesis it is evident that, in the majority of spermatocytes, each vesicle contains either a pair of autosomes in chiasmate association or else the univalent X-chromosome. Evidently vesicle formation is preceded by some pairing process distinct from that at zygotene and one which may be initiated at the pre-meiotic mitosis. Occasional cysts are found containing homologous autosomes which are not joined by a chiasma. Presumably, zygotene pairing failed in these cases, due perhaps to an incomplete pre-zygotene association, though desynapsis cannot be excluded. But in about

5% of the spermatocytes, homologous autosomes are in separate vesicles or, less often, distinct lobes of the same sac. Such cases indicate a failure or incompleteness of pre-zygotene association leading to a preclusion of zygotene pairing. How ever they arise, univalents auto-orient at first metaphase and divide equationally. But, prior to second metaphase, homologous half-univalents show the same "secondary pairing" as the diassociated members of a half-bivalent. Asynapsis does not therefore lead to non-disjunction.

In *Nautococcus* the prophase nucleus becomes lobed, the lobes separating non-homologues, but distinct vesicles are not delimited. A curious feature which this type shares with some others is that the terminal association between the two chromosomes of a bivalent lapses even before metaphase so that these may shift their position relative to each other. In fact their anaphase separation is almost always asynchronous. When the separation of the chromatids of one chromosome is far in advance, the four homologous chromatids may take up a linear arrangement, the outer chromatids being those which separated first. In this event, the spindle fibres of the two chromatids in each half of the linear array "fuse" and these chromatids then proceed together to their particular pole. Such chromatids are thus already aligned in parallel at telophase and they do not dissociate.

Llaveiella shows the same early lapse of terminal association and asynchronius anaphase separation as *Nautococcus* but there is an added complication. In many coccid bugs, not only can half-chromatid units be resolved optically but they may separate and show degrees of mechanical independence.

Table 53. *Pairing Behaviour of Some Luzula Hybrids.* (Data of NORDENSKIÖLD 1961.)

Parents		Hybrid	Maximum pairing		Fertility
			Type	% observed	
1. Diploid L. campestris 2n = 12 AL	Aneuploid × L. campestris 2n = 16 = 8 AL + 8 BL	2n = 14 = 10 AL + 4 BL	4 AL_II + 2 (AL − 2 BL)_III	76	50% seed set
2. L. campestris 2n = 12 AL	× L. australasica 2n = 24 BL	2n = 18 = 6 AL + 12 BL	6 (AL − 2 BL)_III	71	75% seed set
3. L. campestris 2n = 12 AL	× L. sudetica 2n = 48 CL	2n = 30 = 6 AL = 24 CL	6 (AL − 4 CL)_V	—	Completely sterile

This is the case at meiosis in *Llaveiella*. The lapse of terminal attraction and the separation of "sister" half-chromatids, while the non-sister half-chromatids of a chromosome remain paired along the junction between sister chromatids, gives four pairs of non-sister half-chromatid units per "bivalent". These become variously arranged relative to each other while they retain their original orientation relative to the poles. At first anaphase these half-chromatid units behave independently and asynchronously.

Thus in *Llaveiella taenechina* the chromosome number is $2n = 4 + XX/XO$. At meiosis in the male there are therefore eight half-chromatid autosomal pairs. Presumably, the four half-chromatids derived from one bivalent are closer to each other than they are to the four half-chromatids from the other, non-homologous bivalent. Anyway the unit of orientation at second metaphase is a pair of half-chromatids. These are doubtless homologous but how they are constituted is not clear. The second anaphase separation of these units is also asynchronous and linear chains of four half-chromatids may result. These can be compared with the linear array of chromatids seen at the first division in *Nautococcus* and their anaphase separation is comparable.

Thus, the meiotic products in *Llaveiella* have as many separate half-chromatids as the mother cell has chromosomes. But at second telophase the half-chromatids unravel and are later seen to be paired and jointly coiled. It would appear therefore that a process akin to the "secondary pairing" of interphase occurs at the end of meiosis in *Llaveiella* and, presumably, the chromatid re-constituted from half-chromatids behaves subsequently as though it were essentially a single structure.

The physico-chemical conditions which determine this curious sequence can hardly be guessed at. But it poses genetical as well as cytological problems. The assumption, or conclusion, that the gametic chromosome has a polynemic structure in terms of its genetic component introduces many problems in relation to the structure of the metabolically active gene and in relation to mutation and recombination. Some of the difficulties introduced lose something of their terror if it can be assumed that, in respect of recombination and segregation, the sub-chromatid units behave as one. This assumption certainly seems to be justified in respect of recombination, at least in chromosomes with localised centromeres (see Part Two-section II). But in *Llaveiella*, the mechanical autonomy of the half-chromatid units adds to the difficulties introduced by polynemy in general. Only the complication added in relation to recombination will be considered here.

If crossing-over occurs in *Llaveiella* then the two half-chromatids included in the same spermatid can be genetically identical only if certain conditions are satisfied. Assuming that HUGHES-SCHRADER's description and interpretation are correct and, to our knowledge, they have not been doubted, these conditions are:

(1) "Sister" half-chromatids must be genetically identical.

(2) They proceed to the same pole at first anaphase. Both of these are satisfied by the description but one or other of the following must hold also, namely either:

(3) The half-chromatid pairs at second division must be composed of sister half-chromatids which proceed to the same pole while the homologous half-chromatid pair moves to the opposite one, or else.

(4) The half-chromatid pairs of second division must contain half-chromatids from different homologous chromosomes and orientation and disjunction of these pairs must be directed so that identical half-chromatids from different half-chromatid pairs proceed to the same pole. In other words, the plane of separation at second division must be in the plane of the chiasma and rotation must be avoided.

There is nothing in the description to suggest that the third sequence obtains. It would require that sister half-chromatids be arranged in pairs in the linear array of four half-chromatids at second anaphase—an arrangement different from that concluded for the chromatid-chain at first division in *Nautococcus* and that at second division in *Protortonia*. It may be, of course, that the half-chromatids which subsequently associate in the spermatids are genically different so that the gametes can have a kind of internal heterozygosity. Whether or not these different half-chromatids act together or separately cannot be discussed. But any polynemic structure introduces the possibility that the gene may be a laterally as well as a longitudinally differentiated structure. The possibility of a mosaic development arises only if the half-chromatids of one mitotic generation are related to the chromatids of the next in the same way as the chromatids of one generation are related to the chromosomes of the subsequent one. And they need not be (see also page 45).

The most bizarre meiotic patterns, however, occur in the males of mealy bugs, armoured and palm scales (BROWN 1963). Here, while the female maintains the typical aphid-coccid sequence, three highly irregular types of male meiosis are known:

(1) The diaspidid system—here the paternal chromosomes are eliminated during early embryogenesis after which the males develop as haploids and meiosis is replaced by an essentially mitotic division.

(2) The lecanoid system—with the exception of *Puto*, all the species of lecanoids so far studied possess a distinctive chromosome cycle where the paternal chromosomes become heteropycnotic at the blastula stage in male embryos and remain so throughout development and throughout meiosis. Two meiotic divisions occur: the first is equational for all chromosomes but at the second division the heteropycnotic chromosomes are segregated from the isopycnotic. This segregation occurs without any prior synapsis or even any form of secondary pairing. Finally there is a supression of sperm formation in half of the meiotic products, for only the two nuclei derived from the isopycnotic chromosomes produce normal sperm.

In lecanoids then, unlike the *Llaveiini* considered earlier, not only is synapsis suppressed but so too is secondary pairing. And this is combined with a differentiation of the two haploid sets of chromosomes which show differential cycles of condensation and differential meiotic behaviour. Only maternally-derived chromosomes are transmitted and so the males behave as though they were produced by haploid parthenogenesis. Indeed, since the heteropycnotic set is morphogenetically "inert" (BROWN and NELSON-REES 1961) the males also develop as though they were haploid.

(3) The Comstockiella system.—Here, as in lecanoids, the paternally-derived chromosomes develop positive heteropycnosis in male blastulae. But, with the exception of one chromosome, (D^H), these lose their heteropycnosity during pre-meiosis or early meiosis and pair with their euchromatic homologues. The

time of pairing also varies from pre-prophase to late prophase. No chiasmata have been seen and it is assumed by Brown that crossing-over does not occur. In some cases the homologues can be shown to be double prior to pairing, e. g., *Comstockiella sabalis* and *Ancepaspis tridentata*. Nevertheless there is only a single meiotic division with the pairing partners separating at random and both division products develop into sperm.

The single chromosome which remains heteropycnotic does not usually pair with its euchromatic homologue (D^E). Instead D^H divides equationally and is eliminated either by anaphase lagging or else by telophase ejection. Its daughter halves persist as pycnotic residues during the early phases of spermiogenesis but are excluded from the definitive gametes. The euchromatic homologue D^E also divides equationally and so makes a contribution to both meiotic products.

Since the D^E chromosome divides in the single meiotic division but the pairing homologues do not, it follows that, in the sperm the D^E chromosome is expected to consist of a single chromatid unless there is a compensating reproduction. And this, unchecked, is expected to produce mosaic embryos containing aneuploid sectors. Since this does not obtain it is clear that some form of compensation mechanism must operate. In *Aonida lauri* this appears to take place at pre-meiosis or early meiosis though the details are as yet obscure. In other cases e. g., *Comstockiella sabalis*, it may take place at a post-meiotic stage. In the former event the behaviour of the D-chromosomes is reminiscent of that shown by the X and Y chromosomes in sex-ratio *Drosophila* (see Part Two-IV-2). However, in some species, as in *Ancepaspis tridentata* and *Odonaspis penicellata*, the D-members do not appear to be a fixed pair. Rather the identity of the D-pair varies, though in any one cyst the same pair is involved in all cells.

It will be appreciated, however, that while an "extra" duplication brings the D^E chromosome into line with the others, the net result of the division sequence is the production of sperm nuclei whose chromosomes are double. Indeed, the suppression of one of the meiotic divisions is a common means of avoiding reduction in apomicts. Now, sperm generally have a 1C-DNA level. Consequently, the DNA which male gametic nuclei are obliged to make prior to the first cleavage division must be synthesised while these nuclei are in the egg cytoplasm. But, in the Comstockiella system there is only one meiotic division. If therefore the two components of the chromosomes are chromatids, as Brown believes, and not sub-chromatid units, then the sperms have an unreduced chromosome complement. And this requires a second kind of compensation: the chromosome reproduction which normally precedes first cleavage must be omitted. A study of DNA synthesis during sperm formation in these types would prove most rewarding*.

Finally, it is possible to get both lecanoid and Comstockiella systems operating in the same individual and, in some cases, even in different cysts of the same testis as in *Nicholiella bumeliae*.

An important conclusion which emerges from a study of diffuse centric systems is that certain concepts formulated in connection with localised systems are not strictly applicable or have a modified meaning where the centromere is non-localised.

---

* See note 2 Appendix.

First, the distinction between cross-over and non-crossover segments is made relative to a localised centromere, the segregation of which is invariably pre-reductional; so also, therefore is that of non-crossover segments. Second, terminalisation is, of course, a movement towards the chromosome ends. But, where centric activity is localised, it is also a movement away from the centromere, i. e., towards the end of an arm. But arms have no meaning where the centromere is diffuse. To which end therefore, will chiasmata move in chromosomes with a non-localised centromere, for these do not show localised repulsion? Presumably to the nearer end, but this may not be the case: one end may be different from the other and where two chiasmata form they may move away from each other, i. e., one towards each end, irrespective of their joint position relative to the ends.

In this connection it is interesting to note that bivalents between unequal-sized chromosomes in an inter-racial hybrid of *Cimex* were asymmetrical at metaphase by which time terminalisation was complete (see page 144). This means that the chiasmata moved away from the inequality so that "symmetrical" bivalents with a lateral chiasma were not produced.

However, terminal movement in diffuse systems does provide some basis, consistent with that used in organisms with localised centromeres, for distinguishing cross-over and non-crossover segments, at least where a single chiasma is involved. The non-crossover half-loop is that which, by terminalisation, extends at the expense of the other. Thus, in a rod bivalent with a terminal chiasma derived from a single interstitial one, the cross-over chromatids are on each side of the terminal association whether the centromere is localised or not. Where it is localised, and as we shall see in some cases where it is not (e. g., *Hemiptera*: *Heteroptera*, see page 144), a rod of this kind orients lengthwise on the spindle and separation is across the chiasma. This means that the non-crossover segments show pre-reduction while cross-over regions separate equationally at first division.

But in *Luzula*, in *Puto*, and indeed in all the *Hemiptera-Homoptera*, the rods lie with their long axis at right angles to that of the spindle. Anaphase separation is then post-reductional for the non-crossover regions and the two chromatids which move together at first anaphase are, as we have seen, associated by a half-chiasma. Thus, while the onset of first anaphase in localised systems is determined by the cessation of both terminal attraction and that between homologous chromatids in pairs, a lapse of only the latter is required in the *Luzula-coccid* type of meiosis. However, the former is also finally broken at telophase in coccids and, as we have seen, re-pairing may then be necessary for the second meiotic sequence. This secondary pairing involves a parallel and, presumably, specific alignment of homologous chromatids. In localised systems the complete separation of homologous chromatids is prevented by the regions of association adjacent to the centromere. Should these lapse precociously, as DOWRICK (1953) found in abnormal pollen mother cells of a presumably apomictic *Chrysanthemum*, homologous chromatids fall apart.

The principal differences between the standard sequence in types with localised centromeres and the basic sequence in non-localised types are, therefore:

(1) There is no point of localised repulsion from which terminalisation can be said to proceed.

(2) Bivalents lie in the plane of the equator and non-crossover regions show post-reduction.

(3) There are no localised regions of association or active mobility, and

(4) Homologous chromatids disjoin, partially in *Luzula* and completely in coccids, at first telophase and then re-pair prior to second metaphase.

The co-existence of non-localised centric systems and first division auto-orientation led CASTRO (1950) to the conclusion that the one peculiarity (auto-orientation) depends on the other (diffuse spindle fibre system). There is a number of arguments which militate against this generalisation, the most formid-able of which has been raised by UESHIMA (1963). Despite objections to the contrary (see below) it is now generally accepted that all hemipteran insects have holokinetic chromosomes. UESHIMA succeeded in crossing two "strains" of the heteropteran *Cimex pilosellus* which parasitise two distinct host species. In the males of strain A (host species *Antrozous pallidus*) there were twenty-two autosomes and an XY pair where the X was the largest and the Y the smallest member of the complement. In strain B (host species *Eptisecus fuscus*) only twenty autosomes were present and here the Y was indistinguishable from the autosomes so far as its size was concerned. Clearly these "strains" are in fact cryptic species. Hybrids were made by crossing females of strain A with males of strain B and at meiosis the autosomes formed nine bivalents, eight of which were hetermorphic, and one trivalent. The heteromorphic pairs always con-tained a single chiasma which had completely terminalised by first metaphase and the two short chromatids invariably passed to one pole and the two longer ones to the other. In the trivalents, there were always two terminalised chiasmata by first metaphase and the configuration always orientated such that two of the six chromatids moved to one pole and the remainder to the other.

SUOMALAINEN and HALLKA (1963) also have evidence for first division co-orientation in male *Psyllina* (jumping plant lice) which like the aphids and the coccids belong to the *Sternorrhyncha*. And this type of orientation also charac-terises male *Auchenorrhyncha* (HALLKA 1959).

The situation is somewhat complicated by the claim of PIZA (1958) that in scorpions and in the *Hemiptera* the chromosomes are not holocentric but dicentric, the centromeres occupying terminal positions at opposite ends (di-telocentric). KUSANGI and TANAKA (1959) and KUSANGI (1962) have made similar claims for *Luzula*. The argument in support of these claims is based on two kinds of observation:

(1) The apparent restriction of centric activity to the chromosome ends at meiosis, and

(2) The turning-in of the two extremities of each chromatid at meta-anaphase towards the pole of the spindle, a feature which is particularly pronounced in scorpions.

The SCHRADER's (1961) have recently re-examined the first problem in detail using three species of pentatomid bugs—*Euschistus servus*, *E. tristigmus* and *Solubea pugnax*. In all three species the entire mitotic chromosome shows centric activity with chromosome fibres forming along the entire length of each chromatid so that the chromatids remain parallel as they separate. Moreover, experimentally-induced chromosome fragments continue to divide both at mitosis and meiosis.   But at normal meiosis there is a restriction of centric activity to a limited region of the chromosome ends at both first and second metaphase, so that the chromatids move end-first to the poles.   This restriction does not involve any irreversible change in chromosome structure for, as we have already mentioned, fragments induced by X-irradiation perform a normal meiosis.   In different species of heteropterans this change from the holo- to the telo-kinetic condition involves differences in time of onset and degree of effect. Thus in the X-chromosome of the coreid *Protenor* (SCHRADER 1935) it occurs

Table 54.   *The Occurrence of Neo-centric Systems.*
Those at mitosis are not, in our opinion, equivalent to those at meiosis.

Division	Species	Reference
1. Mitosis	*Haemanthus katherinae* and *albiflos puniceus* *Leucojum aestivum* and *vernum* *Clivia cyrtanthiflora* *Iris aphylla* and *pumila* *Hymenophyllum tunbridgense* *Triturus viridescens*	BAJER and ÖSTERGREN (1961)
2. Meiosis	*Secale cereale* *Zea mays* *Elymus wiegandii* *Lilium formosanum* *Pleurozium schreberi* *Phalaris* hybrids *Bromus* hybrids	KATTERMANN (1939) PRAKKEN and MÜNTZING (1942) ÖSTERGREN and PRAKKEN (1946) REES (1955) RHOADES and VILKOMERSON (1942) VILKOMERSON (1950) ZOHARY (1955) VAARAMA (1954) HAYMAN (1955) WALTERS (1957)

between the first and second meiotic divisions while in *Rhytidolomia senilis* it occurs after a typical holokinetic orientation.   Indeed in this latter species the chromosomes are so large that one can actually follow the shift from holocentric to telocentric orientatation.

This restriction of centric activity to the ends is absent from all the coccids and aphids so far studied and the SCHRADERS have explained this in the following terms.   In the *Homoptera*, as in *Luzula*, the problem of orientation of a diffuse centric system in the presence of unterminalised chiasmata has been overcome by reverting from co- to auto-orientation at first meiosis.   But in the pentatomid bugs, as in many other heteropterans, the temporary restriction of centric activity to the chromosome end functions as an alternative mechanism.

As the SCHRADERS themselves point out, the disproportionate activity of the chromosome ends which leads to them taking precedence in anaphase movement has never been adequately explained though it has been observed in many organisms with a non-localised centromere.   In *Tityus*, as indeed in *Haematopinus* (BAYREUTHER 1955), this movement of ends is evident even during first metaphase

and it is this behaviour which convinces PIZA of the dicentricity of the system not-
withstanding the fact that some X-ray induced interstitial chromosome fragments
in *Tityus* appear to show the same terminal activity (RHOADES and KERR 1949).

Now a number of lines of evidence show that the ends of chromosomes possess
special properties reminiscent of those shown by the centromere. Thus:

(1) The telomere, like the centromere, has a compound structure (LIMA-
DE-FARIA 1956).

(2) Both regions may show a special cycle of division (LIMA-DE-FARIA and
BOSE 1963).

(3) The two regions exhibit a tendency for non-homologous association at
meiotic prophase (see Table 58) both amongst themselves (centromere with
centromere and telo-
mere with telomere) and
with each other (centro-
mere with telomere).

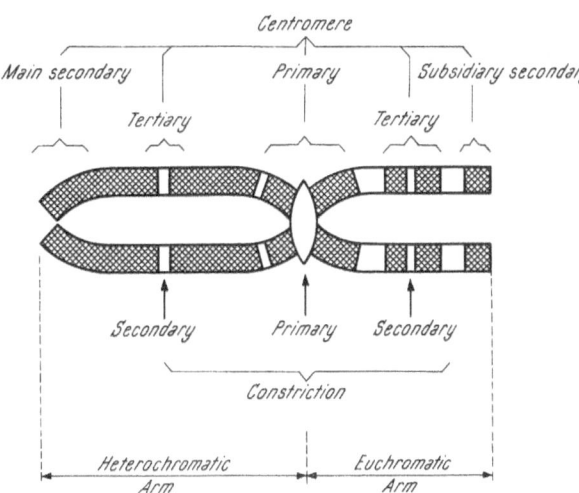

Fig. 130. Structure and centric activity of the A-chromosome in *Pleurozium schreberi* according to Vaarama (1954).

(4) Both regions
show an attraction to
the polar organisers of
the cytoplasm leading
to the phenomenon
of polarised behaviour.
Thus when the chromo-
somes first appear at
the onset of division
they often lie with their
centromeres directed to-
wards the pole of the preceding division having
remained throughout in-
terphase in a Rabl or
polar field, the consequence of centromere-centriole attraction at the preceding
ana-telophase.

Similarly at zygotene-pachytene of first meiotic prophase the chromosome
ends are commonly polarised towards the centriolar apparatus in a bouquet
polarisation (see also page 6).

(5) Like the centromeres, the telomeres can manifest active mobility when
on the meiotic spindle (Table 54 and see Part Two, IV-2). Such active ends
are called neo-centromeres and they have been observed both at mitosis and
meiosis. At meiosis, however, neo-centric activity is long-lasting whereas at
mitosis the activity is short-lived and acts for only a fraction of the mitotic
period. At meiosis, these neo-centromeres always arise in complements with
heterochromatic segments, that is, it is dependent on a specific genotypic constitution, though euchromatic ends may also show such activity. Extra or
supernumerary chromosome fibres develop from the end of the chromosome but
only in the presence of the main centromere. By contrast, at mitosis, neo-centric
regions can appear at any position though only in one of two sister chromatids
at any one time.

The most remarkable neo-centric system on record is that of *Pleurozium* (VAARAMA 1954). At mitosis only the primary centromere is active but at meiosis a variety of sites may show centric activity and exceptionally these may function simultaneously. The A-bivalent, for example, is characterised by the presence of a heterochromatic arm and a predominantly euchromatic one. The commonest site of secondary activity is the distal end of the H-arm and, less commonly, the corresponding end of the E-arm. The secondary constrictions in both arms may also show activity but this is rare (Fig. 130). Finally, by X-ray treatment it was possible to produce fragments of two kinds—"telic" (possessing an end segment) and "atelic" (lacking an end segment). And, in this event, fragments with an end segment behaved normally despite the fact that they lacked a true centromere.

These relations between centromere and telomere suggest that perhaps one should view the end activity of the chromosomes in species with diffuse centromeres in this light rather than arguing for a strict dicentricity.

VII. Meiosis in Species Hybrids

Species are differentiated to different degrees and in different ways. These include differences in gene frequency or type, chromosome morphology, chromosome number or various combinations of these. Consequently when species are crossed, either in nature or by experiment, the course of meiosis will be variably affected. Most species which produce hybrids with regular meiotic pairing and normal fertility appear to be differentiated primarily by genic factors. On the other hand most hybrids derived from parents differing in chromosome structure or number are at least partially sterile (Table 55).

If the parental species have the same number of chromosomes there may be sufficient homology between them to allow them to pair at zygotene. But equality of chromosome number is not sufficient to guarantee pairing, for synapsis is often partially or even completely inhibited in hybrids between species which have identical chromosome numbers. Alternatively pairing may be near normal even when the parents have marked differences in chromosome number as we have seen in hybrids between different species of *Luzula* (see page 137) and as has also been shown in certain lepidopteran hybrids. But in both these examples the centromeres are not localised.

A second feature which tends to characterise species hybrids is a reduction of chiasma frequency (Table 56) and this can occur even following apparently normal pairing. On the other hand in a few cases there may actually be an increase in chiasma frequency in the F_1 hybrid. Even where pairing and chiasma frequency are normal, fertility is not guaranteed for spindle development may be abnormal as may the relations of the chromosomes to it. And sometimes, though meiosis itself is completed, there is a wholesale degeneration of the post meiotic cells so that no normal gametes are formed. WHITE (1946) described such a condition in male F_1 hybrids between the various subspecies of *Triturus cristatus* and *T. marmoratus*.

However, while pre-meiotic failure leading to sterility must be genotypically determined, post-meiotic failure does not always depend on segregation. Thus,

Table 55. *Classification of Meiotic Behaviour in Diploid Hybrids.* (Based on White 1954.)

Type	Example	Reference
(1) Complete pairing and normal behaviour	Euschistus variolaris × E. servus Cimex lectularis × C. columbaris Trimerotropis maritima × T. citrina Syrmaticus soemmeringii × Chrysolophus pictus Pergesa elpenor × P. porcellus	Foot and Strobell 1914 Darlington 1939 Carothers 1941 Yamashina 1943 Bytinski-Salz 1934
(2) Incomplete pairing leading to univalent formation, otherwise normal	Dicranura erminea × D. vinula vinula Cerura bifida × C. furcula Trimerotropis suffusa × Circotettix verruculatus	Federley 1943 Helwig 1955
(3) Complete or incomplete pairing of structurally altered chromosomes, otherwise normal	Lolium perenne × L. temulentum Eyprepocnemis plorans ornatipes × E. p. meridionalis Crepis fulginosa × C. neglecta	Naylor and Rees 1958 John and Lewis 1965 Tobgy 1943
(4) Complete pairing but spindle mechanism and/or anaphase mechanism defective	Chorthippus biguttulus × Ch. bicolor	Klingstedt 1939
(5) Incomplete or variable pairing, spindle mechanism abnormal	Drosophila pseudoobscura ♀ × D. persimilis ♂	Dobzhansky 1934
(6) Little or no pairing, spindle mechanism abnormal	Drosophila persimilis ♀ × D. pseudoobscura ♂	Dobzhansky 1934
(7) Meiosis does not go beyond early first prophase	Gallus gallus × Phasianus colchicus Equus caballus × E. asinus	Yamashina 1943 Trujillo, Stenius, Christian and Ohno 1962

Table 56. *Pairing Properties of Some Newt Hybrids.*
(Data of CALLAN and SPURWAY 1951, LANTZ and CALLAN 1954 and SPURWAY and CALLAN 1960.)

Type	Mean no. Xta/cell.	Max. no. detectable multiples	
		F_1	Back-Cross (Bx)
Triturus cristatus karelinii	42.2	—	—
Triturus cristatus cristatus	37.4	—	—
T. c. karelinii × *T. c. cristatus*	21.2	1	—
Triturus c. carnifex	31.42	—	—
T. c. carnifex × *T. c. cristatus*	16.1 − 21.9	—	—
T. c. carnifex × *T. c. karelinii*	15.1 − 21.2	2	—
Triturus marmoratus	24.25	—	—
T. marmoratus × *T. c. carnifex*	10.8	—	—
B_x − F_1 ♀ (*T. m.* × *T. c. carnifex*) × *T. marmoratus* ♂	7.5 − 21.3	—	3
T. vulgaris meridionalis	22.2	—	—
T. helveticus helveticus	24.3	—	—
T. v. meridionalis × *T. h. helveticus*	7.6	3 +	—

genotypic sterility is indicated in newts (though the effect is post-meiotic), for, since the sperm is not a highly metabolically active phase, unbalanced nuclei frequently survive. This is shown, for example, by the fact that F_1 segregational sterility in animals is expressed as F_2 inviability. In this respect they differ from plants. Thus, while the chiasma frequency in the diploid *Primula kewensis* hybrid is lower than that of its parents, pairing is generally good (Table 57).

Table 57. *Chiasma Frequencies in the Parental Species and in the Diploid and Tetraploid Hybrid, Primula kewensis* (x = 9).
Data of UPCOTT 1939.

Form.	No. of nuclei analysed	No. of chiasmata per Configuration					Chiasmata		
		0	1	2	3	4	Total	per nucleus	per bivalent
P. floribunda	10	0	10	70	10	0	180	18.0	2.00
P. verticillata	10	0	11	58	18	3	193	19.3	2.14
P. kewensis (2x)	20	17	82	80	1	0	245	12.3	1.37
P. kewensis (4x) 2 I	10	...	...	...	...	...	0	...	...
II		0	19	95	10	0	239	...	1.93
III + I		0	0	1	3	0	11	...	1.38
IV		0	0	0	4	20	92	...	1.92
Total		0	19	96	17	20	342	34.2	1.90

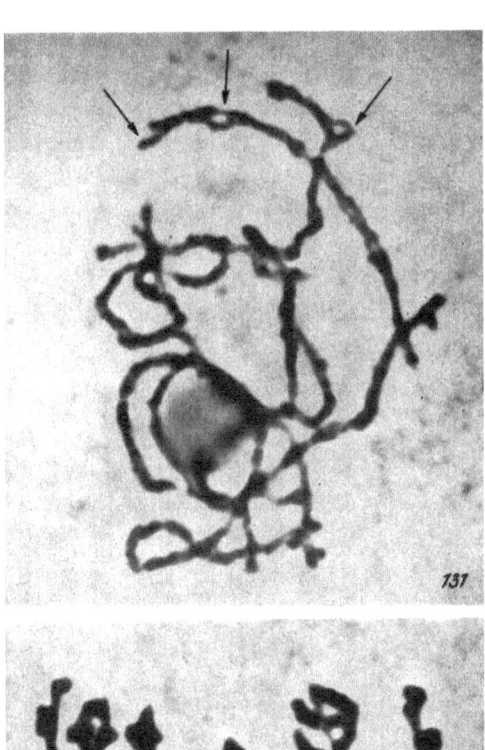

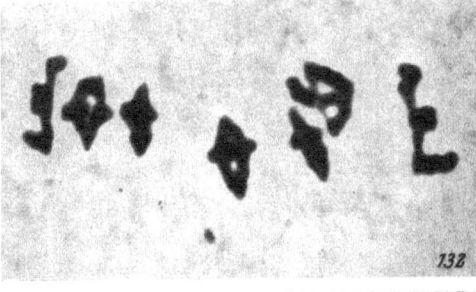

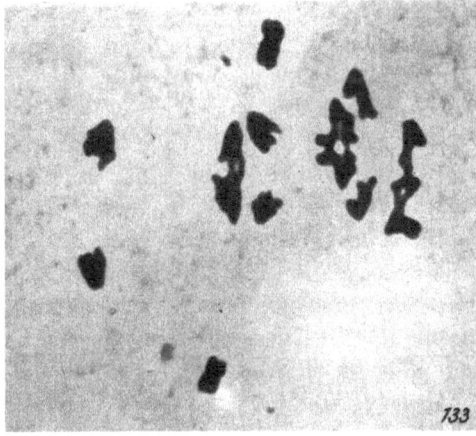

Figs. 131–133. Chromosome behaviour in a *Lolium temulentum* × *Lolium perenne* hybrid. Fig. 131—Pachytene pairing (carmine, ca. × 2000). Note unpaired terminal and interstitial (buckle) duplications (arrows). Figs. 132 and 133—First metaphase and first anaphase to show heteromorphic bivalents. The larger chromosomes are those of *L. temulentum* (carmine, ca. × 1500). Photographs kindly supplied by Dr. HUBERT REES.

Even so, sterility is complete. Despite appearances, this must be segregational and not genotypic sterility. It must be attributed to the recombination of cryptic differences, largely in those parts of the chromosomes which are not commonly subjected to crossing-over. This accomodates the good pairing. In the tetraploid hybrid, however, there is preferential autosyndetic pairing and multiple associations are not as common as the high pairing in the diploid hybrid might lead one to expect. Recombination between the parental genomes is thus suppressed and fertility is restored. Doubling is not expected to have this effect when the F_1 is subject to genotypic sterility.

Where structurally determined differences in size exist between related forms these are maintained in the hybrid between them. This has several interesting consequences. For example, the chromosomes of *Lolium temulentum* (2 n = 14) are about a third longer than those of the related *L. perenne* (2 n = = 14). At pachytene unpaired loops are frequent (Fig. 131) and at metaphase the inequality of chromosome size is manifest by the production of up to 6 asymmetrical bivalents (Figs. 132 and 133). Evidently duplication-deficiency differences exist between the two forms (NAYLOR and REES 1961) as indeed DNA measurements confirm (REES, personal communication). Similarly in *Narcissus* variety "Geranium" (2 n = 17), a commercial hybrid between *N. poeticus* (2 n = 14) and

N. tazetta (2 n = 20), up to 7 asymmetric bivalents may form though usually there are only 1–3 pairs present in any one cell (Figs. 134–135). This example is interesting too because although a first division spindle forms and the chromosomes orient normally on it, there is no first anaphase separation, for sister chromatid attraction does not lapse at first metaphase. Bivalents and undivided univalents thus persist into the second meiotic division though they lose their major coils during the restitution interphase which intervenes between the two meiotic sequences (Figs. 136–137).

There are thus two ways of studying the degree of structural differentiation between chromosome complements. First by a straightforward comparison of

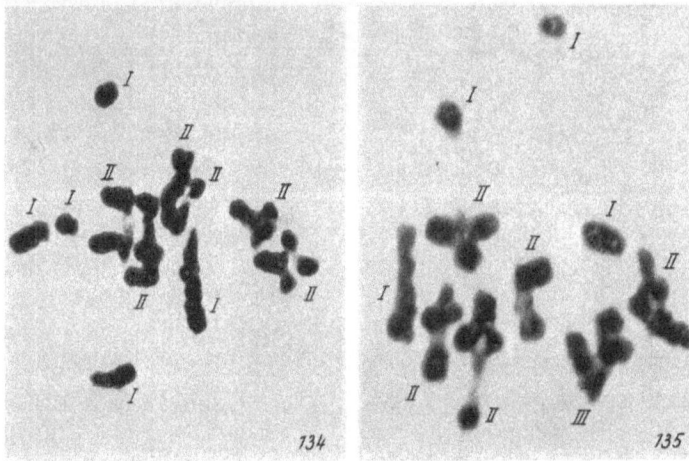

Figs. 134 and 135. Metaphase-I in the commercial hybrid *Narcissus* "Geranium". Note univalents, asymmetric bivalents and in Fig. 135 one trivalent (carmine, ca. × 1000).

the morphology of the chromosomes concerned. Secondly by analysing the pairing behaviour of the chromosomes in hybrids between the two complements. The success of the morphological method depends on the presence of well defined and conspicuous differences involving either chromosome size or else the position of a conspicuous marker like centromere position, secondary constrictions or heterochromatic segments. In relatively few cases therefore can it give unequivocal results. The meiotic behaviour of hybrids is generally held to constitute a much more sensitive test of homology and in favourable cases even small structural differences can be detected. A study of chromosome pairing in hybrids can thus provide valuable information regarding the homology of the chromosome sets in dissimilar taxonomic entities. Indeed it can do more than this—it can even provide evidence of differences in cases where the taxonomist, using his more conventional methods, has failed to find them. Some caution is necessary, however, in assessing the results of the meiotic test. There are five reasons for this:

(1) The work of SEARS and RILEY (see page 120) has shown that in bread wheat, a natural allo-hexaploid, the apparent absence of homoeology between the three ancestral diploid parents is misleading. It depends not on the absence of

structural or genic similarity between the ancestral diploids but rather on the presence of one or a small number of genetic loci which prevents synapsis between homoeologues without affecting the pairing behaviour of homologues.

The implications of this finding are clear—one cannot use absence of pairing in hybrids as unequivocal evidence for absence of homology. But this caution applies primarily to the negative side of the situation. When chromosome pairing does occur, even when it fails to occur consistently, there is good reason in many

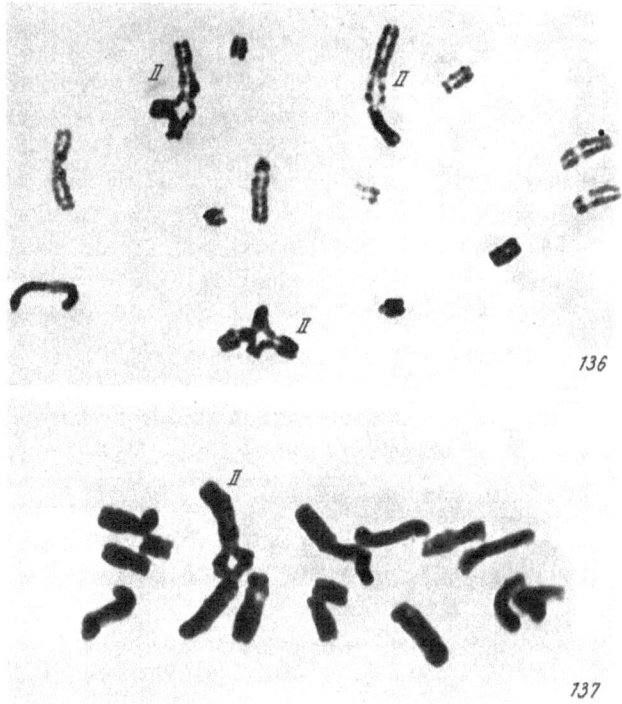

Figs. 136 and 137. Metaphase-II in the commercial hybrid *Narcissus* "Geranium". Chiasmata are even clearer at second division owing to the disappearance of the major coils (carmine, ca. ×1000).

cases to believe that the chromosomes which pair are at least partially homologous. Likewise in many apomictic types meiosis on the male and female sides is often very different. This cannot be due to structural differences nor, indeed, to genotypic ones. But it must be due to differential activity of the genotype comparable with that which determines differentiation itself (LEWIS and JOHN 1963a).

(2) Under certain circumstances chromosomes may show subsidiary associations of various kinds (Table 58). Some of these appear to be due to genuine homology or at least to homoeology (see page 119). But a large number appear unquestionably to be non-homologous. Where pairing is observed in hybrids (or indeed in other instances too, see section V-1), one has therefore to eliminate the possibility of such non-homologous associations before concluding that pairing can be taken as evidence for either homology or homoeology.

The most thorough analysis yet made of subsidiary chromosome associations was paradoxically also the first. This was carried out in 1933 by McClintock who studied the problem in *Zea mays* and came to the following conclusions:

(i) When unbalanced chromosome complements (monosomics, trisomics, deficiency heterozygotes) are present in maize, non-homologous associations involving the chromosomes responsible for the unbalance frequently take place. The same is true when structurally dissimilar complements are present as in individuals heterozygous for an inversion or a translocation. Non-homologous associations are also found occasionally in normal diploids.

(ii) Non-homologous associations appear to be just as intimate as those between homologous parts of chromosomes. And associations of non-homologous parts can proceed at the same time as homologous association: indeed it may even compete with it.

(iii) Associations tend to be in two's whether or not the parts associated are homologous. There are, however, two common exceptions to this; centromere associations and heterochromatic-knob-associations. These, however, do not persist beyond diakinesis. But ends of chromosomes tend to associate in advance of other parts of chromosomes whether the ends are homologous or not.

(iv) Where unpaired chromosomes are present they commonly show some form of folding so bringing non-homologous parts into intimate contact. Such foldbacks are found in single B-chromosomes, in univalents of monosomic plants and, as we have seen in the meiosis of *Chorthippus*, in the X-chromosome of XO species. Many of the foldbacks suggest that the two ends of a chromosome have a tendency to become associated regardless of the way in which the rest of the chromosome behaves (Fig. 138). This means that folding need not always be symmetrical. By diakinesis however, the univalent commonly appears as an open rod. Non-homologous foldbacks were also common in plants heterozygous for deficiencies (Figs. 138*c* and *d*) and in haploids (see section V-1) where foldback pairing between chromosomes frequently persisted to first metaphase giving loose associations simulating "bivalents". Finally foldback pairing also took place in parts of bivalent chromosomes not paired linearly with their homologues in normal diploids (Fig. 138a).

(v) In a paracentric inversion of the short arm of chromosome 8 no loop configurations were present in many cells, the short arm pairing straight. In addition there were many gradations between perfect rods and perfect loops but these two kinds of configuration predominated (see also section IV-1a).

What emerges quite clearly from this work is that in numerically unbalanced or structurally heterozygous complements pairing is not always by homology.

(3) White (1961) has claimed that the occurrence in hybrids of multiples maintained solely by terminal associations may not be indicative of interchange differences as has been held in the past. Rather it depends on an homology between short terminal segments of otherwise non-homologous chromosomes; an homology which is not expressed under normal pairing conditions. But where, as in many hybrids, synapsis is reduced this end-homology finds expression in the formation of multiples which are held together by terminal associations. It is in these terms that he re-interprets the claim, made after a study of meiosis

Table 58. *Subsidiary Systems of*

Type			Example	
			Animals	Plants
a)	In haploids (see pg. 103)		No known example	Antirrhinum, Rye, Barley, Maize, Oenothera
b)	In polyhaploids (see pg. 107) assocn. of { side-to-side univalents { end-to-end		No known example	*Triticum aestivum*
c)	In diploids			
	(I)	Heterochromatic associations (see also VII below)	*Diptera* *Orthoptera*	*Cicer arietinum* *Lycopersicon esculentum*
	(II)	Centromere associations	Lampbrush chromosomes of ♀ newts *Diptera* *Orthoptera*	*Oxalis dispar* (telocentrics only) *Trillium hagae* (3x) and *kamtschaticum* (2x) *Lycopersicon esculentum*
	(III)	Telomere fusions	Lampbrush chromosomes of ♀ newts	
	(IV)	Reflected fusions, i.e. fusions of primary gene products (loop matrices)	Lampbrush chromosomes of ♀ newts	
	(V)	Undefined (see pg. 162)	*Gastrimargus africanus*	
	(VI)	Secondary pairing in chrms. with diffuse centromeres (see pg. 137)	Coccids Aphids	*Luzula purpurea*
	(VII)	Sex chromosomes (see pgs. 185 and 193)	X_{yp} and X_nY *Coleoptera*	No known example
			Diptera *Neuroptera*	No known example
	(VIII)	Inversion heterozygotes	Grasshoppers (pericentric)	Maize (paracentric)
d)	In Allopolyploids-secondary association (see pg. 131)		No known example	*Triticum aestivum* *Ornithogalum, Prunus*

Chromosome Association at Meiosis.

Nature	Cause
Pachytene to Anaphase-I	Mainly non-homologous and due either to het. assocn. (type [c] I below) or foldback pairing.
Post-pachytene to Anaphase-I	s-s by homology e-e by non-homology
Pachytene to diakinesis or in some cases to M I or even M II producing a s-s alignment of bivalents	Non-homologous
Diplotene to M I sometimes persisting to M II. In *Trillium* it has been shown to be random, though assocns. including larger chromosomes occur more often. Also shown more readily in the 3x species	In most cases probably non-homologous, sometimes involving heterochromatic material. But in newts most associations involve homologous centromeres
Diplotene to metaphase I	Both homologous and non-homologous fusions may occur
Diplotene	Usually homologous involving products of allelic pairs
Metaphase I to metaphase II	Presumably non-homology
Late telophase-I to metaphase-II	Chromatid homologous
Pachytene to Anaphase-I, common association to an RNA-body	Non-homologous
Pachytene to metaphase-I. Heterochromatic fusion (see I above)	Non-homologous
Pachytene pairing apparently normal in inverted segments	Non-homologous
Metaphase-I, sometimes persisting to Methaphase-II	Homoeology

in F_1 inter-racial and interspecific hybrids of the genus *Triturus* (Table 58), that different species and subspecies of newts are differentiated from one another by a series of interchanges.

WHITE supports his claim by observations of two kinds:

(i) In grasshoppers, as we have seen (section IV-1 a), pericentric inversions of acrocentric chromosomes can lead, when heterozygous, to the production of

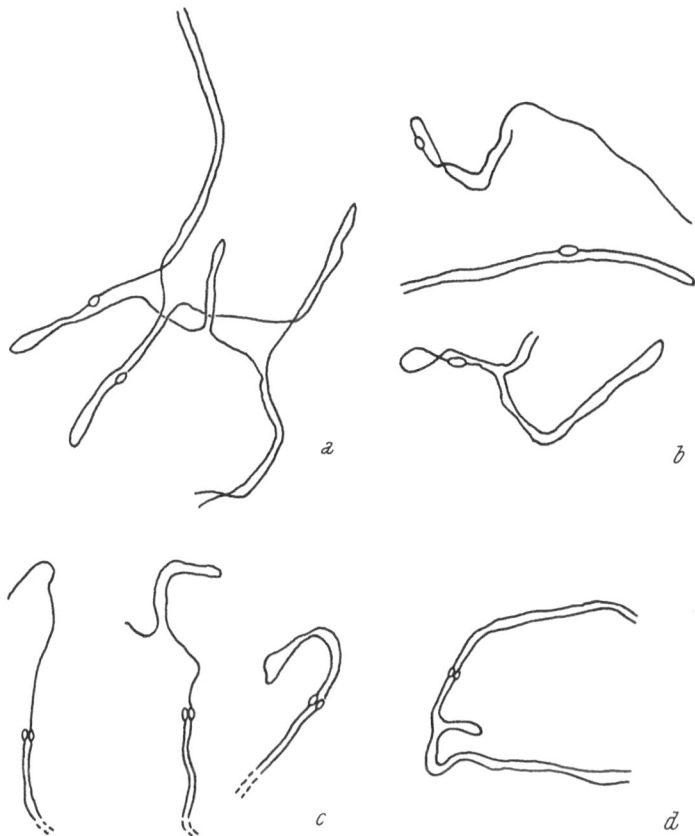

Fig. 138. Foldback pairing in *Zea mays* (a)—two homologues in a diploid showing normal linear pairing near the ends and internal fold-back pairing interstitially; (b)—univalents in monosomic-10; (c)—chromosome-10 bivalents heterozygous for a deficiency of the short-arm; (d)—chromosome-3 bivalent heterozygous for an interstitial deficiency (McCLINTOCK 1933).

heteromorphic bivalents having one acro- and one meta-centric element. Individuals of *Trimerotropis gracilis*, *Moraba scurra* and *Austroicetes interioris* which are heterozygous for two or more such inversions occasionally form chains of four between two such heteromorphic pairs, a terminal association forming between the short arms of the acrocentric members of each heteromorph. In *T. gracilis* and *M. scurra* this happens with a frequency of about 1 in 10,000 first metaphases. In *A. interioris* WHITE claims it is "considerably commoner" but no actual values are given. This terminal pairing is facilitated, according to WHITE, because true synapsis is suppressed in the region of the inversion where non-homologous pseudosynapsis occurs (see page 70).

(ii) *Moraba scurra* exists in two races, an Eastern one with $2n = 15$ ♂ and a Western one with $2n = 17$ ♂. The Eastern race has a pair of large metacentrics replaced in the Western by two equivalent pairs of acrocentrics. WHITE

Table 59. *Summary of* WHITE's *Study on Interpopulation and Interracial Hybrids of Moraba scurra.*

Note that the constitution of the various chromosome types referred to is as follows:

Element	Type			
AB	Fused (F) A⟋○⟍B		Broken (Br) A ○ ○ B	
CD	Standard (St) C⟋○⟍D		Blundell (Bl) ○ C D	
EF	Standard (St) E⟋○⟍F		Tidbinbilla (Tid) ○ E F	

Individual type	Constitution			Xa frequency
	AB	CD	EF	
1. 15-chromosome race	F/F	St/St	St/St	9.496 ± 0.078
		St/Bl	St/St	8.584 ± 0.016
		Bl/Bl	St/St	9.000 ± 0.011
2. 17-chromosome race	Br/Br	St/St	St/St	9.971 ± 0.408
		St/Bl	St/St	9.488 ± 0.157
		Bl/Bl	St/St	9.308 ± 0.200
3. Inter-population hybrids of the 15-race				
a) Yass × Murrumbateman (12 miles apart)	F/F	St/St	St/St	9.98 ± 0.001
b) Hall × Woodonga (150 miles apart)	F/F	St/Bl	St/St	8.51 ± 0.05
				7.98 ± 0.07
				7.98 ± 0.10
4. Interracial hybrids (extremes only given here)				
a) Kyeamba ♀ (17-race) × Collector ♂ (15-race)	Br/F	St/Bl	St/St	7.70 ± 0.07
b) Wallendbeen ♀ (17-race) × Murrumbateman (15-race)	Br/F	St/Bl	St/St	10.19 ± 0.06

(1957) carried out a series of crosses both between members of different races and between different populations of the Eastern races (Table 59).

In the Hall × Woodonga hybrids the chiasma frequency is below average. This is reflected, in part, by a majority of the metacentric bivalents having a chiasma in one arm only and, in part, by asynapsis of one of the four small and indistinguishable chromosome pairs. These events coincide with the occurrence of a series of multiple associations involving an end-to-end union of "heterologous" chromosomes.

ROTHFELS and TESHIMA (1964) have argued along the same lines as WHITE with regard to the occasional multiples which are found in the grasshopper *Chloealtis conspersa*. These were previously attributed to spontaneous interchange in a small minority of sperm mother cells. The re-interpretation was based on the finding that multiple formation occurs more frequently in individuals which are heteromorphic for chromosomes 6 or 8. Heteromorphism in respect of these chromosomes involves the replacement of a normal end of one homologue by a heterochromatic segment of unknown origin. Increased multiple formation depends principally on the more frequent association between the

Table 60. *Occurrence of Multiple Associations in 100 Cells of a Spontaneous Mutant Individual of Triturus helveticus.*
(Data of MANCINO and SCALI 1961.)

Cells with univalents only		Cells with multiples						
		III			IV		VI	
2 I	8 I	1 I	3 I	5 I	No. univalents	2 I	No univalents	With univalents
7	1	1	1	1	2	1	1	0

normal member of the heteromorphic pair and the ends of other specific chromosomes. Associations of this kind were regarded as chiasmate and, therefore, specific. Consequently, they were taken as indicative of end-homology between otherwise heterologous chromosomes. In this organism, multiple formation was not correlated with univalent formation. But the competitive efficiency of a normal end for pairing with heterologous chromosomes appears to be enhanced by the deletion and replacement of the corresponding end of its homologue.

Clearly WHITE's argument, if admitted as satisfactory, is capable of extension to all cases where multiple associations of a terminal nature have been used as evidence for translocation differences between complements brought together in a hybrid. Indeed WHITE has even re-interpreted his own earlier claim of interchange heterozygosity in *Metrioptera brachyptera* (WHITE 1940) in terms of this hypothesis of duplicate end segments.

This explanation, however, omits a consideration of a number of pertinent points:

(i) The multiple associations observed in the two pericentric inversion heterozygotes cited occur so rarely that one cannot rule out spontaneous interchange. In *A. interioris* the precise frequency is not given but in our experience a frequency of 1 in a 100 cells is far from uncommon for spontaneous chromosome mutation. Moreover cases are already on record where terminal interchange of relatively short chromosome segments leads to the production of a multiple in which the component members are associated solely by terminal contact. Thus X-ray induced interchanges in the grasshopper *Gesonula punctifrons* are of exactly the same form as those figured by WHITE (RAY-CHAUDHURI 1961). So too are quadrivalents in tetrasomic cells of *Pyrgomorpha kraussi* (see Fig. 113). Even interchanges involving long segments are expected to give multiples with terminal associations where chiasmata tend to be distally localised, *Oenothera* provides a classic case. Indeed WHITE and MORLEY (1955) themselves have shown that the chiasma frequency of distal regions increases in the presence of heterozygous pericentric inversions (see page 70).

Again, HELWIG (1955) reported multiples, of the type found by WHITE, in hybrids between *Circotettix verruculatus* and *Trimerotropis suffusa*. But, as he points out, they sometimes occur as frequently in normal individuals of both species despite the fact that there is extensive failure of pairing only in the F_1 hybrid.

Table 61. *Pseudo-multivalent Formation in Diploid Tomato.*
(Data of GRÖBER 1961.)

No. pseudomultivalents per nucleus						Total nuclei analysed	% nuclei with pseudomultivalents
0	1	2	3	4	5		
295	32	12	7	2	2	350	15.7

(ii) WHITE claims that the pairing homologies in interpopulation hybrids of *M. scurra* are too numerous to be accounted for in terms of interchange differences. However, this claim is based on the assumption that one of the four small chromosomes, designated chromosome 4 by WHITE, is involved in three of the five multiple associations. But the four small chromosomes of *M. scurra* are not easily distinguished. Indeed WHITE himself admits that in cells containing small univalents it cannot be established whether the same small pair is involved in different cells. That the same one is involved is suggested only by the fact that two asynaptic pairs have not been observed in the same cell. Moreover in the Kyeamba × Collector hybrids the chiasma frequency is in fact below that in the Hall × Woodonga cross, yet no multiples form here. On the other hand, it is far from clear why multiple associations should be expected to be associated with low chiasma frequencies if, as WHITE believes, the "extra" terminal associations themselves depend on chiasma formation. Nor is it clear how frequent the multiple associations are.

(iii) CALLAN has come out in favour of WHITE's hypothesis of reduplicated and translocated telomeres (see footnote to WHITE 1961, page 320). He gives two reasons for this. First, he has failed to find "trivalents" with interstitial chiasmata in both arms of the common chromosome and second, multiples cannot be detected in the lampbrush chromosomes of F_1 *T. cristatus* hybrids. Let us note, then, that multiples are found in the sex with the *higher* chiasma frequency (see Table 6, page 26). We, however, regard as more important the difference between the sexes in respect of chiasma localisation (see Table 5, page 25). The occurrence of distal chiasmata in ♂ newts would surely account for the absence of markedly interstitial chiasmata in the ♂ F_1's, while the absence of multiples in the ♀ F₁'s is also expected if the interchanges involve terminal segments only, since in females chiasmata tend to form in less distal positions.

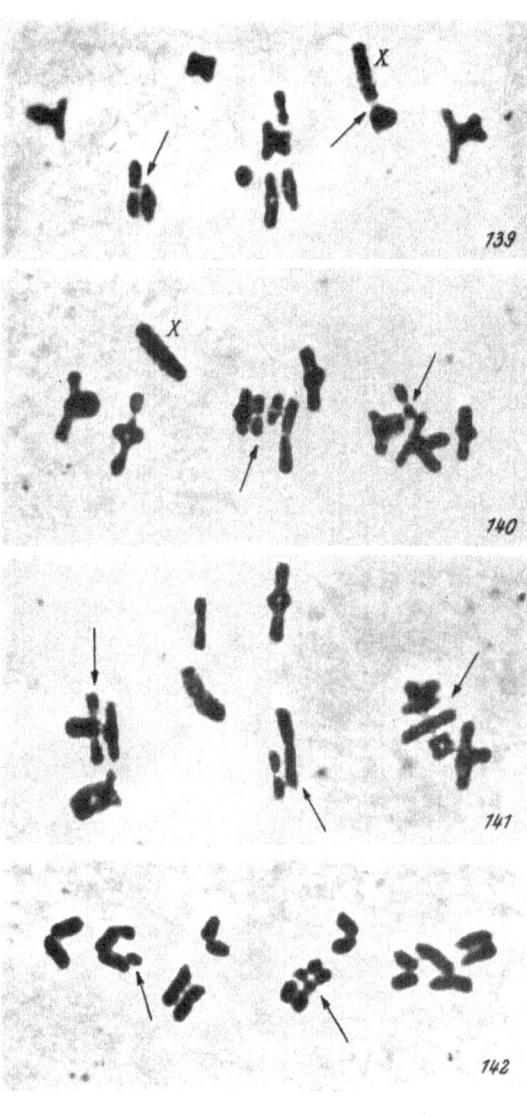

It is also difficult to appreciate how so many newt telomeres can be regarded as homologous when, as CALLAN points out "we can recognise many if not indeed most of the chromosome end regions."

Finally WHITE (1946) himself has described F₁ hybrids between *T. cristatus* × *T. marmoratus* where, despite the presence of extensive univalent formation, no multiples were formed. MANCINO (1961) found "trivalents" in a male hybrid of *T. italicus* ♀ × *T. vulgaris* ♂ which he also interpreted in terms of WHITE's hypothesis. Yet this same author, in conjunction with SCALI (1961), also reported on a specimen of ♂ *T. helveticus* where multiples of three, four and six, consisting of terminally associated members, occurred in nuclei both with and without univalents (Table 60). And these were interpreted as interchange multiples.

A second and, in our opinion, a more cogent argument has been put forward by WHITE. He points out that the co-existence of cells with 23 or 24 strictly terminal chiasmata and cells with multiples of four in the same hybrid individual provide evidence in favour of end duplications. He further points out that Fig. 5a in LANTZ and CALLAN (1954) shows 18

Figs. 139–142. Secondary association in male meiotic cells of *Gastrimargus africanus* (2n = 22 + X) at first (139–141) and second (142) metaphase (orcein-phase contrast, ca. ×1500). Note associations are formed between similar and different sized chromosomes. Fig. 139—One close side-by-side association (arrow) and one end-to-centromere association of a bivalent with the unpaired X (arrow). Fig. 140—One (s-s) association of four bivalents and one association of three involving centromere-to-centromere- and centromere-to-end approximation (arrow). Fig. 141—Side-to-side associations of two and four. Fig. 142—An (11 + X) cell with side-by-side associations of 2 and 3 respectively (arrows).

terminal chiasmata but, since two chains of six are observed in some cells, a maximum of 16 terminal chiasmata can be expected in the absence of end duplications. But this argument, as WHITE appreciates, depends on two factors. First, the ability to recognise strictly terminal chiasmata and second, the generally-accepted view that terminalisation is arrested by a change in homology.

In a study of a hybrid obtained by pollinating *Secale cereale* with X-rayed pollen of *Secale montanum*, PRICE (1958) observed cells with multiples of six and multiples of four together with bivalents. The multiple of six (or it's "break-down" products) can be obser-ved in untreated hybrids; the association of four was attribu-ted to X-ray induced interchange. The chromosomes involved in this interchange behaved in various ways. Neither univalents nor as-sociations of three involving them were observed and 2 II versus 1 IV were found in 259 : 51 cells. A distinction between complete and incomplete terminalisation in multiples of four was made only with regard to "frying-pan" associations and the correspon-ding Y-shaped configurations. If terminalisation can proceed bey-ond a change in homology, three of the four chromosomes in the multiple could become associated terminally (at the junction of the Y or the ♀) following chiasma for-mation in an interstitial segment and in a pairing segment distal to it. PRICE claims to have seen triple chiasmata of this kind (formed by the terminalisation of two chias-mata) in 2 out of 7 Y-shaped and 5 out of 22 "frying-pan" multi-ples (see also MARQUARDT 1948).

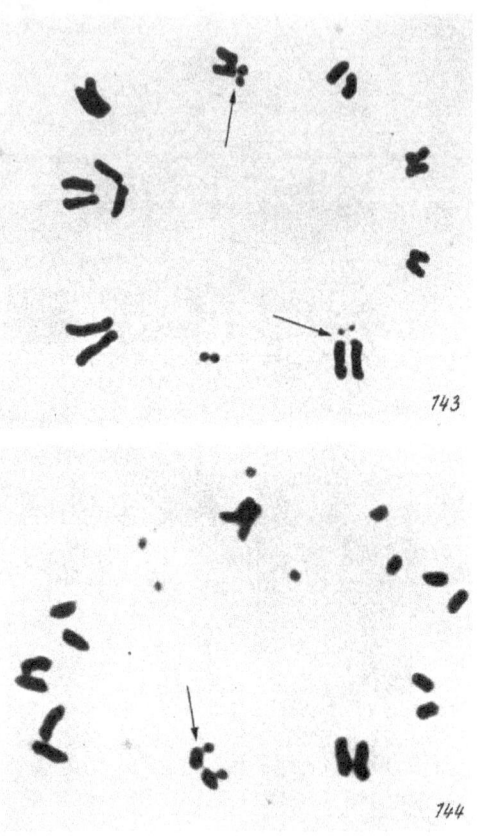

Figs. 143 and 144. Secondary association at second division in *Schistocerca paranensis* (orcein, ca. ×1500).

Comparable quadruple chiasmata were described by DARLINGTON and GAIRD-NER (1937) in interchange hybrids of *Campanula*. But here they were interpreted as a consequence of chiasma formation in short duplicated end segments. In other words they were accounted for on the basis subsequently suggested by WHITE for *Moraba*. Thus, if one accepts PRICE's explanation, WHITE's argument regard-ing the expected frequency of terminal associations in interchange multiples is invalidated. But if one accepts the explanation of DARLINGTON and GAIRDNER, then we have support for WHITE's contention and support which comes from an undoubted interchange heterozygote!

(iv) WHITE has not ruled out the possibility of secondary pairing and asso-ciation, more particularly telomere association of a non-homologous type. It is true that this type of pairing generally lapses by first metaphase so that its persistence in White's case would seem to argue against this. But non-homologous associations can persist; they do so in haploids (see page 103) and they may do so in diploids (Matsuura and Kurabayashi 1951). Again, in diploid tomato (2 n = 2 x = 24) Gröber (1961) has shown that in a proportion of cells the terminal macro-chromomeres and the centromeres associate non-specifically in early first prophase. Both regions are heterochromatic, more especially the centromere. Persistence of such association leads to the bivalents maintaining contact up to first metaphase and so pseudo-multivalents consisting of a side-to-side alignment of two bivalents are produced (Table 61). Half-bivalents may show a similar parallel alignment at second metaphase. In grasshoppers both centric and non-centric ends are not infrequently heterochromatic so that a basis for secondary association certainly exists. We have recently found loose secondary associations at extraordinarily high frequencies in *Gastrimargus africanus* (Figs. 139–142), both at first and at second meta-anaphase. But these associations are not present at diplotene although the chromosome ends are markedly heteropycnotic at this time. Henderson (private communication) has seen the same phenomenon in occasional individuals of *Schistocerca gregaria* and we have seen it in occasional cells of *Schistocerca paranensis* (Figs. 143–144).

▲ *E. plorans ornatipes*

▲ *E. plorans meridionalis*

Fig. 145. Distribution of *Eyprepocnemis plorans ornatipes* and *Eyprepocnemis plorans meridionalis* (Dirsh 1958).

All four points suggest that a much more thorough analysis of the situation needs to be made before concluding, as White has done, that many or all of the chromosome ends in a complement may prove to be homologous and owe their origin to a translocation of short terminal duplications. Recently we have had an opportunity of analysing the meiotic behaviour in some F_1 grasshopper hybrids belonging to the genus *Eyprepocnemis*. Our results have a bearing on this problem.

Nineteen species are commonly recognised in the genus *Eyprepocnemis* but the majority are not sharply differentiated (Dirsh 1958). One of these,

Table 62. *Chiasma Frequencies of the two Subspecies of Eyprepocnemis plorans.*

Subspecies	Chiasmata per bivalent			Total Xta (5 individuals, 20 cell per individual)	$\bar{x}$	P
	1	2	3			
E. p. ornatipes	873	221	6	1333	13.33	0.05 − 0.01
E. p. meridionalis	820	265	15	1395	13.95	

Eyprepocnemis plorans, is divided, with difficulty, into four sub-species largely on the basis of colouration and the overlapping pattern of certain morphometrical differences. Our hybrids were produced from two of these, *E. plorans ornatipes*

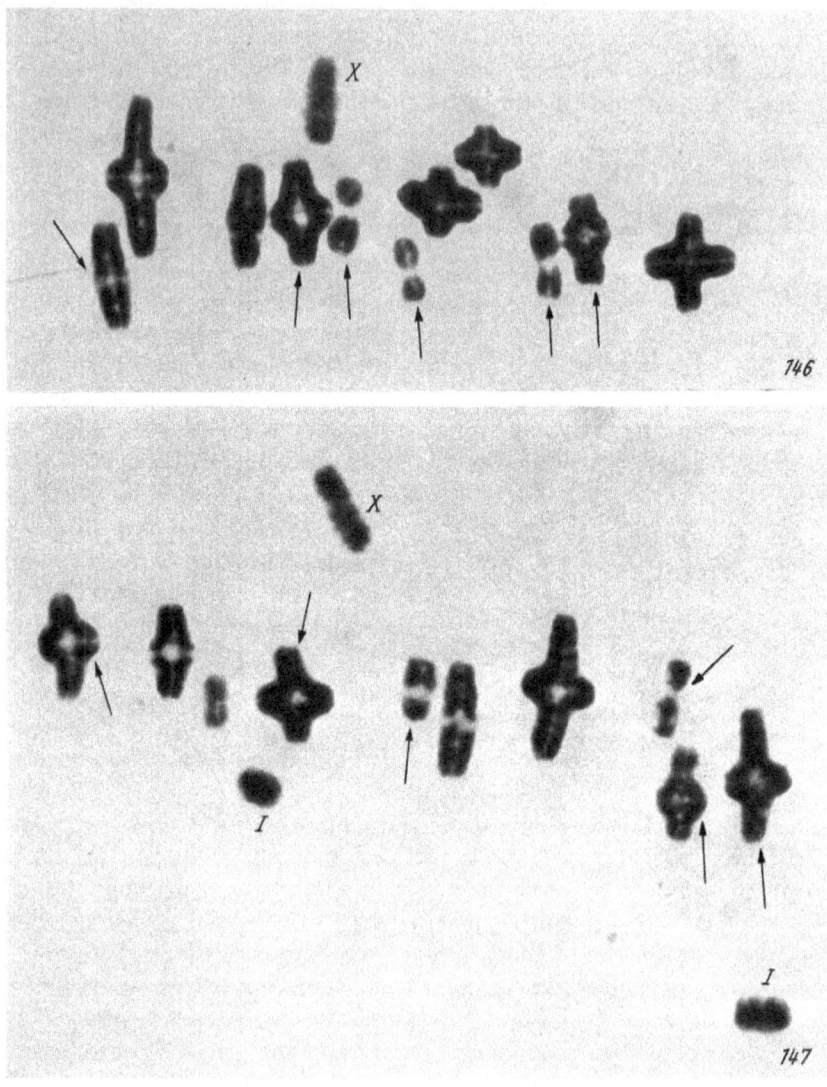

Figs. 146 and 147. Metaphase-I of meiosis in F$_1$ hybrids between *Eyprepocnemis plorans ornatipe s* and *Eyprepocnemis plorans meridionalis* (orcein, ca. × 2000. Fig. 146—11 II + X, six of the bivalents are distinctly heteromorphic (arrows). Fig. 147—10 II + 2 I + X, the two autosomal univalents are unequal in size and five of the bivalents (arrows) are heteromorphic.

and *E. p. meridionalis*. *Ornatipes* is distributed across Africa in a narrow belt north of the equator, while *meridionalis* forms a similar belt south east of the equator (Fig. 145). The two subspecies were crossed reciprocally by Mr. PHILLIP HUNTER-JONES of the Anti-Locust Research Centre, London and through his kindness we have had the opportunity of examining males from parental, F$_1$ and

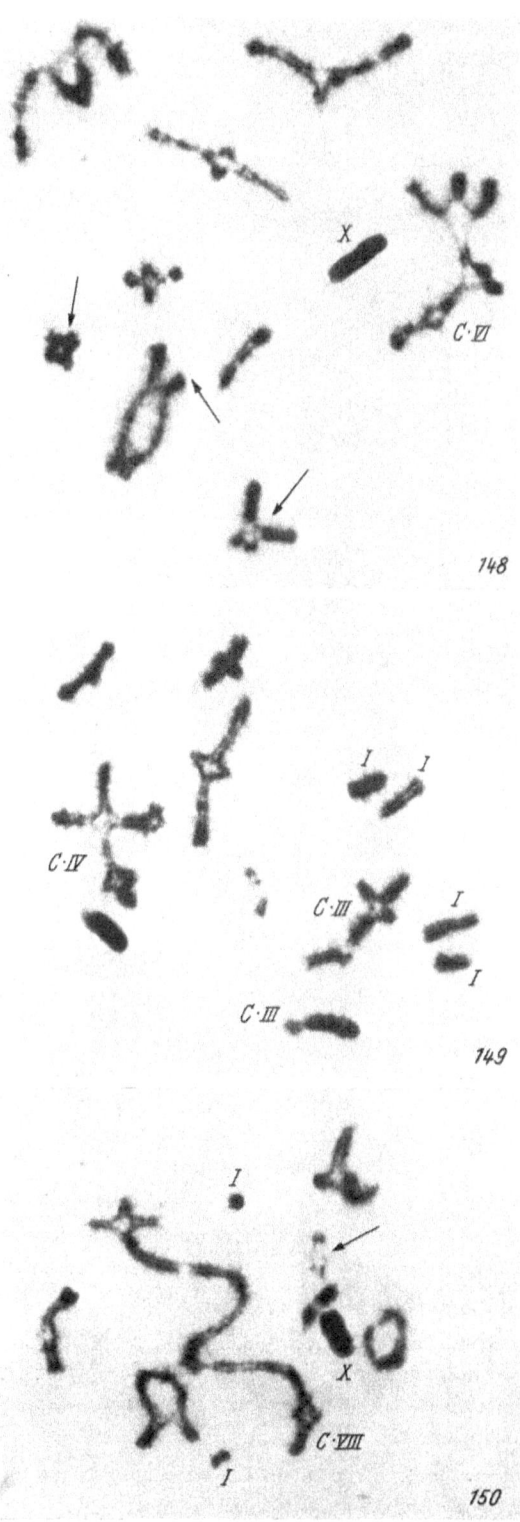

backross generations (John and Lewis 1965).

Both subspecies have a characteristic acridoid complement of 23 acrocentric chromosomes in the male sex and a comparison of mitotic and meiotic morphology in the parental stocks reveals no readily apparent difference between the chromosomes in the two complements. Both have 2 long, 6 medium and 3 short autosome pairs together with an unpaired X. Despite this apparent lack of difference the two subspecies differ significantly in chiasma frequency (Table 62).

Meiosis in the F_1 is characterised by the presence of:

(i) Heteromorphic bivalents in all three size groups (Figs. 146–153).

(ii) Multiple associations ranging from chains of three with an associated univalent to chains of ten (Figs. 148–153). A majority of these multiples are open chains in which the members are associated terminally or nearly so. We have occasionally found rings of four and frying-pan associations. We also have occasional but indisputable evidence for the presence of successive interstitial chiasmata (e.g., Fig. 149) which, together with the evidence from the heteromorphic bivalents,

Figs. 148–150. Diplotene in F_1 hybrids between *Eyprepocnemis plorans ornatipes* and *Eyprepocnemis plorans meridionalis*. Heteromorphic bivalents are arrowed (orcein, ca. × 1500).

must mean that interchanges are involved in producing some of the multiples. And we have no reason to doubt that many of them owe their presence to interchange differences between the two parental forms.

(iii) One or more pairs of univalents, some of which are unquestionably potential members of multiple associations (Figs. 149 and 153).

As one might expect from these findings the F_1 is sterile and no F_2 has ever been produced. We are undoubtedly dealing here with a case of sibling speciation in which both forms are chromosome homozygotes but are nevertheless homozygous for structurally different combinations of chromosome segments. And the significant point in the present connection is that most of the multiples which we find are of the type which WHITE claims as evidence against the occurrence of interchange.

(4) We have already drawn attention to the danger of using the bridge and fragment technique to assess the presence of paracentric inversions (see page 67). Yet this has been used extensively in species hybrids as a measure of inversion differences between the parental forms. An alternative explanation for the occurrence of bridges and fragments in such hybrids is chromosome breakage for, as we have seen earlier (see page 32), foreign cytoplasm has the capacity for inducing breakage.

(5) If MATTHEY (1963) is correct in his interpretation of meiosis in female hybrids from the cross *Acomys cahirinus* ♀ × *A. minous* ♂ it would appear that a particular cytoplasm may be involved in the fusion as well as the fragmentation of chromosomes. Gametes of ♀ *A. cahirinus* invariably contain 19 chromosomes, one of which is an X-chromosome; the 18 autosomes consist of 16 metacentric

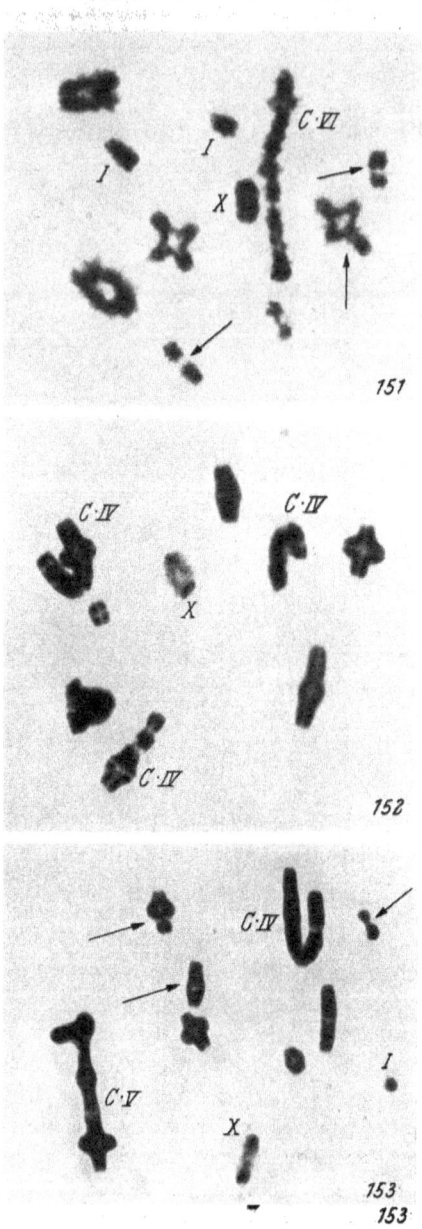

Figs. 151–153. Diakinesis (151) and first metaphase (152–153) in F_1 hybrids between *Eyprepocnemis plorans ornatipes* and *Eyprepocnemis plorans meridionalis* showing various combinations of univalents, heteromorphic bivalents (arrows) and multiple associations (orcein, ca. ×1500).

and two telocentric elements. Males of *A. minous*, on the other hand, are polymorphic (Table 63); indeed a given male may be mosaic in respect of karyotype so that, as far as the autosomes are concerned, at least 4 different male gametes are possible—14 M + 4 T, 15 M + 4 T, 13 M + 5 T and 14 M + 5 T. All the male hybrids obtained from crossing females of *A. cahirinus* to males of *A. minous* have autosome complements, which are directly predictable from the listed gametic types, namely 31 M + 6 T, 30 M + 6 T, 29 M + 7 T and 30 M + 7 T.

Table 63. *Summary of* Matthey's *Observations on Acomys.*

Type	Sex	2 n	No. autosomes present	
			Metacentrics (M)	Acrocentrics (A)
1. *A. cahirinus*	♀ (XX)	38	32	4
	♂ (XY)	38	32	4
2. *A. minous*	♀$_1$ (XX)	38	28	8
		40	30	8
	♀$_2$ (XX)	40	28	10
	♂$_1$ (XY)	39	29	8
		38	28	8
	♂$_2$ (XY)	40	30	8
		39	27	10
		40	28	10
3. F$_1$ hybrids *A. cahirinus* ♀ × *A. minous* ♂	♂$_1$ (XY)	38	30	6
		38	29	7
	♂$_2$ (XY)	39	31	6
Note autosome constitution of expected progeny will be:-		39	30	7
		38	29	7
♀ gam. + ♂ gam. ⟶ F$_1$				
16 M + 2 Å + { 14 M + 4 Å ⟶ 30 M + 6 Å	♀$_1$ (XX)	39	32	5 *
15 M + 4 Å ⟶ 31 M + 6 Å		39	33	4 *
13 M + 5 Å ⟶ 29 M + 7 Å		38	31	5 *
14 M + 5 Å ⟶ 30 M + 7 Å		38	32	4 *

* All these represent unexpected autosome classes.

The female hybrids, however, cannot be so derived. Two of the four observed types can be explained by assuming a simple fusion of two telocentrics so giving a metacentric element. Thus:

$$30 \text{ M} + 7 \text{ T} \to 31 \text{ M} + 5 \text{ T}$$

and,

$$31 \text{ M} + 6 \text{ T} \to 32 \text{ M} + 4 \text{ T}$$

But the other two (32 M + 5 T and 33 M + 4 T) are not so simply derived. Here Matthey has to invoke non-disjunction of a telocentric element coupled in the latter case with a pericentric inversion which transforms a T into an

M-element. Despite this MATTHEY concludes that centric fusion involves nothing more than a simple approximation and fusion of the centromere regions of two telocentrics. Let us note that this conclusion is based on observations made exclusively from somatic mitoses, indeed from somatic mitoses in a colchicine treated cell lineage. Let us also note that no frequency data are given for the various karyotypes observed. One has no idea therefore of the true extent of the variation within the soma and no evidence at all that the same variation exists within the germ line. And until this is established there is little factual basis for comparing this process with genuine centric fusion.

VIII. Meiotic Behaviour of Special Chromosomes

1. Sex-chromosome Systems

The commonest system of genotypically controlled sex determination depends on the segregation of a difference at meiosis. This system may operate at the haploid level, as in many algae and bryophytes, or at the diploid level. From the standpoint of the meiotic system, the sporophytic phase in types with genotypically controlled haploid dioecism is comparable with the heterogametic sex of types showing diploid dioecism (LEWIS 1961).

The difference, the segregation of which ultimately determines sex, may be a simple mendelian one. In this case special segregational problems do not arise. But, as we have seen, in the basic meiotic sequence, the segregation of two chromosomes, or parts of them, depends on the co-orientation of their centromeres at metaphase of the first division. This, in turn, depends on their association, this on chiasma formation and this on pairing. Pairing is quantitatively limited and qualitatively specific; it depends on homology. Thus, lack of homology means failure of specific pairing; failure of pairing means failure of chiasma formation; failure of chiasma formation means failure of metaphase association and so failure of co-orientation and segregation.

Of course, success at the early stages does not ensure the success of the later ones. Homologous chromosomes may fail to pair; paired chromosomes may not form chiasmata; and so on.

Now, a chiasma is the visible consequence of the genetical event known as crossing-over (see Part Two—section II). And in many genetic systems there are chromosomes or chromosome segments which, for one reason or other, must be protected from recombination. This is the case where sex is determined by a chromosme-borne difference more extensive than a single mendelian one. How then is the segregation of these differences ensured if crossing-over is precluded?

The commonest mechanism for reconciling the rival demands of segregation and protection from recombination depends on linkage between the critical, so-called, differential segments, in which crossing-over does not occur, and pairing segments which show crossing-over and whose segregation is not directly involved in sex determination. The chromosomes concerned are labelled X and Y when the male is heterogametic, the Y being confined to the male. With female heterogamety, on the other hand, it is customary to refer to Z ($\equiv$ X) and W ($\equiv$ Y) chromosomes, the latter being exclusive to the female.

This state of affairs (XY♂, XX♀ or ZZ♀, ZW♂) can be regarded as the basic sex chromosome mechanism and so long as the differential segment is in one piece and linked with a pairing segment special meiotic problems do not arise. However, the pairing of the pairing segment is doubtless impeded by the associated differential segment and, if this proves to be a serious impediment, some system of pre-zygotene association may be favoured (see below).

Be this as it may, it is the derived sex-chromosome systems which introduce the greatest segregational problems and they require considerable modifications of the meiotic system to function efficiently. Change has been in various directions. In some cases, the X and Y have become completely differential, i. e., no crossing-over occurs between them at all. In other cases the Y component has been lost altogether. Increase in number may occur also. Thus the number of X-chromosomes, the number of Y-chromosomes, or the number of both may increase. In these cases crossing-over may or may not be retained or retained only in part. And systems with many X-chromosomes may have no Y-chromosomes.

Thus, the segregational problems introduced by the advent of these derived mechanisms, and whose solution must be largely antecendent to the origin of the mechanisms, are of different kinds.

In XO and X_nO systems the variable and irregular behaviour which characterises univalents must be avoided (see page 54). In the former the regular segregation of the X-chromosome to either spindle pole is sufficient but in the latter all the members of the X complex must be concerted in their movement. Multiple sex chromosome mechanisms require the co-ordinated orientation of more than two chromosomes. And, where crossing-over has been eliminated, a means of communication other than that afforded by a chiasmate association must be found.

Thus, from the point of view of the meiotic problems they create, sex chromosome systems can be considered in the following groups:

1. Chiasmate systems:

    a) Basic XY system.
    b) Derived XY systems.
    c) Multiple systems — X_nY, XY_n, X_nY_n.

2. Non-chiasmate systems:

    a) Absence of partner(s).
      i) Simple — XO.
      ii) Compound — X_nO.
    b) Differential partner(s).
      i) Simple — XY.
      ii) Compound — X_nY, XY_n, X_nY_n.

Simple Chiasmate Systems

Let us begin with the basic XY chiasmate system.

This is the commonest single system. It has been found at the haploid level in bryophytes and at the diploid level in animals and flowering plants. Some claims have also been made for the gymnosperms but the evidence is conflicting.

This system, as we mentioned earlier, does not introduce any special meiotic problems and so it will be considered only briefly.

So far as we are aware, the separation of the X and Y is invariably pre-reductional in plants, e. g. *Melandrium* and hop (Fig. 154). In animals however, consistent pre-reduction, consistent post-reduction and variable frequencies of both have been claimed for particular species. It will be appreciated that the factors which determine pre- versus post-reduction in simple chiasmate systems are quite different from those which determine a similar difference in non-chiasmate systems including the XO types.

In the basic system the difference depends on the location of the pairing segment relative to the differential segment and the centromere. Invariable

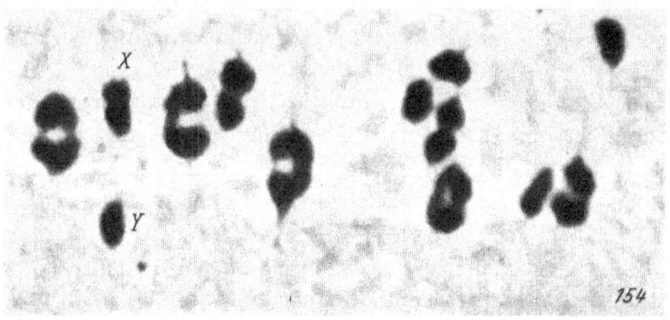

Fig. 154. *Humulus lupulus* (2n = 18 + XY; carmine-phase contrast, ca. × 2000) at first meta-anaphase.

pre-reduction means complete linkage between the differential segment and the centromere which must, therefore, either terminate the differential segment or else be included in it. It means also that the pairing segment is terminal. Invariable post-reduction, on the other hand, must mean that the pairing segment separates the centromere and the differential segment and that there is no crossing-over in the other arm of the chromosome. This arm cannot, however, be differential in the same sense. If it were, consistent two-strand (reciprocal) double crossing-over in the interstitial segment would be necessitated. This would give pre-reduction and it has been claimed for *Drosophila* (DARLINGTON 1934). But where crossing-over can occur on either side of a pairing-segment-included centromere, pre- or post-reduction can obtain.

It will be appreciated that the differential segments in diploid systems have a rather different status from their counter-parts in haploid systems for in the latter there is only one kind of diploid and this is heterozygous for the difference determining the sex of the haplophase.

The chromosomes of mammals have attracted considerable attention recently and the majority of them have been found to have an XX/XY mechanism. Now, where post-reduction can be shown, there is certain evidence for chiasma formation but, as discussed above, the absence of post-reduction need not mean an absence of crossing-over. There is, however, considerable disagreement as to the relative frequency of post-reduction in mammals or, indeed, whether it occurs at all in some cases (Table 64). WAHRMAN and RITTE (1963) have recently

Table 64.   *Pre- and Post-reduction in XY mammals.*

Organism	Percentage		Total cells analysed	Reference
	Pre-red.	Post-red.		
Rat	90	10	—	Koller and Darlington 1934
	99.75	0.25	400	Ohno, Kaplan and Kinosita 1956, 1958
	100	0	—	Makino 1942
Golden hamster	81.6	18.4	253 MI	Koller 1938
	77.5	21.5	111 AI	
	100	0	—	Matthey 1952
Chinese hamster	55	45	64	Pontecorvo 1943
	100	0	—	Yerganian 1959
Man	90	10	212	Koller 1937
	100	0	64	Sachs 1954
Mouse	100	0	—	Makino 1941
	88	12	119	Slizynski 1955

produced convincing evidence for pre- and post-reduction in *Apodemus mysta-cinus*, a species in which it is also possible to demonstrate chiasmata in the XY pair at diplotene (Fig. 155). In the related species *A. sylvaticus* a large proportion of secondary spermatocytes contain a heteromorphic (XY) half-bivalent indicative of post-reduction. On the other hand, Ohno and Weiler (1961) found that in *Mesocricetus auratus* the segregation of the X from the Y was, in every instance, post-reduc-tional (see also page 189).

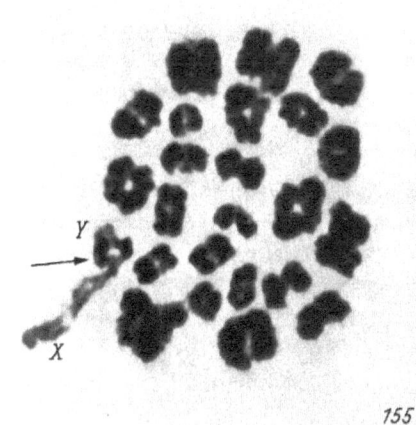

Fig. 155. A chiasmate association between the X and Y chromosomes of *Apodemus mystacinus* (2n = 48) Photo-graph (ca. ×2500) kindly supplied by Dr. J. Wahrman (Wahrman and Ritte, unpublished).

While the basic sex chromosome system is undoubtedly a chiasmate XY-mechanism, all instances of such a mechanism cannot be considered as ancestral for the group in which they are found. This is because neo-XY systems can develop from either XO or non-chiasmate XY systems following interchange with the autosomes. Neo-XY chiasmate mechanisms are known, for example, in a number of grasshoppers (White 1954, Saez 1957, Saez and Diaz 1960, Mesa 1961) and beetles (Smith 1952, 1953, Virkki 1959, 1962) and in at least one phasmid—*Isagora schraderi* (HUGHES-SCHRADER 1947). But, mechanically speaking, all these systems are comparable with the basic XY mechanism and need not detain us further.

Multiple Chiasmate Systems

The commonest system of this kind depends on the controlled segregation of three chromosomes. Such a mechanism determining haploid dioecism has been recorded in the liverwort genus *Frullania* many species of which have been found to possess an $X_1 X_2 Y$ mechanism. Multiple sytems involving three chromosomes have been described in higher plants also where they control diploid dioecism. But here, as in *Humulus* and *Rumex*, the mechanism is $XY_1 Y_2$ (Figs. 156 and 157). In animals, on the other hand, diploid dioecism controlled by both $X_1 X_2 Y$ and $XY_1 Y_2$ mechanisms are known.

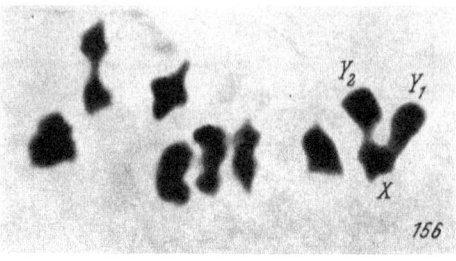

Fig. 156. *Humulus japonicus* ($2n = 14 + X Y_1 Y_2$; carmine, ca. × 2000 at first metaphase).

In all but one of the known cases (see below), the trivalent produced is a linear chain in which the genic complex represented by a single chromosome is centrally placed. Clearly, therefore, these systems depend on the segregation of the central chromosome from those at the ends of the chain. And this depends on convergent orientation (Figs. 156, 157 and 158). In fact, the mechanical problems involved are similar to those which obtain in interchange heterozygotes, namely, the multiple must form regularly, be maintained until after prophase and it must disjoin regularly. In these connections both the structural properties of the chromosomes concerned and genotypically controlled aspects of chromosome behaviour are expected to be important (see page 82).

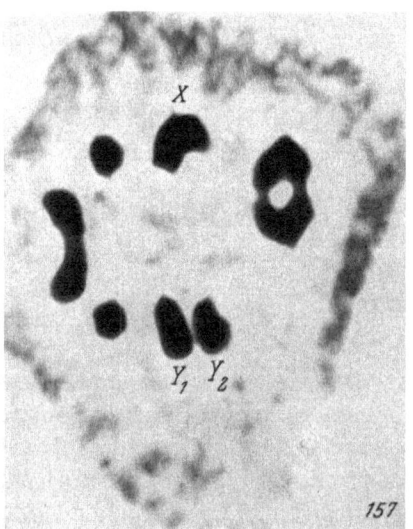

Fig. 157. *Rumex hastatulus* ($2n = 6 + X Y_1 Y_2$; carmine-phase contrast, ca. × 2000) to show segregation of the sex chromosomes at first anaphase.

Many of the investigations on multiple systems have been on a purely qualitative plane and there is little quantitative data on the regularity or otherwise of the behaviour of the sex chromosomes. Some of the available data on multiple formation and disjunction in $X_1 X_2 Y$ mantids is given in Table 65. It would appear that the orientation of the multiple is sequential. First, the Y-chromosome co-orients with one or other of the X-chromosomes, the remaining X being indifferent at this time (cf. RICKARTS 1964). Later this X-chromosome also orients relative to the Y (CALLAN and JACOBS 1957). Moreover, in *Mantis religiosa* its orientation is delayed relative to that of the autosomal bivalents, but in the mecopteran *Boreus brumalis* its anaphase disjunction precedes that of the autosomes. Indeed, in this species it occurs during the pre-metaphase stretch which is of minor extent here compared with that in mantids and phasmids (COOPER 1951).

In mantids this stretch has a considerable influence on the orientation of the sex multiple, the frequency of mal-orientation being greater before the stretch than it is at metaphase (WHITE 1941, CALLAN and JACOBS 1957). Thus in *Stagmomantis carolina* over 50% malorientation was observed at pre-metaphase but only 3% at metaphase (HUGHES-SCHRADER 1943). More detailed information for a species of *Hierodula* is given in Table 66. Some of the re-orientations which occur during the pre-metaphase stretch (e. g., linear to convergent) must involve a reversal in the direction of centromere orientation. The means of achieving such regularity is not clear. But convergent orientation is expected to be the most stable arrangement

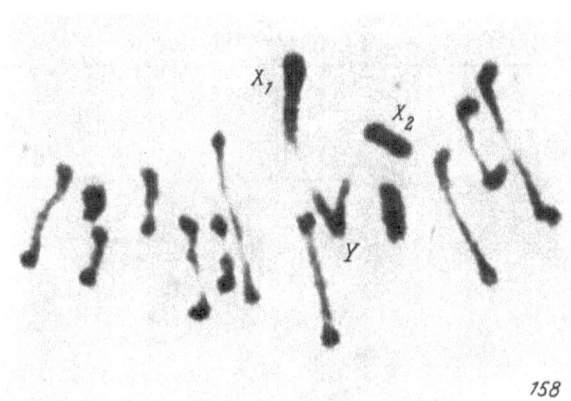

Fig. 158. *Sphodromantis gastrica* (2n = 12 + X₁ X₂ Y) at first metaphase to show convergent orientation of the sex trivalent. Photograph kindly supplied by Dr. S. A. HENDERSON.

for a wide range of chains of three chromosomes and, as BAUER, DIETZ and RÖBBELEN (1961) showed from a study of living material, the most frequent modes of orientation in both bivalents and multiple chromosome associations are those which have the highest stability.

A curious safety mechanism has been described in mantids, for cells with univalents, whether of auto- or allo-somal origin, are blocked at first metaphase, their nuclei become pycnotic and they eventually degenerate. This state of affairs has been described in *Mantis religiosa* (CALLAN and JACOBS 1957), species of *Orthodera* and in *Rhabdomantis pulchella* (WHITE 1962). The situation is particularly remarkable when one considers that the ancestral sex mechanism in mantids depended on the regular behaviour of an unpaired X-chromosome in the male. To our knowledge the condition has not been described elsewhere. Indeed it would not appear to be necessary in all cases. Thus in "abundant material of 9 species" of X₁ X₂ Y morabine grasshoppers, inefficiency owing to failure of sex-multiple formation or orientation has never been found (WHITE 1962). Again, SMITH (1955) claims the invariable formation and disjunction of the XY₁ Y₂ multiples in *Rumex hastatulus*.

The mantids provide also the exception to the rule that the multiple formed in compound chiasmate systems is a linear chain. The unique condition was described by WHITE (1962) in *Rhabdomantis pulchella* where the two X-chromo-

somes pair with the same arm of the Y. Thus, at metaphase, the three sex
chromosome are held terminally, or nearly so, by a triple chiasma. In other
words the pairing arms in this mantid are associated in the same manner as the
legs in the Arms of Man. Malorientation of this multiple is seen at premetaphase
but this state of affairs is soon corrected. This is in contrast to the variable
behaviour exhibited by multivalents of comparable structure in triploids and
trisomics. The basis for this regular behaviour in *Rhabdomantis* is even less
obvious than in the case of the linear sex multiples. But two possibilities suggest

Table 65. *Incidence of Meiotic Abnormalities Involving the Sex-chromosomes of four
Mantids with an $X_1 X_2 Y$ Mechanism. The Morphology of these Chromosomes in Spho-
dromantis viridis Did not allow the Identification of the X-univalent.*
(Compiled from WHITE 1941, CALLAN and JACOBS 1957.)

Species	Normal Orientation	Mal-oriented	Failure of Pairing[1]			% Abnormal Cells	Total Cells
			$X_1Y + X_2$	$X_2Y + X_1$	$X_?Y + X_?$		
1. *Paratenodera sinensis*	535	6	11	3	2	3.95	557
2. *Sphodromantis* sp.	684	25	24	4	16	9.16	753
3. *Sphodromantis viridis*	481	9	—	—	15	4.75	505
4. *Mantis religiosa*	1206[2]	(< 2%)	42	19	2	4.96	1269

[1] Cells with univalents appear to be blocked at first metaphase (see text)
[2] Includes 27 cells with autosomal univalents.

themselves. First, the two X-chromosomes cannot be distinguished and their
pairing arms are longer than that of the Y-chromosome. This means that while
the $X_1 — Y$ and $X_2 — Y$ inter-centric distances are equal, the $X_1 — X_2$ inter-
centric distance is longer. And these inter-centric segments provide the means
of centric communication. Thus the Y centromere is in equal communication
with those of the two X-chromosomes which are not in equal communication
with each other. The second possibility is that the Y-centromere is in some sense
stronger than those of the X-chromosomes individually. The fact of misdivision
and subsequent isochromosome formation (see page 54) provides unequivocal
evidence that the centromere is a multiple structure composed of supplementary,
as opposed to complementary, parts. And observations on the structure of the
centromere support this conclusion (LIMA-DE-FARIA 1954). The possibility exists
therefore that quantitative differences may exist between the centromeres of
different chromosomes, as we have seen them to exist between the two members
of a dicentric system (see page 95), and that this may determine differences in
activity. In fact the X-chromosome of the marsupial *Potorous tridactylis* is conspicuous by virtue of its extended centric constriction and this species has a
multiple XY_1Y_2 system (SHARMAN and BARBER 1952).

Of course, differences in centric activity may be determined by things other than structural differences in the centromeres themselves. Differences "in state" may be involved comparable with those which can be inferred between the centromeres of mitosis and meiosis. A change (or difference) in centric activity has been shown also to depend on a difference in the nature of the associated chromatin. This effect has been found in the case of dicentric chromosomes produced by inversion crossing-over in *Drosophila* (see page 69) and in *Sciara* where the anomalous behaviour of the sex-chromosomes is concerned (CROUSE 1961).

Indeed, in the mantids themselves and elsewhere also, the movement of the sex-chromosomes at the pre-meiotic spermatogonial mitosis may be delayed relative to that of the autosomes. This too indicates a difference in centric-organisation and/or activity.

Table 66. *Orientation of the* $X_1 X_2 Y$ *Sex-chromosome Multiple before and after the Completion of the Pre-metaphase Stretch in a Species of Hierodula.*
(Data of INAMDAR 1949.)

Stage	Normal Orientation	Malorientation		% Total Malorientation
		Linear	L-shaped	
Premetaphase	42	15	7	31.6
Metaphase	149	3	1	2.6

XO-systems

This simplified system presumably arises following the progressive erosion and subsequent loss of the Y-chromosome (DARLINGTON 1939, 1958) but some, at least, of the material it originally contained may be translocated to the X-chromosome or the autosomes during the process. But whatever genic adjustments are necessary, this simplified mechanism can be entertained only in meiotic systems which are pre-adapted to handle univalents. It is true that spontaneous univalents do, sometimes, behave in a regular manner (suggesting, in fact, that a pre-adaptation obtains) but typically their movements are erratic compared with those of hereditary univalents like the unpaired X in the heterogametic sex of XX/XO species (see page 10). Only in the yam, *Dioscorea sinuata*, has such a system been claimed among plants but it is a common one among animals. In most XO systems pre-reduction is the rule but invariable post-reduction is not uncommon and division of the X at either first or second division has been recorded for the same individual in a few cases.

Pre-reduction of the X

The behaviour of the unpaired X-chromosome at the first meiotic division has been studied in detail from fixed and living material in the grasshopper *Melanoplus differentialis* (NICKLAS 1961). Here the X is virtually telocentric and positively heteropycnotic during prophase in the male. But, by prometaphase only the centric end, which is visibly double by this time, is overstained, the rest of the chromosome being negatively heteropycnotic (compare Fig. 159). Following its attachment to the spindle the X moves towards a pole but not within 2–3 μ of it. Only rarely does it remain in this position until the end of anaphase. Typically, it traverses the spindle and approaches the opposite pole as closely as it did the first. And it may make as many as eight pole-to-pole trips, the last occurring as late as autosome congression. The centromere leads in all these movements so re-orientation precedes each trip (Fig. 160). Each trip is usually from one pole to the other but in 4% of the cells re-orientation was observed at the equator. The trips are usually made at uniform velocity but

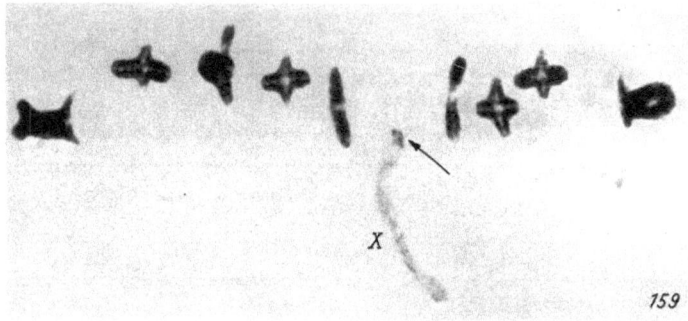

Fig. 159. First metaphase in *Pyrgomorpha kraussi* showing the X-chromosome negatively heteropycnotic except at its centric end (orcein, ca. × 1500).

retardation in the equational regions was observed in about 5% of the cells. The velocity between trips, which appear to traverse the same arc of the spindle, may, however, vary by a factor of ten. The body of the X-chromosome in this species lies in the cytoplasm while its centromere is attached at the periphery of the spindle (compare below). Typically, the attachment-fibres of sister half-centromeres are syntelically arranged towards the destined pole of a given trip; that is they are both directed to the same pole. But the sister half-centromeres of X-chromosomes in the equatorial region may show amphitelic orientation, being directed to different poles (BAUER, DIETZ and RÖBBELEN 1961). SHIMAKURA (1957) has described similar movements in grasshoppers of two other genera, namely *Chrysachroan* and *Chloealtis* and univalent X-chromosomes in interchange progeny of X-irradiated *Tipula oleracea* also behave in this manner (BAUER, DIETZ and RÖBBELEN 1961). Indeed in *Tipula* it was possible to show that X-univalents with amphitelic centromeres remained in the equator during the anaphase movement and only after did they pass, still undivided, to a common pole. Moreover, during their movement the sister half-centromeres are still clearly connected to both poles. Perhaps this property accounts for the fact that occasionally in *Chorthippus* we have found the X to divide after the autosomes have completed their anaphase separation (Fig. 161).

In many *Orthoptera*, however, it has been found that unpaired autosomes lag and otherwise misbehave at first anaphase, even though the behaviour of the X is almost invariably pre-reductional. This holds in hybrids where failure of

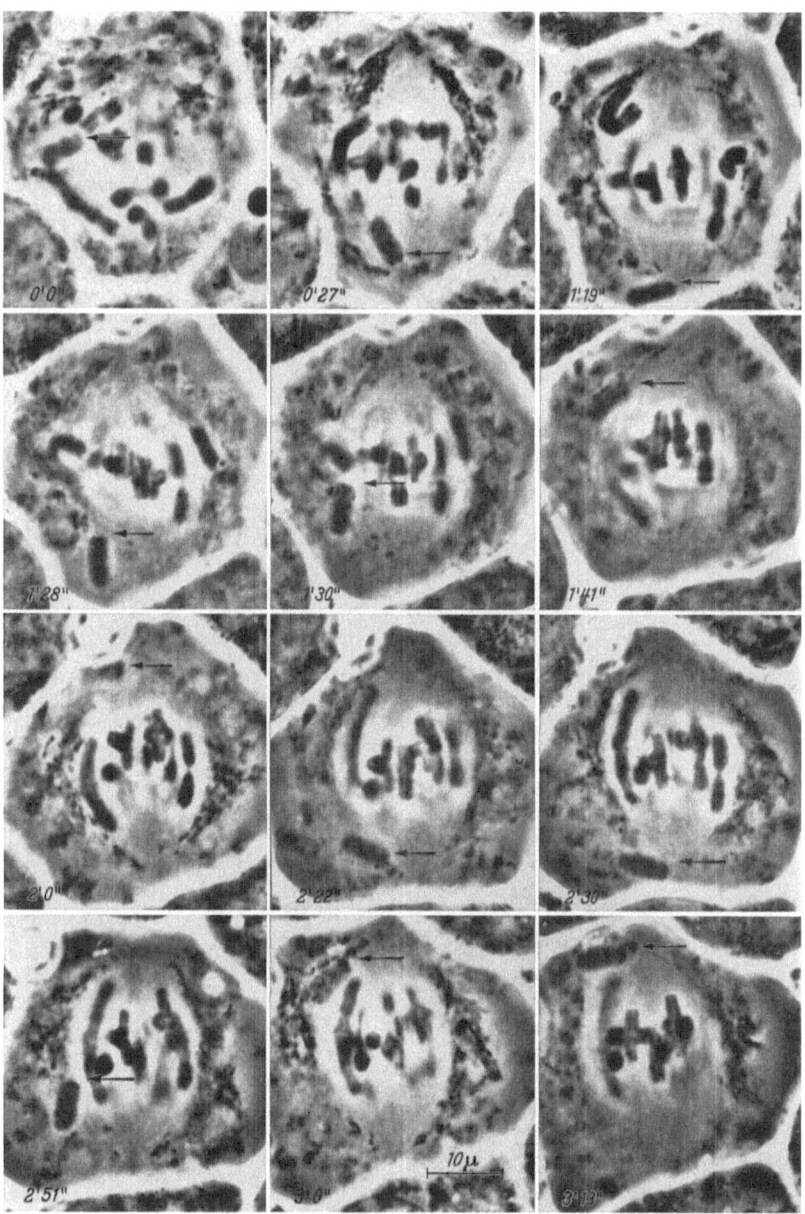

Fig. 160 a.

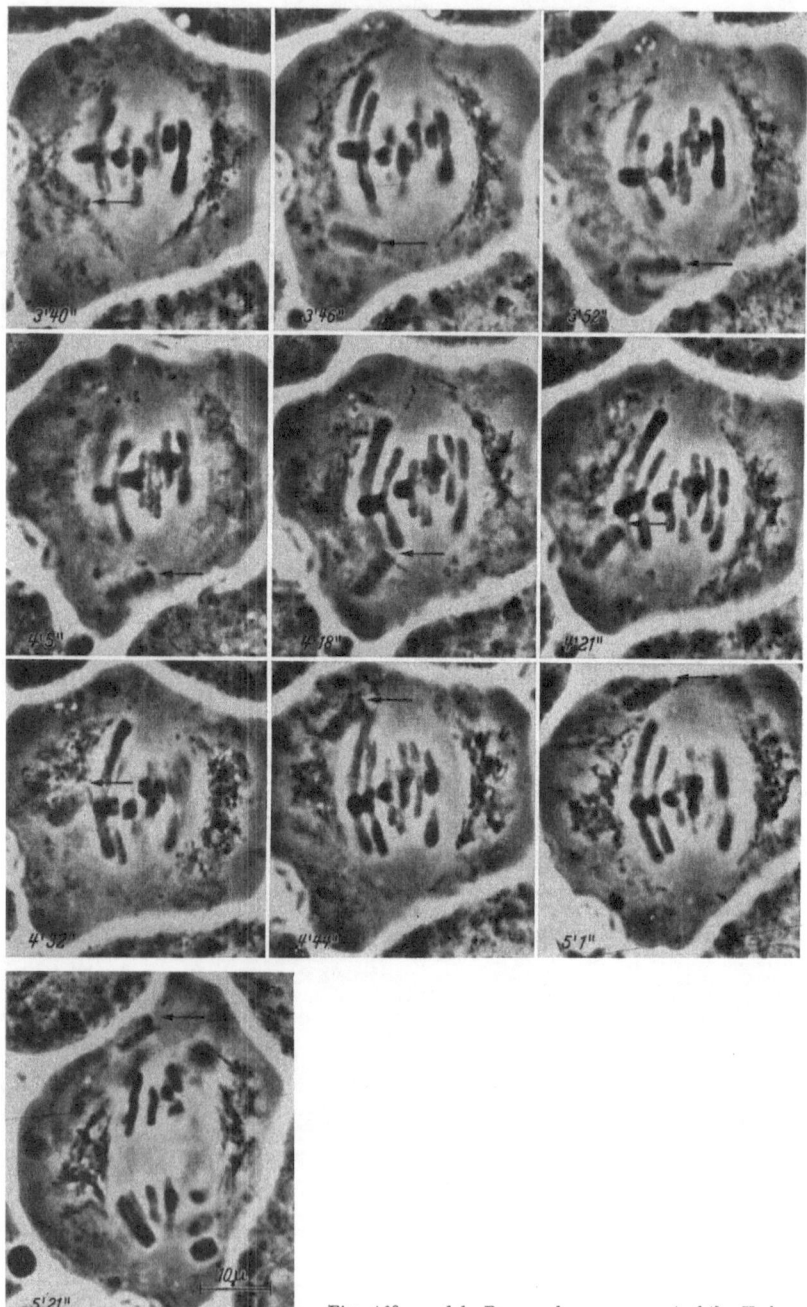

Figs. 160 a and b. Pre-anaphase movement of the X-chromosome in *Melanoplus differentialis* (NICKLAS 1961).

autosome pairing follows non-homology or partial homology (John and Lewis 1965), in aneuploid cells (Lewis and John 1959) and spontaneous asynaptics (John and Henderson 1962). Here, again therefore, we must infer a difference between the autosomes and the sex-chromosome, a difference which presumably resides in the centromere or in its vicinity.

Whether the *Melanoplus* sequence is typical cannot yet be judged but in some orthopterans at least, e. g., *Pyrgomorpha*, *Romalea* and *Eyprepocnemis* the X

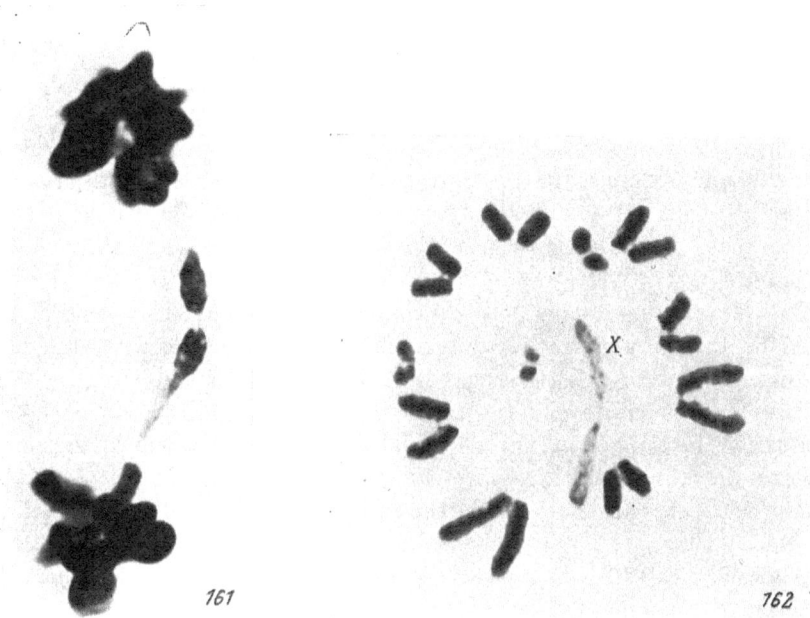

Fig. 161. Equational separation of the X-univalent at first-anaphase in *Chorthippus brunneus* (orcein, ca. ×1500).

Fig. 162. Polar view of second metaphase (11 + X) in *Eyprepocnemis plorans* showing the negatively hetero-pycnotic X and the smallest autosomal half-bivalent lying in the centre of a hollow spindle (compare with Fig. 187; orcein, ca. ×2000).

appears to occupy the centre of a hollow spindle at first metaphase in the male, a position it may even maintain at second metaphase (Fig. 162).

The behaviour of the X-univalent in the mantid *Humbertiella indica* (Hughes-Schrader 1948) is also different from that in *Melanoplus*. This species has nine pairs of autosomes and the unpaired X of the male is enormous in comparison to them. Meiosis is claimed to be non-chiasmate and there is no pre-metaphase stretch. Now, the centromere of the X is impassive during bivalent co-orientation; it comes to lie in the long axis of the spindle and it is moved towards the spindle equator and periphery. It reaches the periphery of the equator with its centromere still inactive and unoriented at a time when bivalent congression is complete. It is then "swept into the cytoplasm . . . over a considerable distance" and only then does the centromere of the metacentric-X orient. It does so towards the nearest pole and definite spindle-fibres of chromosomal origin can be seen to extend polewards. Its anaphase movement is delayed but it is eventually incorporated into one of the telophase nuclei as a more-or-less distinct vesicle.

Thus, the behaviour of the X-chromosome in this species, prior to its orientation, is similar to that of acentric fragments for these too, like the long arms of chromosomes, are expelled from the spindle. This shows that only an active centromere is capable of maintaining an association with the spindle. And, again, we have clear indications of a difference between the behaviour of autosomal and allosomal centromeres. As an aside we may note that the behaviour shown by the auto- and allo-somes in *Humbertiella* is the reverse of that seen in the harlequin lobe of *Brachystethus* (see page 63) where the sex-chromosomes behave normally and the autosomes are extruded. But the extrusion in the latter is due to the extreme lengthening of the chromosome spindle-fibres while in the former it results from a delay in their organisation.

Yet a third type of behaviour seems to characterise the XO species of stone flies, e. g., *Perla maxima*, *P. bipunctata* and *Acroneuria jezoensis* (ITOH 1933, MATTHEY and AUBERT 1947). In these the large metacentric X-chromosome in the male lies in one half of the spindle with its centromere directed towards the equator through which it passes during its reductional first division segregation!

On a priori grounds, the polar movement of an X-univalent could be determined by the essential one-sidedness which has been claimed for the meiotic centromeres of autosomes (ÖSTERGREN 1947). This would permit, or at least facilitate, movement to either pole. But we have a considerable amount of information to show that the allosomal centromeres behave differently from those of the autosomes. And occasionally, as we have seen, the X in *Tipula oleracea* at least, can show amphitelic orientation and yet move undivided to one pole.

A second condition which could determine polar movement, but this time to a particular pole, is mechanical spindle asymmetry similar to that which can be inferred in PMC's of *Rosa canina* and elsewhere (see section VIII–2 and LEWIS and JOHN 1963a). But the pole-to-pole trips in *Melanoplus* generally proceed at uniform velocity so that there is no indication of spindle asymmetry here. Such an asymmetry could be claimed for *Humbertiella* where the spindle fibres of the X-chromosome extend to only the pole of destination. But this condition is consequent upon orientation and does not precede it.

A fixed spindle asymmetry would not, of course, be tenable in XO systems with female heterogamy because three of the meiotic products abort. All the gametes would therefore be of one kind and the heterozygous sex would be homogametic! There is evidence, however, that a reversible mechanical asymmetry is exploited in the psychid moth *Talaeporia tubulosa* which is XX/XO with female heterogamy (SEILER 1921). An adaptable sex ratio is especially important for this species because the females are immobile.

Post-reduction of the X

In systems with a localised centromere, pre-reduction of the X-chromosome is the rule (but see below). This occurs also in some XO species with non-localised centromeres, e. g., aphids, but many with this type of centric organisation show post-reduction for the X-chromosome. We will describe the behaviour as seen in the coccid bugs (HUGHES-SCHRADER 1948).

In *Llaveia*, the sister-chromatids in the unpaired X of the male form spindle fibres directed towards opposite poles and they move apart at first anaphase. In this behaviour they are mimicked by any unpaired autosomes (see page 140). At second division the X-chromosome does not develop spindle fibres but it is caught-up in those from an autosomal chromatid—presumably any autosome— and is moved with it to one pole.

Many coccids, as intimated in section VI–2, have anomalous meiotic systems some aspects of which seem to be related to, though not conditional upon, their non-localised centromeres. But the behaviour of the X-chromosome in these species is essentially similar to that described above for *Llaveia* though the chromatid autonomy shown in types like *Llaveiella* introduces additional anomalies. Thus, in *Llaveiella*, the constituent chromatids of the X-chromosome may be enclosed in separate nuclear vesicles at diakinesis. X-chromatids from separate vesicles orient independently at first division and each of them divides! Subsequently, half-chromatids, one from each sister chromatid, associate and they are moved to the same pole at second anaphase.

So far as we are aware, consistent post reduction has been found only in one group with localised centromeres. The condition has been known since the beginning of the century and the original observations were confirmed and extended by SMITH and MAXWELL (1953). The five species concerned represent four genera in three of the five tribes which comprise the *Lampyridae* (*Coleoptera*) of North America. In these beetles the X divides equationally at first division when the anaphase movement of the half-bivalents is virtually complete. The bi-partite structure of the X is visible some time before its division but its double nature is not obvious in other XO beetles where it shows pre-reduction.

Post-reductional segregation of the X-chromosome has recently been described in the XO males of the damselfly, *Enallagma cyathigerum* (VAN BRINK and KIAUTA 1964). The status of the centric system in this group is uncertain but, in view of the rarity of X-post-reduction in types with localised centromeres, the behaviour of the unpaired sex chromosome suggests non-localised centric activity.

Variable Pre- and Post-reduction

With the exception of uncommon atypical cells, the behaviour of the X, whether pre- or post-reductional, is characteristic of the species or even much larger groups. In some types, however, the first division may be either equational or reductional for the X-chromosome. Thus in *Oswaldocruzia filiformis*, an intestinal endoparasite of the toad, the unpaired X comes to lie near the equator with the five autosomal bivalents. In some cells it subsequently moves undivided to one pole so that complementary products with and without an X-chromosome are produced in equal numbers. In other cells of the same individual, however, the X divides at first division so that there are, in all, three types of first division products but only two kinds of meiotic products (JOHN 1957a). Variation similar to that found in *Oswaldocruzia* was reported by NIGON (1949) in the rare males he observed in laboratory cultures of the normally free-living and hermaphrodite nematodes *Rhabditis elegans* and *Rh. dolichura*.

YO system

A remarkable, indeed so far as we are aware unique, condition has been described in the creeping vole, *Microtus oregoni* (OHNO, JAINCHILL and STENIUS 1963) where the male germ line is YO although both male embryos and the male soma are XY. Indeed both sexes are gonosomic mosaics of the type:

♂ 2 n = 18 soma (XY)/17 germ line (YO) and

♀ 2 n = 17 soma (XO)/18 germ line (XX), thus:

Cell type		Sex	
		♂	♀
Zygote and embryo		XY	XO
Adult	Soma	XY	XO
	Germ line	YO	XX
Gametes		Y or O	X

The single Y in the male germ line is heteropycnotic during the first prophase. At first anaphase the Y moves undivided to one pole and then divides at second anaphase in a manner paralleling the XO males of grasshoppers.

A comparable condition may well exist in *Ellobius lutescens* (MATTHEY 1962).

X_nO systems

The efficiency of a simple XO system depends on the movement of an unpaired X chromosome to either spindle pole at either first or second division. But in systems involving two or more X-chromosomes they must all behave as one and proceed to the same pole. This necessitates some means of communication between the X-chromosomes or else some external force capable of giving co-ordinated movement without direct communication. As in the simple XO systems, so in the multiple X_nO ones, the X component may show either pre- or post-reduction.

Pre-reduction

A multiple X_1X_2O system is probably the ancestral one in spiders and the meiotic sequence in species of *Tegenaria* has been studied in detail by REVELL (1947). In *T. atrica*, for example, the two X-chromosomes are positively heteropycnotic and associated in the pre-meiotic nuclei. As leptotene develops the centric, pre-condensed ends of the autosomes become concentrated towards one side of the nucleus together with the sex-chromosomes. This polarisation is in relation to the centrioles which lie close together at this stage. When sister centrioles move apart at diplotene, the majority of the autosomes, together with the two sex chromosomes, remain polarised towards one centriole while the other autosomal bivalents move with the second centriole. The diakinesis nucleus is thus bipolar and both centromeres of each autosomal bivalent are directed towards the same centriole. Following the organisation of a spindle, this monosyntelic system gives way sequentially to disyntelic co-orientation. But the

X-chromosomes do not change their orientation (Fig. 163). At metaphase they are negatively heteropycnotic, aligned in parallel and they lie in one half of the spindle. In fact, they remain near the centriole towards which they were polarised at prophase and they become included in the daughter nucleus that forms at that end of the spindle.

REVELL described a similar sequence in other species of *Tegenaria* which differed from *T. atrica* in having not two but three X-chromosomes. Conditions in the indian spider, *Stegodyphus specificus* ($2n = 28 + X_1X_2O$) appear to be similar but the two X-chromosomes tend to have a more equatorial position at first metaphase (SHARMA and SINGH 1957). The details of the sequence differ, however, in *Aranea* where the joint polarisation of the sex-chromosomes does not appear to be established until after the spindle has been formed (PÄTAU 1948). In a study of various Indian spiders, SHARMA, JANDE and TANDON (1959) showed that the X-chromosomes in females of *Senelops radiatus* ($2n = 26 + X_1X_1X_2X_2X_3X_3$) formed bivalents.

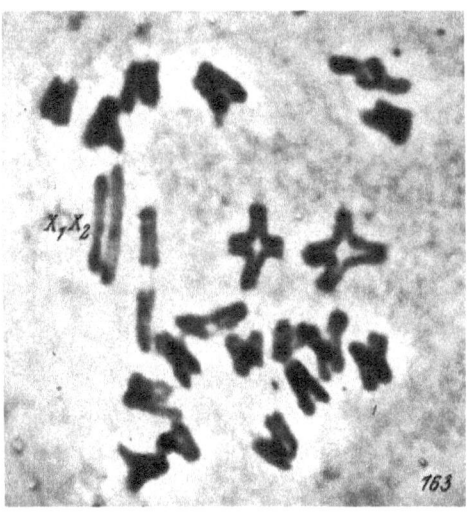

Fig. 163. Sequential metaphase-I congression from a bipolar diakinesis in the male of *Tegenaria atrica* ($2n = 20 + X_1 X_2 0$). The two X-chromosomes retain their original polarisation and lie parallel (orcein-phase contrast, ca. × 2000).

Several species of stone flies in the genera *Perla*, *Isoperla* and *Isogenus* are X_1X_2O in the male. At meiosis, these two X-chromosomes like those of spiders, are aligned in parallel. But the X-complex behaves in the same way as the single X of XO species of stone flies (see page 179). Its members lie in one half of the spindle with their centromeres directed towards the equator through which they pass during first anaphase. The same pattern of "backward" anaphase movement also characterises the $X_1X_2X_3$ species of *Perloides* though here the three sex-chromosomes of the male form a zig-zag configuration in one half of the spindle (MATTHEY and AUBERT 1947).

Yet another variant has been described in X_nO species of cyprid ostracods (Table 67). Here the several X-chromosomes—two in *Notodromonas monacha*, three in *Scottia browniana*, *Platycypris baueri* and *Physocypria kliei*—form a non-chiasmate X-aggregate. The basis of this association is not clear in all four cases but in *Notodromonas* the heterochromatic sex-chromosomes come together in the region of their centromeres at pre-diakinesis. The X-aggregate moves *en bloc* to one pole at first anaphase after lagging behind the separating half bivalents. And in the 3 X forms, this collective movement occurs despite the fact that not all the three centromeres orient to a common pole. During interkinesis the aggregate disbands and all the sex chromosomes divide quite independently at second anaphase.

Table 67. *Sex chromosome Systems in Cyprid Ostracods.*
(Data of DIETZ 1955, 1958.)
All show male heterogamy and are pre-reductional in type.

Sex Mechanism		Species	Metaphase-I Behaviour
XY		*Cyclocypris laevis*	Chiasmate $X^M Y^T$ bivalent
XO		*Cyclocypris globosa*	X^M-univalent
$X_n O$	2 X	*Notodromonas monacha*	Non-chiasmate $X_1 T X_2 T$ aggregate
	3 X	*Scottia browniana* *Platycypris baueri* *Physocypria kliei*	Non-chiasmate $X_1 T X_2 T X_3 T$ aggregate
$X_n Y$	3 X	*Cypris dietzi, whitei,* *fodiens* and *compacta*	The 3 telocentric X's from a non-Xte sex aggregate but one of them also forms a Xte association with the single Y^T
	3 X	*Cypria exsculpta*	The 3 telocentric X's form a non-Xte sex aggregate while the single Y^T remains as a sex univalent
	4 X	*Cypria exsculpta* and *ophthalmica*	Three of the X^T's form a non-Xte sex aggregate. The fourth X^T and the single Y^M form an independent bivalent
	6 X	*Heterocypris incongruens*	The six X^T's form a non-Xte sex aggregate but one of them also shares a Xte association with the single Y^T
XY_n	3 – 6 Y	*Cyclocypris ovum*	All the sex chromosomes ($3-6\ Y^T$ and $1\ X^M$) form univalents
	4 – 7 Y	*Cyclocypris ovum*	One of the Y^T chromosomes and the single X^M forms- a chiasmate bivalent, the remaining $3-6\ Y$'s all remain as univalents

Note: M = metacentric, T = telocentric.

Finally, species having diffuse centromeres and a multiple $X_n O$ system are known in which X-chromosomes also show pre-reduction. Thus in the coccid

Matsucoccus, the six univalent X-chromosomes of the male form a ring⁻or plate at first metaphase which is surrounded by an outer ring formed by the more slowly condensing autosomal bivalents. The X-chromosomes then move as a group to one pole at first anaphase and each one divides equationally at second anaphase.

Post-reduction

Other species which resemble *Matsucoccus* in having diffuse centromeres and a multiple X_nO sex chromosome mechanism differ from it in that the sex-chromosomes divide equationally at first division. This is true of the species of *Dysdercus* (Fig. 164) though other X_1X_2O heteropterans like species of *Anisops* show pre-reduction (JANDE 1959). Post-reduction is found also in nematodes. Thus the two X-chromosomes in *Belascaris triquetra* and *Ganguleterakis spinosa*, the five in *Ascaris lumbricoides*, the six in *Toxascaris canis* and the eight in *Contracaecum incurvum*, all divide equationally at first anaphase (see WHITE 1954).

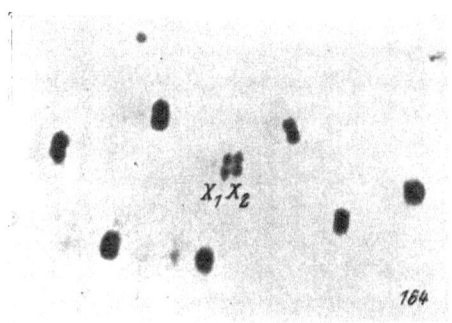

X_1X_2

164

Fig. 164. Metaphase-I in *Dysdercus koenigii* ($2n = 14 + X_1 X_2 0$). The X-univalents move the centre of the spindle and divide equationally. The two non-homologous X-chromatids then associate and move together at anaphase-II (carmine, ca. × 1500).

In some of the Heteroptera a very intimate association of the non-homologous X-chromatids produced by the equational first division is established at anaphase-I or earlier. In this respect certain X_1X_2O species resemble those with an XY system (see page 190) and for this reason have sometimes been confused with them (see BATTAGLIA 1956).

Non-chiasmate XY Systems

To some extent, the problems presented by a mechanism of this kind are similar to those proposed by multiple X_nO systems. Thus, in both, some means of communication other than chiasmata must obtain, directly or indirectly, between the sex-chromosomes. The two mechanisms differ, however, in that, having communicated, the sex-chromosomes in one case have to move together (X_nO) whilst in the other they must move apart. And in the latter event they may again separate at either the first or the second division.

Pre-reduction

The males of all the known Neuroptera Planipennia are XY and in most of them the two sex chromosomes are unpaired at first metaphase. Even so they are co-oriented at a distance, one sex member lying in each half of the spindle in a "distance conjugation" (Fig. 165). The stages of first prophase have been studied by SUOMALAINEN (1952) who finds that the sex-chromosomes are positively heteropycnotic throughout first prophase. At zygotene they often lie widely separate but by pachytene they are mostly situated near to or even touching one another in a non-chiasmate association. In one case, *Chrysopa abbreviata*, SUOMALAINEN demonstrated a small nucleolus connecting them (his Fig. 72 and see page 186). HIRAI (1956) also describes a nucleolar association at prophase in *Myrmeleon formicarius* and *Palpares pardus asani* and a similar observation has been made by MAKINO and KANO (1947). By diakinesis, however, the X and Y are seen as univalents lying at varying distances from each other. It would appear, therefore, that we can rule out precocious disjunction following co-orientation of associated chromosomes as a cause of distance pairing, though this certainly operates in the X_1X_2Y mecopteran *Boreus brumalis* (COOPER 1951). Here, the three sex chromosomes are associated at prophase and, doubtless, by chiasmata. There is, however, a pre-metaphase stretch in this species and, though this is a very mild one in comparison with the extreme stretch of mantids and phasmids, the sex chromosomes are separated by it. In consequence they may already be about 4–5 µ apart when the autosomal bivalents congress. Precocious disjunction of X and Y (so-called Xy_p, see below) has been recorded also in beetles, e. g. *Zopherus haldemani* (SMITH 1952).

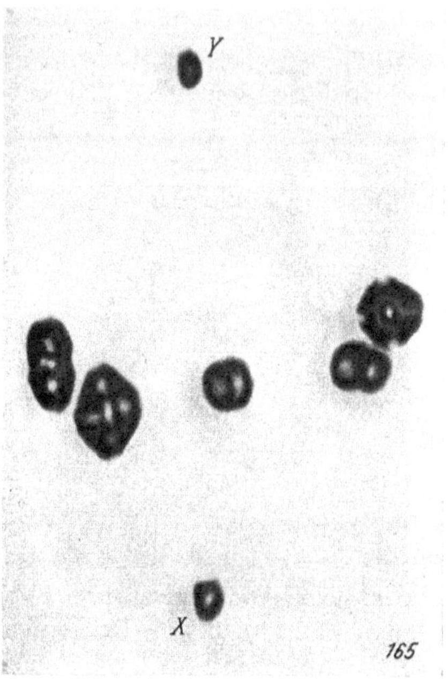

What may be a genuine case of distance pairing, but in a multiple system, is described later (see page 191). It would appear, however, that while chiasma formation has frequently been dispensed with, pairing has not. We, for example, discovered a novel non-chiasmate pairing system in beetles (JOHN and LEWIS

Fig. 165. Distance co-orientation of X and Y univalents in the neuropteran *Chrysopa sp.* (2n = 10 + XY). Photograph (ca. ×2500) kindly supplied by Dr. J. WAHRMAN (WAHRMAN and FRIEDLANDER, unpublished).

1960). In ladybirds and other beetles with an Xy_p (sex parachute) mechanism, the two sex-chromosomes of the male are positively heteropycnotic and associated both in the pre-meiotic and the early meiotic prophase nuclei. The autosomes are characterised by blocks of centric heterochromatin but for the most part these remain separate.

At pachytene the sex bivalent stains less deeply than at earlier stages. It is thus less stained than the autosomal heterochromatin but more stained than the euchromatic regions and it takes the form of a signet ring. At this time a nucleolus can be seen associated with the sex-bivalent. If this is present at an earlier stage, we have not been able to see it. It may be, therefore, that this structure is not organised until pachytene when the sex chromosomes lose some of their stainability. By metaphase-I the sex-chromosomes and autosomes stain to the same extent, as they do for the remainder of the sequence.

Preparations made with stains, like Feulgen, which stain only DNA reveal the typical parachute appearance of the sex bivalent at metaphase in which the small y appears to be associated at each end with the large, metacentric X. This condition has been described by various authors in about twenty different families of the group and it led to the suggestion that two terminalised chiasmata were responsible for the association. Aceto-carmine preparations however, show that the nucleolus is still present at metaphase and this is really responsible for the Xy_p association at this time (Fig. 166). In fact the nucleolus persists until anaphase and it moves to the same pole as the X-chromosome.

Thus Xy_p beetles resemble the bugs in that the association of the sex chromosomes before and during early prophase depends in part on the non-specific stickiness of positively heteropycnotic heterochromatin supplemented by the development of a nucleolus between them (see Figs. 2, 14 and 15 of UESHIMA 1963): specific pairing and chiasma formation are not involved. They differ from most bugs in that the Xy_p bivalent is maintained until anaphase and undergoes pre-reduction. And this in turn depends on the persistence of the nucleolus which they organise. In most bugs on the other hand the nucleolus does not persist, the X and Y separate before first metaphase and divide equationally at first-anaphase (see below).

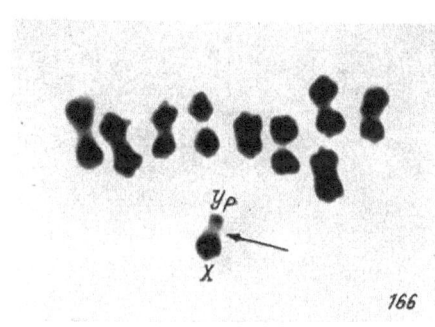

Fig. 166. Nucleolar-associated Xyp bivalent in *Tenebrio molitor* (2n = 18 + Xyp; carmine, ca. ×3000).

In at least three species of tingid bugs, however—*Bredenbachius consanguineus, Monathia globulifera* and *Tingis benglana*—the X and Y undergo pre-reduction (JANDE 1960). During leptotene and pachytene the two sex chromosomes appear as separate heteropycnotic bodies. In the "diffuse" stage which follows pachytene these two bodies are associated with a common nucleolus and by now both X and the Y are bipartite. By diakinesis the nucleolus has disappeared and the X and the Y are again quite separate. But during the formation of the first metaphase plate the sex-chromosomes re-associate end-to-end as a pseudo-bivalent which undergoes co-orientation, as do autosomal bivalents.

The formation of sex pseudo-bivalents has been described in at least two other XY-types. In the mycetophlid *Apolipthisa subincana* (FAHMY 1949) the pre-condensed sex chromosomes emerge unpaired at the onset of meiosis though they frequently lie in parallel alignment (somatic pairing). They remain unpaired until post-pachytene but then tend to move closer together and by first metaphase are paired over a short region of their length in about 85% of the cells. The nature of this association is variable. The X and Y may be paired over both arms or over only a short segment of one arm or by their centric regions alone. In the remaining 15% of the cells the sex-chromosomes are present as univalents at metaphase-I. But even these eventually succeed in pairing during anaphase-I and after lagging on the spindle the X and Y disjoin at first anaphase, each then dividing at second anaphase.

The second instance of pseudo-bivalent formation occurs in chrysomelid beetles of the genus *Alagoasa* (VIRKKI 1961). Here the sex chromosomes occur in a variety of forms but are always much larger than the ten autosome pairs. In three species—*A. extrema*, *A. ceracollis* and *A. acutangula*—the X and Y are present at diakinesis as univalent chromosomes both of which are clearly double. They do not pair until pro-metaphase and even then associate only by their terminal regions forming a loose pseudo-bivalent. In *A. bipunctata* and an unidentified species (*Alagoasa* sp. 1) the X and Y chromosomes do not pair at all. But they come to lie close together at first metaphase and following this brief contact separate to opposite poles.

But to return to the question of the persistent nucleolus in beetles. So far as we are aware, chromocenters formed by the fusion of positively heteropycnotic segments always break-up at the later division stages unless the regions con-

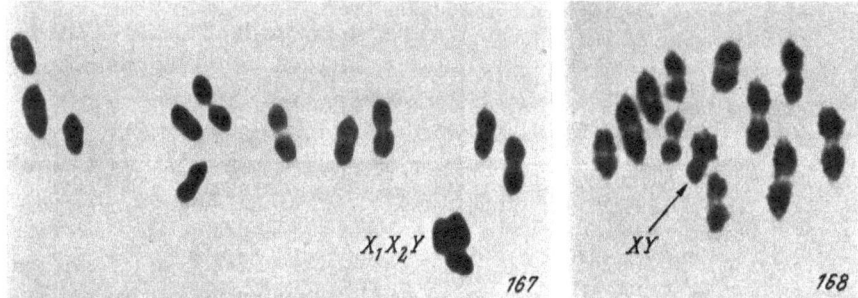

Figs. 167 and 168. The two types of male found in *Forficula auricularia* (orcein, ca. ×2250). Fig. 167—2n = 22 + X₁ X₂ Y. Fig. 168—2n = 22 + XY. CALLAN (1941) concluded that the Y-chromosome in this species was dicentric but this is a very doubtful interpretation.

cerned are responsible for the organisation of a common nucleolus and, perhaps, even then, only if that nucleolus persists and does not disappear prior to the disruption of the nuclear membrane as it usually does. It may be, of course, that the real difference is not the presence versus the absence of a distinct and discrete nucleolus but whether the materials, which in beetles are organised into a spherical body, remain on the chromosomes or diffuse away from them. This possibility is suggested by CALLAN's (1941) account of meiosis in the earwig *Forficula auricularia* (2n = 22A + XX/XY or X₁ X₁ X₂ X₂/X₁ X₂ Y). As in bugs and Xy$_p$ beetles the sex chromosomes of the earwig are completely differential and positively heteropycnotic in the pre-meiotic nuclei of the male. The species is polymorphic with respect to the sex chromosomes (Figs. 167 and 168) but whether there is one X and one Y or two X's and one Y the sex members associate non-specifically at the telophase of the pre-meiotic mitosis. If they fail to associate at this time in 2n = 25 males a bivalent and univalent condition is found at leptotene and this condition persists to first anaphase. Clearly, initial association is a consequence of the characteristic stickiness of over-condensed heterochromatin and if this fails it cannot be rectified.

At leptotene and zygotene the autosomes organise a nucleolus which behaves in a typical manner and disappears during diplotene. The sex chromosomes do not organise a discrete nucleolus but CALLAN describes them as having a covering of nucleolar material which is Feulgen-negative. This sex-chromosome-associated material is still present at diplotene but it is not figured by CALLAN in his post-diakinesis drawings. It may well be, however, that some, at least, of this material persists and is responsible for the maintenance of the sex-chromosome associations for they show pre-reduction as in the beetles. It will be appreciated that a layer of nucleolar material would be more difficult to detect than a spherical nucleolus and the latter escaped detection for fifty years. Indeed it still evades some (VIRRIKI 1963).

Whatever the details of the situation in the earwig, it is clear that there, as well as in the bugs and Xy_p beetles, initial association of the sex-chromosomes is the result of the stickiness of positively heteropycnotic material.

When we turn to the mammals on the other hand there are many cases where nucleolar material appears to be concerned in bringing the sex chromosomes together at the early stages. The basis for the later, metaphase association of the sex chromosomes in these types is a matter of some dispute and may not be the same in all of them.

The nucleolar material associated with the sex-chromosomes at meiosis in the males of mammals is commonly called the sex vesicle and it has been shown in *Microtus agrestis*, at least, that this structure is organised only in the meiotic cells (SACHS 1953). It may be the only nucleolus present at this stage, as in the mouse, or autosomally-organised nucleoli may be present also, as in the rat.

In *Microtus agrestis* the X and Y chromosomes are very large compared with the autosomes and, like those of bugs and beetles, they are positively heteropycnotic at leptotene. But, although they lie near one another they are not fused into a common chromocentre. In fact each organises its own sex vesicle and their initial association appears to depend on the fusion of these vesicles into one large structure. In this vesicle the X and Y are associated terminally or by a short terminal segment. At metaphase they are associated end-to-end and the first anaphase segregation is invariably reductional.

It would appear, therefore, that the initial approach of the sex-chromosomes is due to nucleolar fusion. But, as we mentioned earlier, their subsequent mode of association in mammals is a subject of dispute. Where a pairing region occurs between the invariably pre-reductional centromere and the differential segment, an equational first division separation of the latter is possible. The realisation of this possibility gives clear evidence for crossing-over and a chiasmate association. Of course, pre-reduction for this region does not exclude crossing-over but in the absence of crossing-over post-reduction is precluded.

So far as we know, invariable post-reduction has been claimed only in one mammal genus, namely the rodent *Apodemus* where it is seen in five species. A sixth form is said to show the opposite behaviour invariably and invariable pre-reduction has been claimed also in the fox, mink, guinea pig, porpoise and a number of bats and insectivores (see WHITE 1954).

In many cases, however, different investigators disagree regarding:

1. The relative lengths of the pairing segments.

2. The positions of the centromeres, and

3. The nature of the association whether by chiasmata or otherwise. Indeed, some investigators have frequently changed their minds.

For example, MATTHEY (1936) claimed that both pre- and post-reduction occurred in *Apodemus sylvaticus*; KOLLER (1941) agreed and gave their relative frequencies as 8 and 92% respectively. But, even so, MATTHEY subsequently doubted his earlier conclusion. Similarly, in the striped or Chinese hamster (*Cricetulus griseus*) PONTECORVO (1943) claimed to have seen both pre- and post-reduction and in 1952 MATTHEY agreed that some post-reduction did obtain. But a year later he claimed there was pre-reduction only, an opinion held also by YERGANIAN (1959). In the golden hamster, *Mesocricetus auratus*, too there is a similar conflict of opinion. KOLLER (1938) who first examined the situation claimed 81.6% pre-reduction and 18.4% post-reduction. MATTHEY (1952) reported that the separation of the sex-chromosomes was always pre-reductional, so too did OHNO and WEILER (1961). EMMONS and HUSTED (1962), on the other hand find both pre- (97.13%) and post-reduction (2.87%).

As we have already pointed out, invariable pre-reduction does not necessarily mean the absence of a pairing segment and crossing-over. This is because pre-reduction is obligatory when the centromeres are in the differential segment, or between it and the pairing segment, whether chiasmata are formed or not. But opinions regarding the existence or otherwise of chiasmata have fluctuated also as the following quotations from the works of the same investigators show.

Of the mouse:

"The X and Y are connected end-to-end through the vesicle. Thus, there appears to be no possibility of the exchange of genetic material between the mouse X and Y during prophase" (OHNO, KAPLAN and KINOSITA 1957) and,

"The end-to-end association is very likely due to a chiasma formed between the extremely short second arms of the X and Y" (OHNO, KAPLAN and KINOSITA 1959).

And of the rat:

"The large X and small Y-chromosomes appear always to be joined end-to-end without any chiasma formation" (OHNO, KAPLAN and KINOSITA 1956) and,

"The chiasma formation between the X and Y, however, appears to be limited to the extremely short second arms which may be euchromatic" (OHNO, KAPLAN and KINOSITA 1958).

Synaptinemal complexes (see page 204) are not found in the sex vesicle of the mouse at a stage when these are developed by the autosomes. This suggests that the association of the X and Y is different in kind from that of the other chromosomes (SOLARI 1964).

Finally OHNO and WEILER (1962) point to an interesting correlation between the size of the Y and its apparent meiotic behaviour. In a majority of mammals, including men, rats, cats, mice and opossums, the Y is one of the smallest chromosomes while the X is among the largest. From the onset, both the X and the Y are positively heteropycnotic and, so it is claimed, they always associate end-to-end so that meiosis is always pre-reductional.

In *Rattus natalensis* (HAMERTON 1958), *Allocricetulus eversmanni* and *A. curtatus* (MATTHEY 1961) and *Cricetus griseus* (OHNO and WEILER 1962), however, the Y is almost as large as the X and the two chromosomes pair side-by-side presumably indicating a substantial pairing segment. Indeed, in *Cricetus griseus* when the major portion of the X and Y are heavily condensed, the distal two-thirds of the long arms of both remain isopycnotic and are associated side-by-side. And OHNO and WEILER were able to show the presence of one, or occasionally two, chiasmata at diakinesis in 21 out of 50 cells analysed *.

---

* See note 3 Appendix.

One of the most intriguing of non-chiasmate XY systems showing pre-reduction is that found in tipulid flies (Bauer 1931, Henderson and Parsons 1963). The X and Y are both heteropycnotic at the onset of meiosis and are associated non-specifically. This association usually lapses during diakinesis. When it fails to do so, it is terminated by a fairly violent pre-stretching of the chromosomes (John 1957 b) so that the X and Y exist as univalents during first metaphase (Fig. 169). But they do not divide at first anaphase. Instead they lag and then segregate, despite the fact that in many instances the component chromatids of each sex univalent become considerably attenuated between the two polar groups. In only one instance, *Tipula maxima*, have they been found

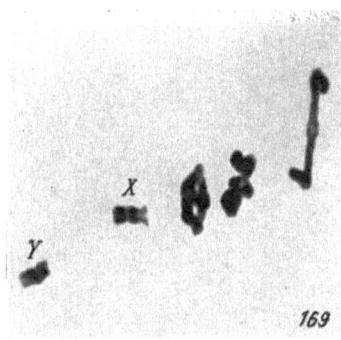

to undergo post-reduction (John 1957 b) and even here it appears to be an irregular case since Bauer (personal communication) and Henderson and Parsons (1963) have found the X and Y to undergo pre-reduction in the specimens they have examined. Where post-reduction does occur the X and Y chromatids undergo non-homologous pairing at second prophase and then segregate at second anaphase.

Fig. 169. Metaphase-I in the male *Tipula maxima* (2n = 6 + XY). The sex-chromosomes are unpaired and form a system of hereditary univalents (orcein, ca. × 1500).

Post-Reduction

As far as we know, no XY system with localised centromeres undergoes regular post-reduction in the absence of chiasmata, but where diffuse centromeres are present the situation may be otherwise. Thus the majority of the XY-Heteroptera show post-reduction. The precise details vary somewhat, we shall consider the case of *Dicranocephalus agilis* (2n = 10A + 2m + XY). In the male meiosis of this species, the early prophase nucleus includes a positively heteropycnotic body and a nucleolus organised by the autosomes. This body represents the associated X and Y chromosomes. Their condensed condition and later behaviour leaves one in no doubt that their association is non-chiasmate and even non-specific (Lewis and Scudder 1958). It is, in fact, usual for adjacent positively-heteropycnotic segments to associate in this way, whether they are allo- or auto-somal, for they, like nucleoli, are mutually adhesive.

By pro-metaphase I the chiasmate, autosomal bivalents stain as deeply as the sex-chromosomes which are still associated. The latter move to the spindle equator and may show a temporary co-orientation. But they soon fall apart and the X and Y auto-orient, side by side, on the periphery of a hollow spindle (Fig. 170). The co-oriented autosomal bivalents occupy the periphery also.

Now, the sex-chromosomes become negatively heteropycnotic relative to the autosomes, a condition they retain throughout the remainder of the sequence. And they divide equationally at the first division, though their chromatids separate slightly later than the autosomal half-bivalents.

Since the spindle is narrower at the poles than at the equator, the X and Y chromatids approach each other as they proceed to the poles. This approach is facilitated by their prior side-by-side auto-orientation and this, in turn, reflects their non-chiasmate prophase association. When they touch at late anaphase the negatively heteropycnotic sex-chromatids associate. There is no interphase and the sex "pseudo-bivalent" co-orients at second division, this time in the centre of the spindle (Fig. 171).

As we intimated earlier, this association of X and Y chromatids can be compared with that between non-homologous X-chromatids in multiple X_nO systems (see page 184). In the latter, however, the members of the pseudo-bivalent pass to the same pole while in the former they pass to opposite poles. How this difference is determined is completely obscure but, paradoxically, post-reduction results from both events.

The complement of *D. agilis* includes also a pair of small, so-called m-chromosomes. These, like the sex-chromosomes are completely differential but the mechanism of their reduction is different. They are not over-stained at prophase and they are not associated up to diakinesis. But, by metaphase they have moved together and now co-orient in the centre of the first-division spindle (Fig. 170). At this stage they, like the sex-chromosomes are understained. It is interesting to note therefore that while the differential sex-chromosomes show post-reduction, the similar m-chromosomes are pre-reduced.

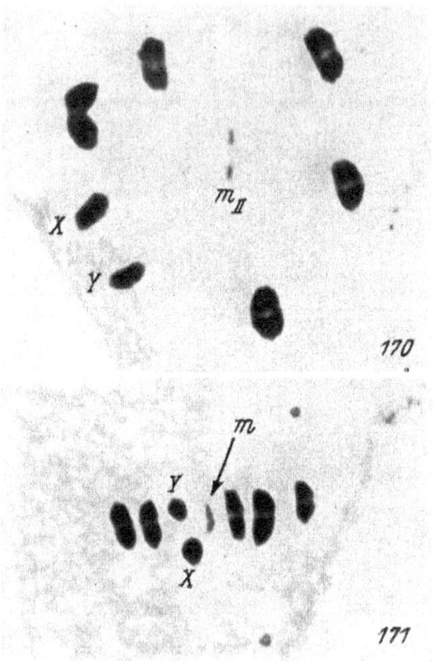

The spindle-positions occupied by the sex and m-chromosomes tend to be constant for a given species or even

Figs. 170 and 171. First and second metaphase of meiosis in the male of *Dicranocephalus agilis* (2n = 10 A + 2m + XY); orcein, ca. × 2000). At metaphase-I (Fig. 170) the m-bivalent is placed in the centre of the hollow spindle while the univalent X and Y chromosomes lie next to one another at the periphery. At metaphase-II (Fig. 171) the m and the pseudo-XY bivalent are both placed centrally.

a larger group (see MANNA 1962) though in some cases they may show some variation even within individuals.

Non-chiasmate Multiple XY Systems

The mechanisms which have been held to operate in simple, non-chiasmate XY systems have also been described for comparable multiple (X_nY and XY_n) systems. These too, therefore, may show either pre- or post-reduction. Indeed, as we have already seen in the case of *Forficula*, simple and multiple systems may coexist within a given species and the mechanism in both types is essentially the same.

Pre-reduction

Distance pairing has been claimed for the cricket *Eneoptera surinamensis* (PIZA 1946). This species is peculiar in having only three pairs of autosomes and, in the male, three sex-chromosomes, XY_1Y_2. At first metaphase, or what is first metaphase for the autosomal bivalents, the X lies in one half of the spindle with its centromere directed towards the nearer pole. The two Y-chromosomes lie free from each other but both are in the other half of the spindle again with

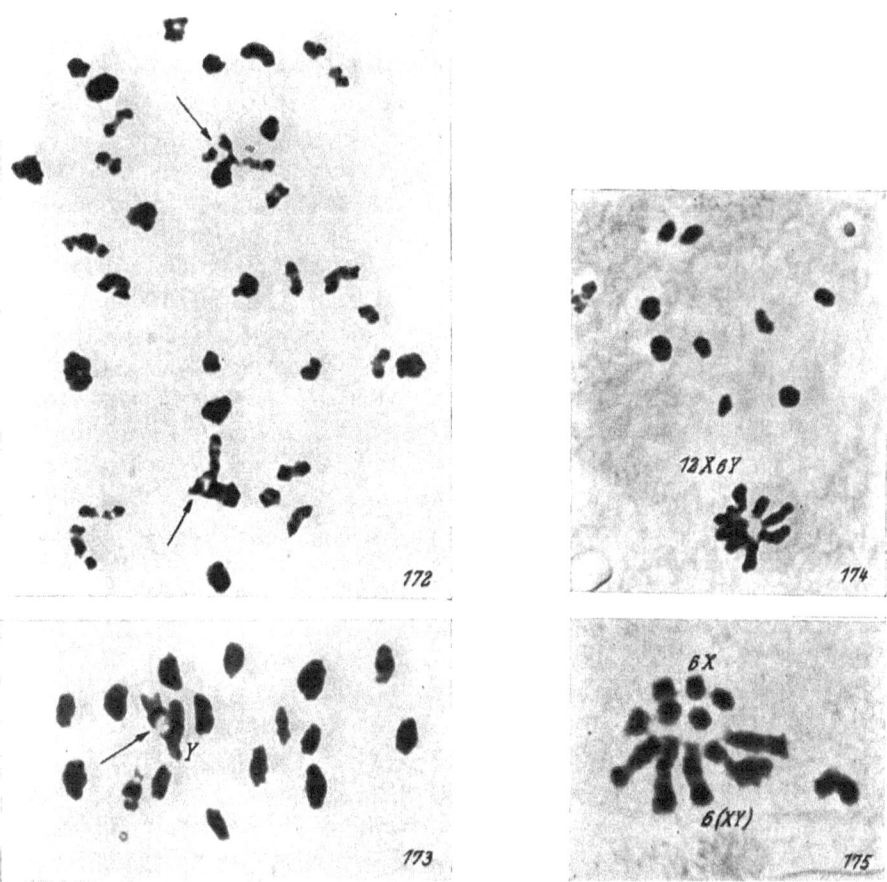

Figs. 172 and 173. Meiosis in the male of *Blaps mucronata* ($2n = 32 + X_1X_2X_3Y$). Fig. 172—Two cells at dia-kinesis. The "frying-pan" shaped sex multiple is in close association with an autosomal bivalent in each case. Fig. 173—metaphase-I—the Y-chromosome is co-oriented with the three X-chromosomes and the nucleolus is quite distinct (carmine-phase contrast, ca. ×2000).

Figs. 174 and 175. *Blaps cribosa* ($2n = 18 + 12 \times 6 Y$). Fig. 174—A complete cell (ca. ×2500). Bivalents are formed between each of the six Y-chromosomes and 6 of the 12 X-chromosomes, the remaining 6 X's being asso-ciated by means of a nucleolus with the bivalents. In Fig. 175 one of the XY bivalents has become detached and lies to the right of the main group. Photographs kindly supplied by Dr. J. WAHRMAN (WAHRMAN and FREUND, unpublished).

their centromeres oriented. PIZA has figured diakinesis and at this stage there are three univalents.

The Neuroptera are generally XY in the male, and, as will be recalled (see page 184), distance pairing has been claimed in them. But in *Plethosmylus decoratus* an $X_1 X_2 Y$ mechanism has been described (HIRAI 1956). In this species the three sex-chromosomes are associated with the nucleolus at pachytene. By late diakinesis they "lie distantly far apart, but connected by fine threads with one another."

A quite different, highly peculiar and, at present, inexplicable situation was described by PAYNE in the mole cricket *Gryllotalpa hexadactyla* and the observations have been confirmed by WHITE (1951). At first metaphase in the males of this species there is an unpaired chromosome which doubtless corresponds to the single X-chromosome of congeneric XO species. But there is, in addition, an unequal bivalent. And while this has no connection with the univalent X, the larger of its two members proceeds to the same pole as the X at first anaphase. The most logical explanation is that this represents an $X_1 X_2 Y$ system since DIETZ (1958) has described an equivalent situation in *Cyclocypris ovum* (Table 67). Here two kinds of sex systems are known. In the one the single metacentric X and the 3–6 telocentric Y-chromosomes all form univalents, yet nevertheless show pre-reduction. But in the second, one of the 4–7 telocentric Y-chromosomes forms a chiasmate bivalent with the single metacentric X while the 3–6 remaining Y's all behave as univalents.

In the Coleoptera, neo-XY systems have developed, presumably as a result of X/autosome exchange. This may have been necessitated by the erosion and loss of the original y_p for it presumably gives a mechanically more stable meiotic system. But a few XO beetles are known. Be this as it may, the neo-XY mechanisms, unlike the basic Xy_p system, are chiasmate. But the same process has arisen in the numerically tetraploid types with multiple $X_n Y$ mechanisms. Mechanically speaking these are a combination of the other two. Thus, in such types, it would appear that the Y-chromosome forms a chiasmate connection with the most recently evolved X-chromosome while the other X-chromosome or chromosomes are held to the chiasmate XY pair by the nucleolus. We first observed this situation in *Blaps mucronata* ($X_1 X_2 X_3 Y$ Figs. 172–173) before, in fact, we found that the nucleolus plays a comparable role in the basic Xy_p system of the group (LEWIS and JOHN 1957, JOHN and LEWIS 1960). And equivalent observations on other $X_n Y_n$ beetles have been made by WAHRMAN (Figs. 174–175) and GUÉNIN (unpublished communication).

DIETZ (1958) has described $X_n Y$ systems similar in principle to those of beetles, in cyprid ostracods (Table 67). Thus in *Cypris dietzi, whitei, fodiens* and *compacta* ($X_1 X_2 X_3 Y$) and *Heterocypris incongruens* ($X_1 X_2 X_3 X_4 X_5 X_6 Y$), the several telocentric X-chromosomes form a non-chiasmate sex-aggregate (compare $X_n O$ cyprids, see page 182) but one of its members is joined by chiasmate union to the single metacentric Y. The nature of the X-aggregate is not known but persistent nucleoli tend to characterise the entire group (DIETZ 1958). An even more remarkable condition is found in *Cypria exsculpta*. In $X_1 X_2 X_3 Y$ forms the 3 telocentric X-chromosomes again form a sex-aggregate but the single Y remains univalent and yet undergoes pre-reductional segregation. The $X_1 X_2 X_3 X_4 Y$ form of *C. exsculpta* involves three of the X-chromosomes in a sex-aggregate with the fourth X and the single Y forming an independent bivalent. This system also occurs but regularly in *Cypria opthalmica*.

Post-reduction

Multiple non-chiasmate $X_n Y$ systems are quite common in the Heteroptera (Table 68) where as many as five X-chromosomes may occur consistently in the male. The many X-chromosomes probably arise following fragmentation for an

Table 68. *Multiple Sex-chromosome Systems in the Heteroptera.*
Sex Multiples Have Been Recorded in 12 of the 29 Families Studied Cytologically.
(Compiled from MANNA 1958, 1962.)

Family	Species	(2n) and Sex Type
1. *Pentatomidae*	1. *Oechalia pacifica*	$2n = 14 = 11 + X_1X_2Y$ or $10 + X_1X_2X_3Y$
	2. *Macropygium reticulare*	$2n = 15 = 12 + X_1X_2Y$
	3. *Dinidor rufocinctus*	$2n = 21 = 18 + X_1X_2Y$
	4. *Thyanta calceata*	$2n = 27 = 24 + X_1X_2Y$
2. *Lygaeidae*	1. *Oxycarenus laetus* and *hyalinipennis*	$2n = 17 = 14 + X_1X_2Y$
	2. *Aphanus japonicus*	$2n = 17 = 12 + X_1X_2X_3X_4Y$
3. *Pyrrhocoridae*	1. *Physopelta schlanbuschi*	$2n = 17 = 14 + X_1X_2Y$
	2. *Dysdercus koenigii*	$2n = 16 = 14 + X_1X_2O$
	3. *Pyrrhopeplus posthumus*	$2n = 24 = 22 + X_1X_2O$
4. *Reduviidae*	1. *Onocephalus impudicus*	$2n = 23 = 20 + X_1X_2Y$
	2. *Ectomocoris cordiger* and *E. ochropterus*	$2n = 23 = 20 + X_1X_2Y$
	3. *Triatoma* sp. 1	$2n = 23 = 20 + X_1X_2Y$
	sp. 2	$2n = 25 = 22 + X_1X_2Y$
	4. *Pygolampis foeda*	$2n = 25 = 22 + X_1X_2Y$
	5. *Pnirontis modesta*	$2n = 25 = 20 + X_1X_2X_3X_4Y$
	6. *Acholla multispinosa*	$2n = 26 = 20 + X_1X_2X_3X_4X_5Y$
	7. *Arilus cristatus*	$2n = 26 = 22 + X_1X_2X_3Y$
	8. *Coranus fuscipennis*	$2n = 27 = 24 + X_1X_2Y$
	9. *Fitchia spinulosa*	$2n = 27 = 24 + X_1X_2Y$
	10. *Rocconata armulicornis*	$2n = 27 = 24 + X_1X_2Y$
	11. *Sinea* sp.	$2n = 28 = 24 + X_1X_2X_3Y$
	12. *Pselliopus cinctus*	$2n = 28 = 24 + X_1X_2X_3Y$
	13. *Velinus nodipes*	$2n = 28 = 24 + X_1X_2X_3Y$
	14. *Pasiropsis* sp.	$2n = 29 = 24 + X_1X_2X_3X_4Y$
	15. *Sinea rileyi*	$2n = 30 = 24 + X_1X_2X_3X_4X_5Y$
5. *Miridae*	1. *Dryophilocoris flavoquadrimaculatus*	$2n = 37 = 34 + X_1X_2Y$
6. *Nepidae*	1. *Nepa cinerea*	$2n = 33 = 28 + X_1X_2X_3X_4Y$
	2. *Laccotrephes griseus* and *L. maculatus*	$2n = 41 = 38 + X_1X_2Y$
	3. *Ranatra elongata, R. filiformis* and *R. sordidula*	$2n = 42 = 38 + X_1X_2X_3Y$
	4. *Ranatra linearis* and *R. elongata*	$2n = 43 = 38 + X_1X_2X_3X_4Y$
	5. *Laccotrephes maculatus* and *L. rubra*	$2n = 43 = 38 + X_1X_2X_3X_4Y$
7. *Cydnidae*	1. *Stibaropus molginus*	$2n = 31 = 28 + X_1X_2Y$

Family	Species	(2n) and Sex Type
8. *Dsyodiidae*	1. *Dysodius lunata*	$2n = 31 = 28 + X_1X_2Y$
9. *Mesoveliidae*	1. *Mesovelia furcata*	$2n = 35 = 30 + X_1X_2X_3X_4Y$
10. *Coreidae*	1. *Haploprocta sulicornis*	$2n = 18 = 16 + X_1X_2O$
	2. *Gonocerus* sp.	$2n = 19 = 17 + X_1X_2O$
	3. *Holopterma alata*	$2n = 20 = 18 + X_1X_2O$
	4. *Pachycephalus* sp.	$2n = 22 = 20 + X_1X_2O$
	5. *Centrocoris* sp.	$2n = 22 = 20 + X_1X_2O$
	6. *Coreus* sp.	$2n = 22 = 20 + X_1X_2O$
	7. *Enoplops* sp.	$2n = 22 = 20 + X_1X_2O$
	8. *Coreus marginatus*	$2n = 23 = 20 + X_1X_2X_3O$
	9. *Acanthocoris* sp.	$2n = 24 = 22 + X_1X_2O$
	10. *Petillia* sp.	$2n = 28 = 26 + X_1X_2O$
11. *Cimicidae*	1. *Cimex rotundatus*	$2n = 31 = 28 + X_1X_2Y$
	2. *Cimex columbaris*	$2n = 29 = 26 + X_1X_2Y$
	3. *Cimex lectularius*	$2n = 29 + = 26 + X_1X_2Y + $ $0 - 12$ extra X's
12. *Nethridae*	1. *Golgulus oculatus*	$2n = 35 = 30 + X_1X_2X_3X_4Y$

increase in their number is not achieved at the expense of the autosomes (cf. X_nY-Coleoptera). This interpretation is consistent with the observation that, while autosomal associations are chiasmate, the sex-chromosomes in multiple systems do not form chiasmata in the male (cf. X_nY Coleoptera).

The general meiotic sequence in X_nY-Heteroptera is similar to that in non-chiasmate XY species. It would appear, however, that co-orientation at second division need not be preceded by an intimate association similar to that described earlier in *Dicranocephalus* and X_1X_2O-Heteroptera. Thus, the orientation is similar to the distance pairing described in *Eneoptera* though the sex-chromosomes of the bugs are closer together even if they do not touch. Even so the many X-chromosomes, which usually vary in size, behave regularly.

A rather special situation has been described in *Cimex lecticularis* (DARLINGTON 1939). Related species of *Cimex* have two X-chromosomes and a single Y. So too do some individuals of *C. lecticularis*. Others, however, may have as many as thirteen extra chromosomes and these are X-like in their behaviour . But when the number of extra "X"-chromosomes is high, some of them separate reductionally at first anaphase. At second metaphase the X-chromosomes typically form a plate opposed to the single Y-chromosome and disjunction is regular. But, non-disjunction—the movement of some X-chromosomes with the Y—is seen again when the number of supernumerary X's is high.

The converse situation—a variable number of Y-like chromosomes—has been observed in several species of *Acanthocephala*, a genus of coreid bugs. In these, the Y-chromosomes, which may number up to five, form a chain as opposed to a plate but even so separation is usually disjunctional (WILSON 1910). Thus in *Cimex* and in *Acanthocephala* one finds chromosomes which, owing to their

behaviour, can be regarded as sex-chromosomes. But, in view of their expendi-
bility and numerical variation within the breeding group, they could as well
be considered as B-chromosomes. And it is to the meiotic behaviour of such
chromosomes that we now turn.

2. B-chromosomes

Earlier we distinguished between what DARLINGTON has called hereditary
and spontaneous univalents. A distinction has been made also between A-chromosomes and what have been variously called accessory, supernumerary or B-chromosomes. The former designates the members of the standard complement but
in this section we are concerned with the latter.

Table 69. *Variation in Number of B-chromosomes Between Plants in Five Species
of Grasses.*
(Compiled from BOSEMARK 1957.)

Species	No. of B's per Plant and no. of Plants							% Plants with B's	Total Plants
	0	1	2	3	4	5	6		
1. *Briza media*	54	2	2	...	...	...	...	6.9	58
2. *Poa trivialis*	23	...	...	2	1	...	...	11.5	26
3. *Holcus lanatus*	60	12	1	...	...	...	...	17.8	73
4. *Alopecurus pratensis*	44	11	7	5	2	...	1	37.1	70
5. *Festuca arundinacea*	10	6	4	2	...	...	...	54.5	22

Polysomy is not an uncommon condition. In many cases the extra chromosomes are unaltered members of the basic complement or else they are so little
altered that their origin from, and homology with, the members of the standard
complement is not concealed. In other cases, however, the extra chromosomes,
which must presumably be derived ultimately from the basic complement, give
no indication of this origin either by appearance or behaviour. Supernumerary
chromosomes of this kind have other peculiar properties. They are found in some,
usually a small minority, of the members of a breeding group (Table 69). Their
number tends also to vary within as well as between individuals. Further, while
phenotypic effects have been demonstrated (JACKSON and NEWMARK 1960,
RUTISHAUSER 1960), it is not usually possible to predict the presence or absence
of B-chromosomes from an examination of the exophenotype.

Chromosomes of this type have been found in plants (see MÜNTZING 1958,
DARLINGTON 1963) and animals (see WHITE 1954). To some extent, the distinction between the hereditary and the spontaneous cuts across that between A-
and B-chromosomes which also conflicts with that between auto- and allo-somes
(e. g., *Clarkia*, *Cimex* see pages 195, 197).

From a mechanical point of view, the behaviour of B-chromosomes at mitosis
is more anomalous and, therefore, more interesting than their conduct at meiosis—
a curious reversal of the usual state of affairs. A detailed discussion of their

mitotic behaviour is outside the scope of the present article. We have, however, discussed this aspect in some detail elsewhere and considered it in relation to mechanical spindle asymmetry and to communications within and between chromosomes (LEWIS and JOHN 1963a).

B-chromosomes are usually smaller and more heterochromatic than the A-chromosomes of the same complement (Figs. 176–178) but this is not always the case. In *Clarkia*, for example, the supernumary chromosomes can be distinguished only when they occur as univalents (LEWIS, H. 1951). But a property shared by all B-chromosomes is their failure to pair specifically with A-chromosomes. However, A- and B-chromosomes do sometimes form non-chiasmate associations. In the case of *Dactylis glomerata* such sticky associations have been found in conjunction with failure of metaphase pairing between homologues (SHAH 1962). This observation may be significant in relation to the situation discussed on page 199.

Homologous B-chromosomes usually pair with each other though *Plantago serraria* appears to be exceptional in this respect (FRÖST 1959). The amount of pairing between a given pair of homologous chromosomes is, of course, expected to vary with the

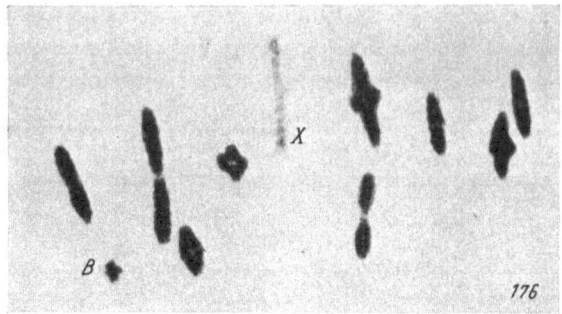

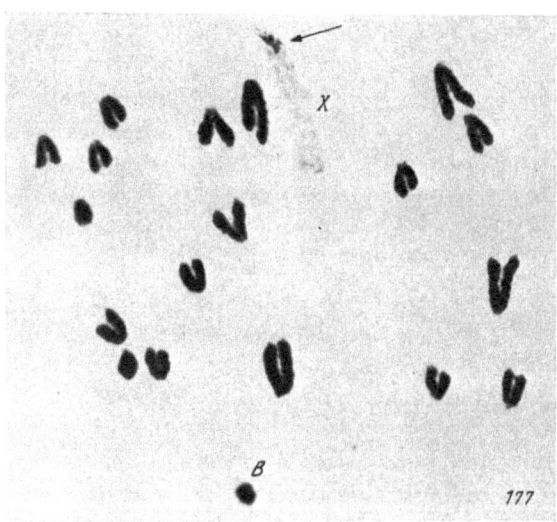

Figs. 176 and 177. B-chromosomes in *Pyrgomorpha kraussi* (2n = 9 II + X + B) at first metaphase (Fig. 176) and first anaphase (Fig. 177) of meiosis (orcein, ca. ×1500).

genotype as has been shown to be the case in rye (see HÅKANSSON 1957). But this could not be demonstrated in *Dactylis* (SHAH 1962). Where more than two homologous B-chromosomes are present, multivalent associations can be formed (Table 70). In some cases internal pairing between the two arms of a B-chromosome have been observed. This was the case in the plants *Poa trivialis* and *Holcus lanatus* (BOSEMARK 1957) for example, and was also observed in five species of trimerotropine grasshoppers (WHITE 1954). Clearly, iso-chromosomes are involved here, though in some of species, e. g., *Trimerotropsis inconspicua* and *T. sparsa*, the B-chromosome showing internal pairing was not iso-brachial.

Chromosomes whose ends are mirror images of each other but which are not iso-brachial and completely iso-chromosomal can of course result following crossing-over in a pericentric inversion. But in view of the existence of perfect iso-chromosomes it would seem highly likely that the hetero-brachials originate from iso-brachial iso-chromosomes. And the iso-chromosomes themselves are

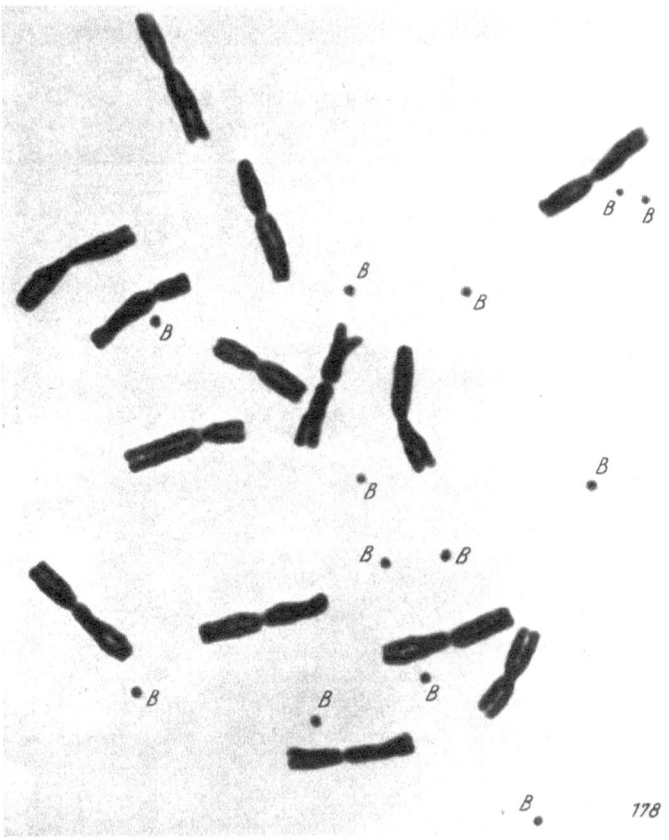

Fig. 178. Mitosis in *Allium cernuum*. The cell shows the fourteen A-chromosomes of the standard complement together with thirteen minute B-chromosomes. These actually varied in number within the plant from 12–20. Photograph kindly supplied by Mr. C. G. Vosa.

doubtless the products of mis-division. In rye, in fact, in addition to the so-called standard B-chromosome which is hetero-brachial, three other kinds of B-chromosomes are known which have been derived from it. Two of these are iso-chromosomes, representing each arm of the standard type, while the third form has originated from the standard following a deletion in its long arm (Müntzing and Lima-de-Faria 1952). In *Festuca* also B-chromosomes derived from other B-chromosomes are known (Bosemark 1956). On the other hand, to judge from their inability to pair at meiosis, an individual may carry supernumerary chromosomes which must have arisen independently from the A-complement (White 1951).

In his study of B-chromosomes in *Dactylis glomerata*, Shah (1963) found two interesting differences between the pairing of the A- and B-chromosomes. Thus, the heterochromatic B-chromosomes of a plant possessing three of them were found to have significantly different chiasma frequencies in two successive years (Table 70, P = 0.001 — 0.01). It is true that the material examined in one year was grown outdoors while that studied in the following year was cultured in a greenhouse. But a similar difference in chiasma frequency was not found in respect of the A-chromosomes. Further, one of the clones studied by SHAH had two B-chromosomes and it was subject to asynapsis. The extent of this anomaly was variable but a genotypic component was indicated. In one year the chiasma

Table 70. *The Pairing of the B-chromosomes in Dactylis glomerata lusitanica*
(Modified from SHAH 1963.)

No. of B's	Stage	Frequency and Type of Association					Xta. per B	Total	
		I	II		III	IV		Cells	Individ.
			Rod	Ring					
2	Diakinesis	0	58	14	...	...	0.60	72	1
	Metaphase	52	278	36	...	...	0.52	340	6[1]
3[2]	Diak. 1959	55	52	3	24	...	0.45	79	1
	Diak. 1960	106	79	18	90	...	0.56	190	1
4	Diakinesis	19	110	18	11	32	0.62	109	2
	Metaphase	1	25	1	1	3	0.56	17	1

[1] The plant scored at diakinesis was scored also at metaphase. The degree of pairing at the earlier stage was significantly higher (P < 0.01). A similar result was obtained in respect of the A-chromosomes. But comparable differences were not found in the plant with three B-chromosomes.

[2] Here, however, the difference between years was significant (see text).

frequency of the B-chromosomes in the asynaptic type was only about 15% of that in normal plants with two B-chromosomes. The chiasma frequency of the A-chromosomes in the same cells, on the other hand, was reduced by less than a half. The heterochromatic nature of the B-chromosomes is probably important in relation to both these differences.

The first anaphase behaviour of unpaired B-chromosomes is often variable— like that of spontaneous univalents. But, in some cases, their behaviour shows considerable uniformity. Thus, in the case of the grasshoppers *Pyrgomorpha kraussi* (LEWIS and JOHN 1959) and *Calliptamus palaestinensis* (NUR 1963) the single B-chromosome showed invariable pre-reduction. In *Holcus lanatus* also, single B-chromosomes, though they lag, do not divide at first anaphase nor do they in *Anthoxanthum aristatum*. This, however, is not the usual behaviour in grasses and an equational division at first anaphase has been found to be the rule in *Zea mays*, *Festuca pratensis*, *F. arundinacea*, *Poa trivialis* and rye (see BOSEMARK 1957). Unpaired B-chromosomes of *Locusta migratoria* on the other hand may behave in either way. Single or unpaired supernumeraries either divide

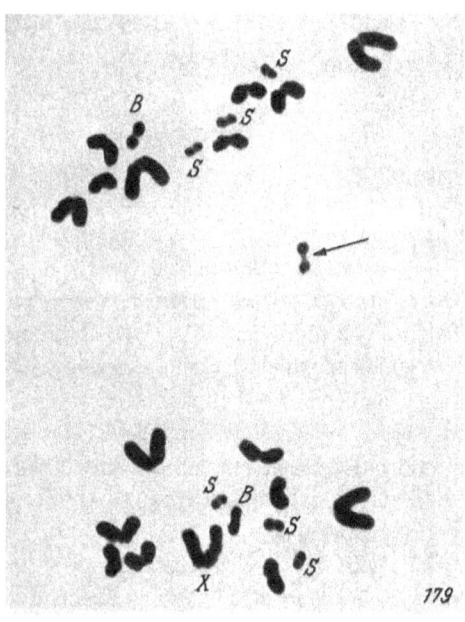

late at first anaphase (Figs. 179–180) in which case they are lost at second division or they divide at second anaphase with no loss (Rees and Jamieson 1954).

B-bivalents usually show normal orientation, congression and disjunction (Figs. 179–180). However, in two grasshoppers, *Trimerotropis sparsa* and *Circotettix thalassinus*, frequent mono-syntelic orientation has been described in the sperm mother cells (White 1954) though even here it does not necessarily characterise all individuals. This led to non-disjunction, one secondary spermatocyte receiving both members of the B-bivalent. A similar result was observed by Catcheside (1950) in PMC's of *Parthenium argentatum*. Here, however, the nature of the association between the heterochromatic B-chromosomes could not be established with certainty. And the non-random segregation seemed to depend on the univalents of lapsed associations having correlated orientation at first metaphase.

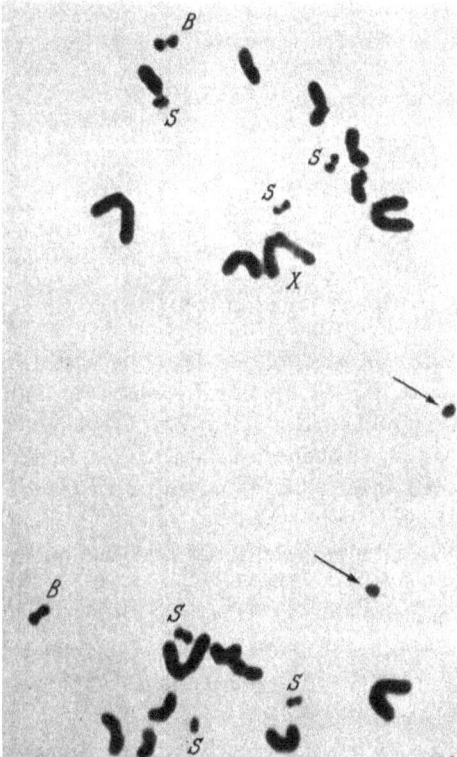

In most species, all four meiotic products are functional on the male side. Consequently, while non-disjunction during meiosis in the male affects the distribution of B-chromosomes it does not affect their average frequency. On the female side, however, three meiotic products generally abort in plants and

Figs. 179–180. B-chromosomes in *Locusta migratoria* (2n = 11 II + X + 3 B) at first anaphase. In Fig. 179 there is a B-chromosome at each pole giving anaphase groups of (11 + X + B) and (11 + B) respectively with one B lagging between the polar groups. In Fig. 180 there is an equivalent situation but the lagging B-chromosome has undergone equational division. In this species the B-chromosome, though small, is slightly larger than the 3 smallest (S) members of the complement (orcein, ca. ×1500).

animals. And polarised non-disjunction in relation to the physiological gradient which determines abortion can lead to the accumulation or elimination of B-chromosomes from the functional gamete. So far as we are aware, only accumulation mechanisms have been described. Thus KAYANO (1956, 1957) found that unpaired B-chromosomes are transmitted preferentially via the egg in *Lilium callosum*. In this species the functional megaspore lies at the micropylar end of the linear tetrad and unpaired B-chromosomes move preferentially to the micropylar pole of the first division spindle. *Trillium grandiflorum* shows the opposite condition for here it is the chalazal megaspore which functions and polarised non-disjunction occurs at first division with the unpaired B-chromosome moving, in this case, to the chalazal end of the spindle (RUTISHAUSER 1960). In *Plantago serraria* also, preferential transmission of B-chromosomes has been described on the female side (FRÖST 1959). It occurs also in rye but here it follows non-disjunction during one of the haploid mitoses which occur during the development of the 8-nucleate embryo-sac. In fact it takes place at the first of these mitoses so that non-disjunction occurs at equivalent stages on both the male and female side in this species (MÜNTZING 1959). Preferential transmission on both male and female sides has also been described in *Collinsia* (DHILLON and GARBER 1962).

In summary we see that B-chromosomes tend to be more variable in their behaviour, both within and between species, than the A-chromosomes. But in some cases, at least, though they misbehave at mitosis and meiosis, their behaviour is more predictable than that of spontaneous univalents especially at particular stages of development.

Part Two

Analytical

I. Chromosome Duplication and Pairing

The early cytologists came to the conclusion that in a majority of organisms the leptotene chromosomes were single and that this singleness persisted at least until pachytene. This, in turn, led to the idea that chromosome duplication occurred during pachytene and not, as in mitotic tissues, during interphase. True there were claims to the contrary. Some workers held the leptotene chromosomes to be visibly double (see RHOADES 1961) but these claims were either treated as suspect or else could be fairly simply explained. Thus, in *Tradescantia* and *Trillium* it is possible to resolve half chromatid units with an ordinary light microscope (NEBEL and RUTTLE 1936) so that the chromosomes at leptotene do show a bipartite structure. But, since the chromosomes at first metaphase are quadripartite, the leptotene threads can still be regarded as single in the sense that the prophase chromosomes of mitosis are double.

Clearly, the visible evidence for structural doubleness has no necessary relationship to the question of chromosome duplication. Indeed, in the diplotene lampbrush phase of female newts the chromosome axis reveals its doubleness only at limited points along its length (CALLAN and LLOYD 1960) while in many plant species the distinction between the two chromatids of each homologue may not be clear before first metaphase.

In recent years it has proved possible to re-examine the question of chromosome duplication through two new avenues of approach:

(1) With the discovery that the main genetic component of the chromosome appeared to be DNA, attempts were made by autoradiographic techniques to determine the precise phase of the meiotic cycle during which DNA synthesis takes place. Although the results vary somewhat a majority of them indicate that DNA synthesis is completed before or during leptotene (Table 71).

Table 71. *DNA Synthesis in the Meiotic Cycle.*

Organism	Time of DNA Synthesis	Reference
1. *Tradescantia*	During leptotene	SWIFT 1950
	Early leptotene	TAYLOR 1953
2. *Lilium*	Before leptotene	TAYLOR 1953 DE 1961
3. *Trillium*	Between pachytene and diplotene	SPARROW *et al.* 1952
4. *Triturus viridescens*	Bulk of synthesis is Completed by or during leptotene but ca. 2 % DNA synthesis takes place during pachytene	WIMBER and PRENSKY 1963
5. *Mouse* ♀	Pre-leptotene (12th — 13th day of foetal life)	RUDKIN and GRIECH 1962 LIMA-DE-FARIA and BORUM 1962
♂	Pre-leptotene	MONESI 1962
6. *Loxa*	During zygotene	ANSLEY 1957
7. *Scutigera*	Earliest stage of the primary spermatocyte	ANSLEY 1954

Some of the variation shown in Table 71 may be spurious. Thus, pin-pointing the onset of meiosis is not easy. It is complicated by the fact that the early stages of first prophase vary considerably in appearance in different organisms and different workers using the same organism do not always delimit stages in the same way. For example, some authors have defined a "premeiotic" spiral stage which has been claimed to constitute the first phase of meiosis. Indeed RHOADES (1961) thinks it may be of widespread occurrence though there is little to support this belief.

(2) Information on the time of chromosome duplication can be obtained also from the pattern of chromosome breakage induced following irradiation at different times in relation to DNA-synthesis and the various stages of division. It is now firmly established that, in the mitotic cycle, irradiation-induced breakage

is of the chromosome type (B″) prior to DNA-synthesis (G_1-phase). After DNA-synthesis (G_2-phase) only chromatid (B′) aberrations are induced but these aberrations are also produced shortly before and during the S-phase as judged by the time of label incorporation (EVANS and SAVAGE 1963, WOLFF and LUIPPOLD 1964).

In *Lilium* bud length is known to be a reliable meiotic indicator. Pollen mother cells can therefore be irradiated at known meiotic and premeiotic stages. With this technique MITRA (1958) claimed that B′ aberrations first appear following irradiation some 12–24 hrs. before the organisation of typical leptotene chromosomes. SAUERLAND had earlier (1956) produced somewhat similar results in *Lilium candidum*.

The evidence for DNA replication prior to pachytene is thus pretty convincing and if DNA replication and chromosome duplication are not the same event the radiation evidence suggests that they either coincide in time or else follow one another in rapid succession.

There are however four qualifications which must be made:

(1) DNA-synthesis extends over a fairly long period of time and all segments of all chromosomes do not replicate synchronously.

(2) Chromosomes which, like the X-chromosome of orthopterans, are completely heterochromatic, replicate later in the meiotic cycle according to LIMA-DE-FARIA (1961), see pg. 263.

(3) The induction of B′-aberrations in what are, on the evidence of label-incorporation (see above), unduplicated segments, may well be spurious for the nature of the breaks produced may reflect the state of replication more accurately than the presence or absence of grains on a photographic emulsion.

(4) The evidence for a concentration of DNA synthesis in the pre-pachytene period has tended to overshadow the reports that at least some DNA synthesis occurs at a later stage. Reports of this kind are not, however, uncommon. Thus, a small rise in thymidine phosphorylation during pachytene has been recorded for suspensions of pollen mother cells in *Lilium* and *Trillium* (HOTTA and STERN 1961). In the former, pachytene incorporation of tritiated thymidine has been described also (PRENSKY 1962) and this phenomenon has been found more recently in the male of the newt, *Triturus viridescens* (WIMBER and PRENSKY 1963). This species has a considerable amount of heterochromatin but this does not seem to be specifically duplicated during the last part of DNA synthesis.

Of one thing we can be sure. If we accept the visual evidence regarding singleness and doubleness of threads as critical evidence for chromosome duplication then we are faced with a paradox. At meiosis claims for the doubleness of the leptotene thread have been used to support the fact that DNA duplication precedes pairing and vice versa. But in mitosis we have comparable claims for the doubleness of the telophase chromatid although DNA replication is known not to occur until mid-interphase. Despite this there is a general statement in one recent introductory book (LEVINE 1962) that the leptotene chromosomes are double when they first appear. Evidently LEVINE has never seen a leptotene nucleus.

At zygotene the visibly-single leptotene threads pair in a highly specific manner. When the process is viewed with the electron microscope it is found that, on pairing, the "single" threads of leptotene form a tripartite ribbon, the so-called synaptinemal complex (MOSES 1958, NEBEL 1959, NEBEL and COULON

1962, Franchi and Mandl 1962). Some third component is thus added to paired chromosomes, a component which is not visible before pairing. In *Drosophila* where crossing-over occurs in females but not in males these synaptinemal complexes are present in oocytes but absent in spermatocytes (Meyer 1963). Presumably therefore they are concerned with crossing-over rather than pairing. At the zygotene stage too there is often visible along the chromosome a series of granular loci called chromomeres which may be sufficiently distinctive to be individually identified. And synapsis of homologues may, as in the lampbrush chromosomes of newts (Callan and Lloyd 1961) be sufficiently exacting to result in a point by point apposition of these chromomeres. Pachytene chromomere patterns do not always coincide in synapsed homologues however (Maguire 1960). A number of theories have been advanced to explain pairing. We have decided not to discuss them for none of them offer a satisfactory explanation of the process. The interested reader is referred to the summaries of Serra (1955) and Rhoades (1961). If he learns nothing else from these he will at least discover how empty hypotheses can be when they are beyond the bounds of experimental testing. It is worth bearing in mind, however, that the chromosome is composed of two kinds of material—genetic and non-genetic. And while pairing requires homology this, in itself, will not ensure pairing, even at meiosis. Thus simple recessive mendelian mutations can cause a genuine failure of pairing between chromosomes which are known to be homologous (see page 282).

Akstein (1962) claims that in the meiosis of mosquitoes there are no traditional leptotene and zygotene stages. He argues that the somatic pairing of homologues which characterises the pre-meiotic mitoses results in homologues showing full pairing at the onset of meiosis. If this interpretation is correct there is no reason why it should not hold in many, perhaps all, Diptera, as indeed Bauer (1931) and Wolf (1941) claimed for the nematocerans they studied.

II. Chiasma Formation

1. General Features

The pairing of homologous chromosomes at zygotene is followed, at the end of pachytene, by their separation. By now, as we have seen, each of the homologues is visibly double and, at one or more corresponding points along their length, two of the four chromatids present in each bivalent criss-cross over one another forming characteristic chiasmata. The origin of these chiasmata has been, indeed for some still is, one of the main subjects of controversy in genetics.

The early cytologists were preoccupied with the question of the origin of chiasmata and spent much time considering two contestant theories—the two-plane or classical theory and the one-plane or chiasmatype theory. This subject has been discussed by many cytologists from Darlington down. We do not intend discussing it at all for it is no longer an open question. The only acceptable theory is that of partial chiasmatypy. Whenever it has been possible to test the correspondence between chiasmata and crossing-over in an unambiguous manner there is a clear parallelism between them both in time and place (see for example Mather 1938, Brown and Zohary 1955, Lewis and John 1963a). Perhaps the most complete data obtained to date comes from a combination of the several studies made on heteromorphic chromosome pairs (Table 72). On the chiasmatype hypothesis paternal and maternal centromeres should segregate at the first meiotic division in forms with a localised centromere. The segregation of a chromosome inequality at second division can only be accounted for, therefore, by postulating crossing-over between the centromere and the region of difference. Hence the finding that second division segregation frequencies are statistically equivalent to first division chiasma frequencies is tantamount to demonstrating that cross-over and chiasma frequencies are equal*.

Genetical data on crossing-over are explicable on the assumption that an exchange of chromatids occurs between the paired chromosomes at a four-strand stage and involves non-sister chromatids (STERN 1931, CREIGHTON and MC-CLINTOCK 1931, BRINK and COOPER 1949, WOLF 1963). It is, however, difficult to define the onset of the four-strand stage cytologically. Certainly the bivalent is 4-stranded by diplotene and it is usually assumed that it becomes 4-stranded during pachytene. If this is so then there is a difference between the time of DNA replication and the time at which the products of replication individualise to form new chromatids. There appears also to be an even greater discrepancy between the X-ray evidence and the observational cytological evidence.

The truth is that we do not understand the relationship between the time of DNA-synthesis and the time of crossing-over. While it might be aesthetically pleasing if crossing-over and DNA replication turned out to be different facets of the same process there is no compelling evidence as yet to show that this is actually so. Thus X-rays can induce breakage with consequent exchange in cells which have completed DNA-synthesis and in those which have yet to make it. Indeed the peak period of radiosensitivity is at metaphase but breakage induced at this time is not observed until the next division (EVANS 1961).

Again, while some radiomimetic chemicals, like maleic hydrazide, certainly produce breakage only if they are present during DNA synthesis others, like 8-ethoxycaffeine, give a pattern of breakage similar to that of X-rays. But we shall return to these points when we deal with the actual mechanics of chiasma production. Before we can do this we need to discuss two relevant questions. First, interference and second, sister-strand exchange.

a) Interference phenomena.—A bivalent very often contains more than one chiasma. There are then several ways in which the successive exchanges of chromatids may be related to one another assuming that these exchanges involve only non-sisters (Fig. 181). The sequences of chiasmata in these instances are said to determine compound crossing-over and if these chiasmata occur at random a 1 : 2 : 1 frequency of 2, 3 and 4 strand types is expected. This, in turn, means that cells having two chiasmata in a segment, like cells having only one, will produce haploid cells a half of which are recombinant in respect of the loci defining the segment. This appears to hold in a number of organisms including maize and *Drosophila*. But in *Trillium Stenobothrus* and *Lilium* (KAYANO 1959) there is some evidence that if a chromatid is involved in one exchange it is less likely to be involved in a second (see also page 72). As a result of this so-called positive chromatid interference the upper limit of recombination may exceed 50% because of a disparity of 2 and 3 strand types and a corresponding increase in the number of 4-strand double-cross-overs.

---

* See note 4 Appendix.

Table 72. *Relationship Between Chiasma Formation and Chromosome Segregation in Marked Chromosomes.*

Nature of Heterozygosity	Species and Reference	Affected Chromosomes	Chiasmata (a)				Segregation (b)				Contingency χ² for 1 D.F. (comparison of a with b) $\chi^2_{[1]} < 4$ Non Sig.
			−	+	Total	% + Xta	A I	A II	Total	% A II Segn.	
	Cricetus aureus KOLLER (1938)[1]	X and Y	206	47	253	18.6	87	24	111	21.6	0.46
Heteromorphic pair	*Lilium formosanum* BROWN and ZOHARY (1955)	A {	183	452	635	70.0	69	172	241	71.4	0.17
			114	120	234	51.3	67	81	148	54.7	0.42
		I {	21	150	171	87.7	15	123	138	89.1	0.15
			72	273	345	79.1	43	149	192	77.6	0.10
	Allium fistulosum ZEN (1961)	S	151	618	769	80.4	61	237	298	79.5	0.037
	Calliptamus palaestinensis NUR (1962)	S	31	379	410	92.3	6	63	69	91.3	0.107
	Lilium maximowiczii NODA (1960)	J_1	490	110	600	18.3	981	219	1200	18.25	0.0019
		J_2	600	0	600	0	1200	0	1200	0	0
Interchange difference	*Disporum sessile* KAYANO (1960)	A } B }	} 1549	} 651	1100 / 1100	} mean 29.6	1987	} 853	1420 / 1420	} mean 30.0	0.12
	Acrida lata KAYANO and NAKAMURA (1960)	I	0	50	50	97.0	2	28	30	93.3	0.60
		II	9	41	50	82.0	3	27	30	90.0	0.94

									Interchange difference
Allium fistulosum ZEN (1961) S	59	208	267	77.9	17	50	67	74.6	0.33
F	138	129	267	48.3	29	25	54	46.3	0.073
Scilla scilloides NODA (1961) a	1128	2820	1974	mean 71.4	278	678	478	mean 70.9	0.097
b			1974				478		
Delphinium ajacis JAIN and BASAK (1963) A	166	134	300	44.7	98	85	183	46.4	0.130
B	130	170	300	56.7	84	99	183	54.1	0.274

[1] 1. The data of KOLLER on Cricetus are included because, although they have been criticised by MATTHEY and by WHITE, a recent reappraisal by EMMONS and HUSTED (see pg. 189) supports them qualitatively.

2. The heteromorphic D-pair in Paris verticellata studied by HAGA (1944) has been omitted from this table. This is because in this species chiasmata are proximally localised (a feature which HAGA would not admit). It is therefore exceptionally difficult material in which to score with accuracy the frequency of cells with a chiasma between the centromere and an inequality (see also pg. 15).

Not only do chromatids show interference, so do chiasmata. Thus as early as 1916 MULLER had demonstrated that, in Drosophila and maize, cross-over gametes with two exchanges, one in each of two small adjacent segments, are produced less frequently than is expected from the product of the incidence of recombination in each of them individually. This he interpreted as a structural hindrance of one exchange on the formation of another in its vicinity and he termed this genetical interference. Subsequently this was shown to be paralleled by a chiasma interference at the cytological level (HALDANE 1931). Chiasma interference is thus quite a distinct phenomenon from chromatid interference. Chiasma interference is the occurrence less often (positive interference) or more often (negative interference) than expected by chance of two or more exchanges in one segment in individual cells at meiosis. Chromatid interference, on the other hand, is the participation less often (positive interference) or more often (negative interference) than expected by chance of the same two chromatids in two or more successive exchanges.

The phenomenon of positive chiasma interference was first interpreted analytically by MATHER (1936). He showed that the relationship between successive chiasmata could be explained if it was assumed that, in every chromosome pair, chiasma formation takes place sequentially such that the first chiasma is formed at a distance, the differential distance (d), which varies around a fixed point which MATHER suggested might be the centromere. Subsequent chiasmata were then held to form at distances varying around a certain interval determined by interference and referred to as the interference distance (i). When the frequency of occurrence of chiasmata at different places along the length of the chromosome are plotted one should therefore obtain a series of overlapping curves of normal distribution. Not until 1963, however, did anyone attempt to make such a plot. Indeed some (HALDANE 1931) claimed this to be an intractible task. HENDERSON, however, has recently carried out an exhaustive study of the distribution of chiasmata in the diplotene bivalents of the male desert locust *Schistocerca gregaria*. This is a species in which all 23 chromosomes are acrocentric and in which not only are the bivalents individually distinguishable but the centric ends are marked by conspicuous heterochro-

Type	Pachytene Configuration	Meiotic products	% recombinants
Reciprocal (2-strand)	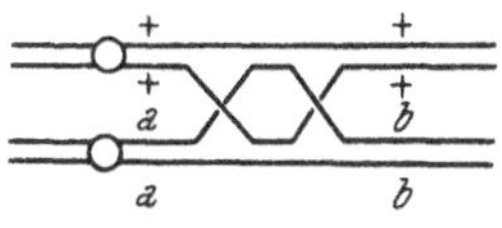	+ + + + a b a b	0
Complementary (4-strand)	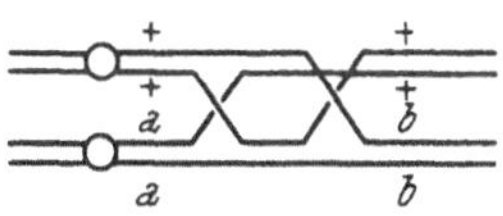	+ b + b a + a +	100
Diagonal (3-strand)	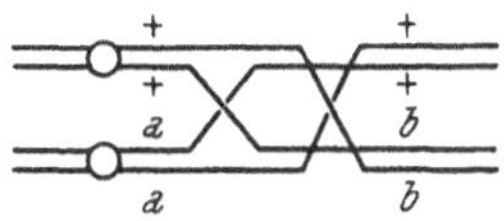	+ b + + a + a b	50
	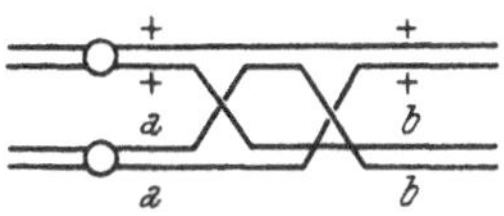	+ + + b a b a +	50
Mean %-recombinants			$\dfrac{200}{4} = 50$

Fig. 181. Relationship of chromatids in successive chiasmata.

matic segments. With this material he finds that chiasma formation usually begins at the distal, non-centric, end so that in this instance the d-distance mean is at zero. The overall i-distance is ca. 7.3 µ of the diplotene length. In some 30–40% of the cases, however, chiasma formation appears to begin at both ends of the chromosome simultanseously and then proceeds towards the middle.

One further feature of positive chiasma interference deserves consideration. In *Drosophila, Zea, Fritillaria chitralensis* and *Uvularia perfoliata* there is no interference across the centromere of metacentrics. But in *Culex pipiens* (PÄTAU 1941, CALLAN and MONTALENTI 1947), *Dicranomyia trinotata* (PÄTAU 1941), *Forficula auricularia* (CALLAN 1949) and *Asellus aquaticus* (VITAGLIANO 1947) it has been claimed that the presence of a chiasma in one arm of a metacentric tends to inhibit the formation of a chiasma in the other. And this it is suggested demonstrates that interference extends across the centromere in these cases. The most comprehensive study of this issue is that of CALLAN and MONTALENTI on *Culex pipiens*. This species has three pairs of mediocentric chromosomes, two of which (M-pairs) are of the same length and about one and a half times as long as the third (m) pair. The proportion of bivalents with a chiasma in each arm is smaller than would be expected on the assumption of statistical independence between the arms (Table 73).

CALLAN and MONTALENTI interpreted their data as implying:

(1) The presence of a chiasma in one arm inhibits the formation of a chiasma in the other.

(2) A proximal chiasma in one arm has a greater inhibitory effect on chiasma formation in the other arm than has a distal one (Table 74) so that the centromere is not effective as an insulator between arms in respect of chiasma interference.

Table 73. *Percentage Frequency of Chiasmata in (M) and (m) Bivalents of Culex pipiens.*
(Data of CALLAN and MONTALENTI 1947.)

Specimen No.	No Chiasma in Either Arm — Univalents only	No Chiasma in One Arm — Bivalents with One Chiasma only	One Chiasma in Each Arm
1	9.09	86.36	4.55
2	32.58	64.25	3.17
3	35.08	63.74	1.17
4	26.55	73.23	0.21
5	36.51	63.49	0.00
6	22.97	70.27	6.76
7	25.00	75.00	0.00
8	19.81	80.18	0.00
9	18.08	80.14	1.77
10	18.77	80.06	1.17

OWEN (1949), on the other hand, has pointed out that these conclusions depend on the assumption that where two chiasmata are formed, each has the same initial probability of formation. He also argues that, since the first chiasma is essential for proper synapsis and disjunction, the initial probability of this first, obligate chiasma might have reached a high value through selection. If this is accepted then the probability of a second chiasma appearing in the other arm will not be the same as the initial probability of the first one being established so that the primary chiasma could not then be said to interfere with the establishment of the second. But spurious interference will be shown.

OWEN's argument does not appear to have been taken seriously by geneticists who customarily accept CALLAN and MONTALENTI's interpretation that a chiasma in one arm interferes not only with the formation of another chiasma in the same arm but also with chiasma formation in the other arm.

Genetic studies on recombination in very short intervals (ca. 0.1 map units) in certain fungi and bacteriophages appears to have revealed a marked negative chiasma interference as far as intra-genic recombination is concerned. The first unambiguous report of this was that of CAVALLI-SFORZA and JINKS (1956) in a paper which has been curiously and persistently ignored by microbial geneticists.

Indeed not only did these authors draw attention to the phenomenon, they also gave an explanation of it—an explanation we may note which was vastly superior to that subsequently proposed.

The detection of such negative interference always demands the adoption of highly selective techniques and one is inevitably led to question the extent to which the result obtained depends on the technique itself. One may also question whether the rare events that these selective techniques reveal bear any direct relationship to the normal process of crossing-over. Indeed in one case the demonstration of apparent high negative interference (EDGAR and STEINBERG 1958) was subsequently admitted to be a spurious consequence of a straight technical error (STEINBERG and EDGAR 1961).

Table 74. *Percentage Distribution of Chiasmata in the M-bivalents of Culex pipiens.* (Data of CALLAN and MONTALENTI 1947.)

Specimen No.	Bivs. with one Xa. only		Bivs. with one Xa. in Each Arm		
	Xa. Proximal	Xa. Distal	Prox./Prox.	Prox./Distal	Dist./Dist.
2	13.02	47.95	0.68	10.95	27.39
4	26.45	48.06	0.32	8.39	16.77
9	21.92	56.68	0.00	6.42	14.97
10	20.35	56.64	0.44	3.54	19.03

Despite misgivings of this kind, the occurrence of multiple or cluster recombinant areas in fungi has, in recent years, produced a storm in the biological tea cup largely as the result of the claims of PRITCHARD and PONTECORVO (see pg. 217). It is worth examining the phenomenon in some detail for it shows how in science, as in politics, given a lead there are always those who will follow.

When first observed, it was thought that the clustering effect would prove a characteristic of recombination in short segments. PERKINS (1962) collated and analysed data from 58,068 tetrads in *Neurospora* relating to gene-marked intervals showing less than 15% recombination. In all cases the non-parental ditypes fell far short of the values predicted from a clustering hypothesis.

There seems stronger grounds for suggesting that negative chiasma interference may prove to be associated with so-called intra-allelic recombination (WESTERGAARD 1964). However, we do well remember that:

(1) High negative interference has yet to be demonstrated for all complex loci, in all organisms and by all techniques. Thus there is little suggestion of it in *Drosophila* even though it is now possible to analyse segments of comparable lenght to those studied in fungi (GREEN 1963). It is true that HEXTER (1963) has data which he has interpreted in support of cluster exchange but, like that of NELSON (1962) in maize, the claim is not convincing.

(2) In T_4 phage, one of the few other organisms where high negative interference has been found, it is now clear that causes other than a novel method of recombination can lead to negative interference. Thus single bursts obtained following single infections with partial heterozygotes containing three or more

closely linked genetic markers show that the frequency of double recombinants is approximately that expected from the single cross-over frequencies observed in the bursts. But single bursts from mixed infections give a seven-fold excess of double recombinants (DOERMANN 1963).

In summary, while one cannot doubt that one may get an observed excess of multiple recombinants within very short regions, it is now clear that:

(1) Such regions have usually to be analysed by special selective techniques involving a small, non-random and perhaps atypical sample of zygotes, and

(2) Whatever the precise cause leading to high negative interference there is no longer any reason to entertain the belief that its production depends on a novel method of recombination as PRITCHARD has claimed.

(b) Sister-strand exchange—So far in our discussion we have assumed that crossing-over is always between non-sister chromatids for this is the only process which can lead to genetically detectable crossing-over and cytologically visible chiasmata. There have, however, been several claims for the occurrence of sister-strand exchange at meiosis and we need now to consider them for if this evidence is to be believed then it forces us to reconsider our ideas and concepts of crossing-over and chromosome replication.

The most compelling claim comes from the work of SCHWARTZ on the cytological consequences of crossing-over in maize plants heterozygous for a ring-rod chromosome 6. We have already considered this work in detail (see page 96) and need only recall that this evidence is, to say the least, highly circumstantial. And even if sister-strand exchange does occur in such heterozygotes it is far from clear that it occurs regularly in normal chromosomes. Thus crossing-over and chiasma formation are frequently irregular in structural heterozygotes (see page 70).

The only other evidence for sister-strand exchange comes from mitotic studies. This evidence is of two kinds:

(i) In *Aspergillus*, yeast and *Drosophila*, homologues are known to undergo a process of exchange in somatic tissues. It is difficult to get precise estimates of mitotic recombination but in *Aspergillus nidulans* PRITCHARD (1963a) assesses that it may be ca. 10^{-3} of that occurring at meiosis.

This process is particularly common and was first discovered in *Drosophila*. This is not surprising when one recalls that somatic pairing, leading to an association of homologues, is developed here to a striking degree. Thus in *Drosophila melanogaster* mitotic recombination may lead to certain genes becoming homozygous in parts of an otherwise heterozygous organism and this, in turn produces a state of somatic mosaicism. In females heterozygous for the X-linked recessives yellow spot (y) and singed spot (sn) the most common kind of mosaicism produced is twin spotting which results from somatic crossing-over between (sn) and the centromere. In fact, the frequency of recovered single cross-overs in this region and those in the (sn) — (y) segment were in the ratio of 110 : 39.5. The corresponding ratio following meiotic recombination was 36 : 20.5 (STERN 1931).

Now, somatic crossing-over in attached X systems will give rise to a twin spot when it involves chromatids which are not attached to a common half-centromere. On the hypothesis that crossing-over can occur between any of the chromatids, twin spotting should be at least as frequent as in the case of unattached X-chromosomes. SCHWARTZ (1954), however, found that the frequency of twin spotting was extremely low in attached-X systems. He concluded from this that chromatids attached to the same half-centromere were more prone to exchange since this type of exchange does not affect genic constitution.

Finally, by ageing the mothers as virgins he obtained a nine-fold increase in the frequency of twin-spotting in the progeny from attached-X females but not in that of unattached-X females. He concluded, therefore, that the increase obtained was due to sister-chromatid exchange in attached X-chromosomes, for this alters the half-centromere connections of the cross-over chromatids. BROWN and WELSHONS (1955) were, however, unable to confirm this effect.

(ii) Further support for the occurrence of sister-strand exchange at mitosis comes from autoradiographic studies and especially those of TAYLOR on *Bellavalia*. By following the distribution of tritium-labelled thymidine TAYLOR (1958) showed that the chromosomes of *Bellavalia*, like those of *Vicia*, behave as though they are 2-stranded before replication. Thus both chromatids of each chromosome are labelled at the first (X_1) c-metaphase after the incorporation of label while after a further cycle of replication in the absence of labelled precursors each X_2-chromosome generally includes one completely labelled and one completely unlabelled daughter chromatid. However, in a high proportion of X_2-cells, sister-chromatids were found which were labelled along only part of their length and in a reciprocal way suggesting that sister-strand exchange had occurred (see also section III-2 b).

Similar results have been observed in all the other types studied to date. These include *Vicia, Crepis, Allium, Cricetus* and an unspecified grasshopper (TAYLOR 1963). It is true that WIMBER (1959) has claimed that the amount of tritium involved in these studies was sufficient to induce enough chromosome breakage to account for the observed exchange frequency. But it is not generally believed (see TAYLOR 1963) that the radiation received from the labelled material is adequate to account for the high frequency of exchange although TANAKA and KOHNO (1961) obtained results which seemed to support WIMBER's claim.

The sum total of our evidence for regular sister strand exchange at meiosis is thus meagre and at mitosis it can be detected only under peculiar conditions.

2. The Mechanism of Crossing-over

The evidence is now quite unequivocal that crossing-over occurs at an effectively four stranded stage and, as such, there have been only two serious contestant theories to account for the mechanism of crossing-over. Either:

(a) Crossing-over is the consequence of breakage and exchange of existing strands, i. e. it is a post-duplication phenomenon so that, at the time of crossing-over, each chromosome of an homologous pair is divided into two sister chromatids, or

(b) Crossing-over is the consequence of the synthesis of new material partly along one and partly along a second of two alternative templates, i. e., it is causally related to the process of duplication itself.

The great difficulty inherent in the first type of mechanism is the elucidation of the precise mechanical forces or chemical agents responsible for the breakage and rejoining of chromatids. The only explanation which has been advanced is that of DARLINGTON (1935) who has related breakage to the torsion which develops in the chromosomes during the first meiotic prophase. As homologous chromosomes pair during zygotene they coil around one another until this relational coiling just balances the force of the underlying spiral organisation which brought it into being. When the relationally coiled chromosomes become double a new state of tension is generated, for each homologue becomes a pair of sister-chromatids which are also coiled relationally about one another. This tension leads to chromatid breakage at various points and also permits the broken ends to rotate. These ends may then rejoin with non-sister ends so leading to the production of a chiasma. This hypothesis offers an explanation for positive interference since the relief of torsion by breakage and exchange will necessarily reduce the force acting on the chromatids in the immediate vicinity of the potential exchange point.

One of the main objections which has been raised against the torsion theory of crossing-over is that it is difficult to understand why a break in one chromatid should be accompanied by a break in an exactly equivalent position in a non-sister chromatid. By contrast it is claimed that this offers no difficulty if it is assumed that crossing-over is the result of an exchange between the two 'new' chromatids during the process of their formation. The first point to make is that while reciprocity may be more intelligible on the second explanation, it is denied by those who postulate it for the explanation of certain other phenomena used to support this same contention of copying-choice (see page 215). And the second point is that such an exchange could, of course, give rise to two-strand doubles only. If recombination is by copying choice therefore it is necessary to postulate the existence of some randomising process to account for the production of three and four-strand doubles. Sister-strand exchange has, in fact, been suggested as the mechanism involved. Thus a double exchange by copying choice associated with a single sister-strand exchange in the region between the two non-sister cross-overs will convert a two-strand double into a three-strand double. Similarly a single sister-strand cross-over in each homologue will result in an apparent four-strand double. But notice that, in both cases, this demands not simply sister-strand exchange: rather it requires sister-strand exchange at particular sites relative to the copy choice switch points. What is more, of all the suggested chromosome models, one which allows equal, reciprocal sister-chromatid exchange without breakage has never been proposed. A copy choice explanation does not, therefore, avoid the difficulties of the breakage hypothesis.

We have already seen that the evidence for sister-strand exchange at meiosis cannot be considered as adequate. Moreover, there is no direct evidence that copy choice occurs at all. It is, in fact, nothing more than a convenient way of explaining aberrant segregation patterns found in certain micro-organisms. Let us then turn to a consideration of these.

A diploid heterozygote, Aa, normally yields a $2:2$ segregation of its alternative hereditary determinants so that half the products of meiosis carry (A) and the other half (a). In yeasts all four meiotic products remain together in one ascus and deviations from the theoretical $2:2$ segregation pattern of monohybrids have been demonstrated many times (WINGE and ROBERTS 1954). Further, quite different genetic mechanisms have been shown to produce the same disturbance of the expected ratio. Abnormal ratios can arise from at least seven causes (Table 75). Most of these require no further comment but one—gene conversion—is sufficiently important to merit a detailed consideration.

The idea of gene conversion was proposed in 1930 by Winkler to account for the exceptional segregation patterns then being reported in fungi and mosses. According to this theory, a recessive allele in a heterozygote can be converted to its dominant allele and vice-versa, and this gives rise to irregular segregation of the four products of meiosis. Thus $4\,(A):0\,(a)$ and $0\,(A):4\,(a)$ tetrads

Table 75. *Causes of Abnormal Tetrad Ratios in Yeasts.*
(Summarised mainly after Winge and Roberts 1954 and Emerson 1956.)

Cause	Ratio $(+:-)$
1. Genic interaction 　a) duplicate genes	$4:0$, $3:1$ or $2:2$ $(\equiv 15:1$ F_2 ratio in diploids$)$
b) complementary genes	$2:2$, $1:3$ or $0:4$ $(\equiv 9:7$ F_2 ratio in diploids$)$
2. Polyploidy and polysomy	$3:1$, $4:0$ or $1:3$
3. Somatic crossing-over	$4:0$ or $0:4$
4. Irregular third ascus division (super-numerary mitosis) followed by differential survival of the products or incorporation of two nuclei within one spore	$4:0$, $3:1$, $1:3$ or $0:4$
5. Spontaneous gene mutation 　a) at somatic reproduction 　b) at meiosis	$4:0$ or $0:4$ $3:1$ or $1:3$
6. Gene conversion	$4:0$, $3:1$, $1:3$ or $0:4$
7. Extra-nuclear heritable factors	$4:0$ or $3:1$

would arise if the (A) or (a) allele were converted prior to chromosome duplication in the meiotic mother cell or if both (A) or both (a) alleles were converted after duplication. On the other hand $3:1$ and $1:3$ segregations would follow if only one of the two (A) or two (a) alleles were converted during or after duplication.

Lindegren (1953) has, in fact, interpreted the irregular segregation patterns in yeasts in terms of Winkler's conversion theory. Let us therefore consider some actual data. Table 76 summarises a study of segregation at thirty-four distinct loci of yeast. The frequency of $0:4$ and $4:0$ segregations is of the order expected were mitotic recombination to be responsible for their occurrence. It is not necessary, therefore, to postulate that such segregations are the consequence of conversion. As for the $3:1$ and $1:3$ segregations, two-thirds of the 34 loci studied had conversion values of 1% or less. But the ad_5—locus showed a conversion value of 15.48% and six other loci—ar_{4-1}, ar_{4-2}, hi_2, ly, thr and ur—showed values between 4 and 6.5%.

The 3 : 1 and 1 : 3 segregations in the 4-spored asci of yeasts correspond to 6 : 2 and 2 : 6 patterns in the 8-spored asci of other ascomycetes. Such segregations have been reported in *Neurospora* (MITCHELL 1955, CASE and GILES 1958), *Sordaria* (OLIVE 1959) and *Ascobolus* (RIZET, LISSOUBA and MOSSEAU 1960) where they have been found in approximately equal numbers. Segregations of 8 : 0 and 0 : 8 are rare or absent but patterns of 5 : 3 and 7 : 1 have been found (KITANI 1962).

The interest of all these cases of presumed conversion stems from the fact that conversion has been held to be due to a copying error during DNA replication. That is, it is explained on a concept of recombination which depends on copy choice. This concept of copy choice stems from the early ideas of

Table 76. *Irregular Segregation at 34 loci in Yeast.*
Wild-type allele given first. The frequency of irregular-segregation is different for different allelic pairs. Two-thirds of the 34 loci studied show 1% or less irregular segregation but several have substantially higher values and in the case of ad_5 it is as high as 15.48%. The frequency is not related to distance from the centromere. (Summarised from Unpublished Data of MORTIMER given in ROMAN 1963.)

Tetrad Type	4 : 0	3 : 1	2 : 2	1 : 3	0 : 4	% Abnormal	Total Tetrads
Frequency	2	117	10,984	104	2	2 %	11,209

BELLING (1931) which were later extended by LEDERBERG (1955). In the extended scheme it is proposed that new genetic strands are laid down alongside the old ones by copying, with the copies reciprocally switching from one parental strand to its homologue at points of exchange. By extension (FREESE 1957) it can be argued that if one section of a strand can be copied twice by non-reciprocal switching, so that the corresponding section of the homologue is not copied at all, then a 3 : 1 ratio will result. This is easiest to imagine if copying is regarded as non-synchronised (Fig. 182). Let us note, in passing, that copy-choice models deal with replication at the chromatid level whereas, as we have seen earlier (see page 45), there are strong indications that the chromatid is not the fundamental unit of replication.

Since the regions of conversion are usually short one must suppose that there is a strong tendency for the 3 : 1 distribution of strands to be corrected within a very short distance either by the back-switching of the original strand, giving the effect of a conversion, or by a compensating switch on the part of the new non-sister strand giving both conversion and reciprocal recombination of outside markers.

In *Sordaria fimicola* about one ascus in 2,000 gives 6 : 2 or 2 : 6 ratios. In the case of the grey locus, KITANI *et al.* (1962) found that the latter type was about five times more frequent than the former. Asci showing 5 : 3 ratios also occur and with about the same frequency as the 6 : 2 segregants. In this case the possibility that the anomalous ratio is due to one or more extra mitotic divisions in the ascus, coupled with a non-survival of some of the products, can be ruled out since three other segregating markers always gave regular

4 : 4 segregation in these same asci. Kitani (1962) has suggested, however, that the 6 : 2 type of segregation has its origin in a DNA duplication process other than the meiotic one. This possibility was suggested by his observation of somatic recombination following one cross which also yielded a large number of aberrant asci. He is of the opinion that somatic recombination, occurring

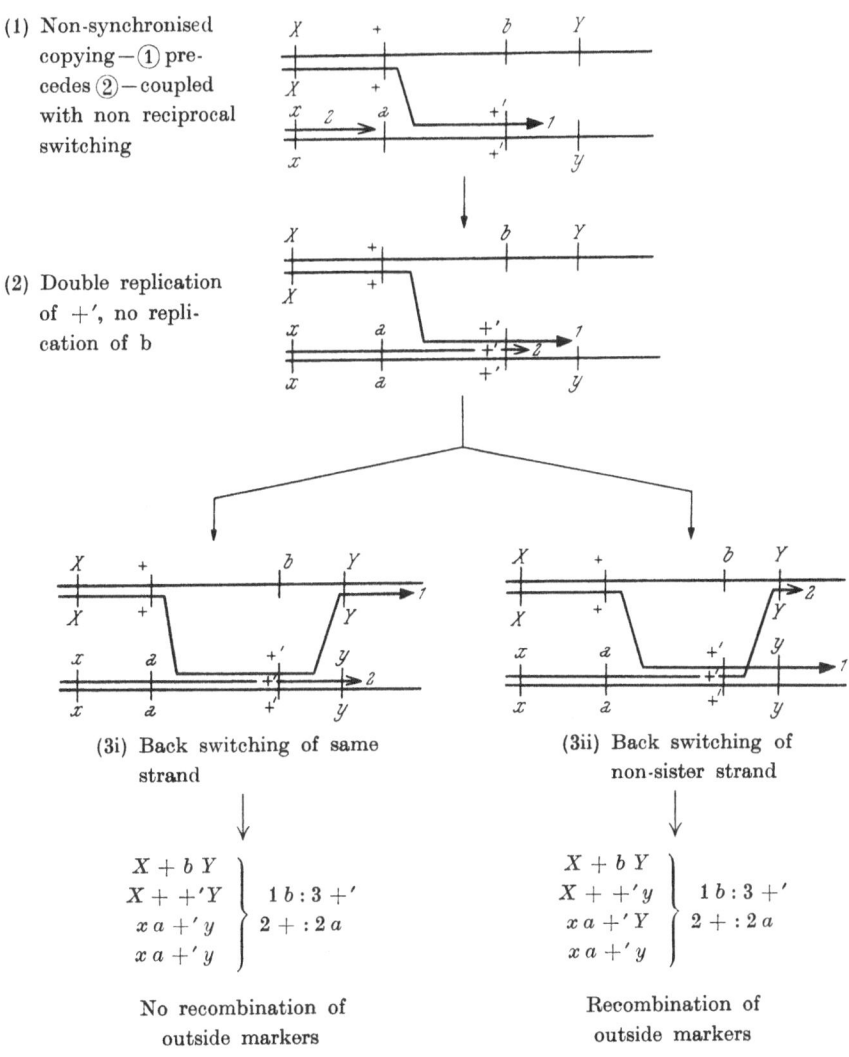

(1) Non-synchronised copying—①precedes②—coupled with non reciprocal switching

(2) Double replication of +', no replication of b

(3i) Back switching of same strand

(3ii) Back switching of non-sister strand

$$
\left.\begin{array}{l}
X + b\ Y \\
X + +'Y \\
x\ a +'y \\
x\ a +'y
\end{array}\right\}\ \begin{array}{l} 1\ b : 3 +' \\ 2 + : 2\ a \end{array}
$$

$$
\left.\begin{array}{l}
X + b\ Y \\
X + +'y \\
x\ a +'Y \\
x\ a +'y
\end{array}\right\}\ \begin{array}{l} 1\ b : 3 +' \\ 2 + : 2\ a \end{array}
$$

No recombination of outside markers

Recombination of outside markers

Fig. 182. Systems of non-reciprocal copy-choice replication.

before perithecia formation, may lead to a few generations of diploid nuclear divisions existing accidentally in the heterokaryotic sector before meiosis. On this basis Kitani has argued that 6 : 2 segregations arise from miscopying of DNA in mitosis while only 5 : 3 types depend on meiotic miscopying. A few of the inter-mutant crosses which gave a low frequency of 6 : 2 asci also gave, though at a lower frequency still, some 7 : 1 asci. These were interpreted (Kitani et al. 1962) as half-chromatid conversions.

The simple truth is that the causal relationship between gene conversion and meiotic recombination, if such exists at all, is far from clear. Nor is it clear how the proposed copy-choice mechanism is related to the duplex DNA molecule, the reproduction of which appears to be semiconservative like, according to some, that of chromatids at mitosis (see page 45).

Much has been made of the question of reciprocity and non-reciprocity in the recombination mechanism. But this question is really a separate issue. Even copy choice is normally regarded as a reciprocal event and only in using it as a basis for gene conversion is it considered to be non-reciprocal. But, whilst the reciprocity of copying is understandable, in order to separate duplex DNA molecules subject to copy choice exchange, breaks must occur in the original templates followed by criss-cross reunion. And these events too, like those on a simple breakage explanation, must be reciprocal.

It has been known for a very long time that, where failure of pairing occurs, no chiasmata form (asynapsis, see Table 95, page 282). It is true that pairing does not in itself lead to chiasma production for cases of desynapsis are also known. But this simply means that while pairing is not sufficient for chiasma formation it is, nevertheless, an indispensable condition. In 1960 PRITCHARD, on the basis of the high negative interference he described in *Aspergillus* (see page 207), proposed that recombination resulted from localised transient pairing during replication which was then followed by conventional pachytene pairing. This, of course, leaves completely unexplained the purpose of the pairing of homologous regions at zygotene-pachytene. To accomodate this PONTECORVO (1959) made the naive suggestion that visible and complete pairing was a mechanical device to ensure segregation, but has nothing to do with crossing-over.

Now, chiasmata have long been known to have two functions which are in a measure distinct. On the one hand they have a genetical function and on the other a mechanical one. It is true that both functions can be dispensed with, so that meiosis is achiamate (see page 20). But, when chiasmata are present they unquestionably serve to maintain the association of homologues after they have ceased to attract one another. And this maintenance of association is necessary to ensure segregation (DARLINGTON 1960). If crossing-over were to occur following limited effective pairing during replication at a pre-zygotene stage and if such crossing-over leads to chiasma formation—a point which PRITCHARD does not deny—then, clearly, there is no need for complete zygotene pairing in order to guarantee segregation (see also page 248).

While some of the earlier studies in microbial genetics were held to support the copy choice model of recombination there has recently been a wholesale return to a breakage-exchange model (see general discussions in EPHRUSSI-TAYLOR 1960, RAVEN 1962). Thus in transformation, recombination of an introduced marker does not require prior replication. Here, therefore, a copy-choice mechanism can be completely excluded and significantly recombination is non-reciprocal. Similarly, when two or more related phage particles carrying different genetic markers are allowed to infect the same bacterium, the recombinants recovered are non-reciprocal although it has been shown (MESELSON and WEIGLE 1961, KELLENBERGER, ZICHICHI and WEIGLE 1961) that recombination has involved the breakage of DNA molecules and the subsequent reunion of the fragments into complete molecules. Thus, not only is copy choice excluded here as the mechanism of recombination, it is also excluded as the cause of non-reciprocal exchange.

We can now with some measure of confidence rule out the possibility that copy choice forms the common basis of all genetic recombination, with sister-strand exchange as a necessary co-factor. There remains the possibility that aberrant events within loci occur at the time of replication but that this has nothing to do with the recombination mechanism which occurs subsequently between any two non-sister members of the four chromatid strands present per bivalent. Let us consider this possibility in more detail. Two sorts of experiments are relevant to its assessment.

Table 77. *Percentage of Isoleucine-independence (IS—representing Allelic Recombinants from is$_a$/is$_b$), with Recombination of Outside Markers, after Various Times of Exposure of Yeast Cells to Sporulation-medium.*
(Data of SHERMAN and ROMAN 1963.)

Hours in Sporulation Medium	Fraction of IS Colonies	Ploidy	No. of Colonies Tested	%-IS with outside Marker Recombination
0	1.9×10^{-5}	100 % 2 n	96	5.2
8	54×10^{-5}	99 % 2 n	114	4.4
20	270×10^{-5}	15 % 2 n	not tested	—
		85 % 1 n	94	70.0

Yeast cells exposed to a sporulation medium normally become committed to meiosis and will undergo sporulation even if transferred to ordinary nutrient medium. However, if the cells are exposed for only a short time they will revert to vegetative growth, provided they are supplied with nutrient medium. Using this technique, SHERMAN and ROMAN (1963) have recently carried out a study of inter-allelic recombination at the isoleucine locus. Cells which were not committed to meiosis and which, consequently, remained diploid after a short exposure to sporulation medium, were found to include those in which inter-allelic recombination had occurred. The time of this recombination was associated with DNA replication as indicated by the incorporation of tritiated thymidine. Inter-allelic recombinants were recovered also, and much more frequently, among the meiotic products. But in these, inter-allelic recombination was correlated with a high frequency of outside marker recombination (Table 77). SHERMAN and ROMAN concluded therefore, that allelic recombination can take place at two different times which correspond with DNA replication and zygotene-pachytene respectively.

The second study which offers some indication that, while events at DNA replication may influence recombination, they do not in themselves bring it about emerges from the studies of LAWRENCE (1961, 1962). In a much neglected pair of publications this author has studied the effect of irradiation at different stages of the meiotic sequence on chiasma frequency in *Lilium longiflorum* and *Tradescantia paludosa* (Fig. 183). In *Lilium* this is possible because of the good correlation between bud length and the stage of meiosis within the anther (see page 258) and here there appear to be three short sensitive stages at which irradiation can produce a modified pattern of chiasma response. The first of these occurs just prior to the start of meiosis at a time when DNA synthesis has

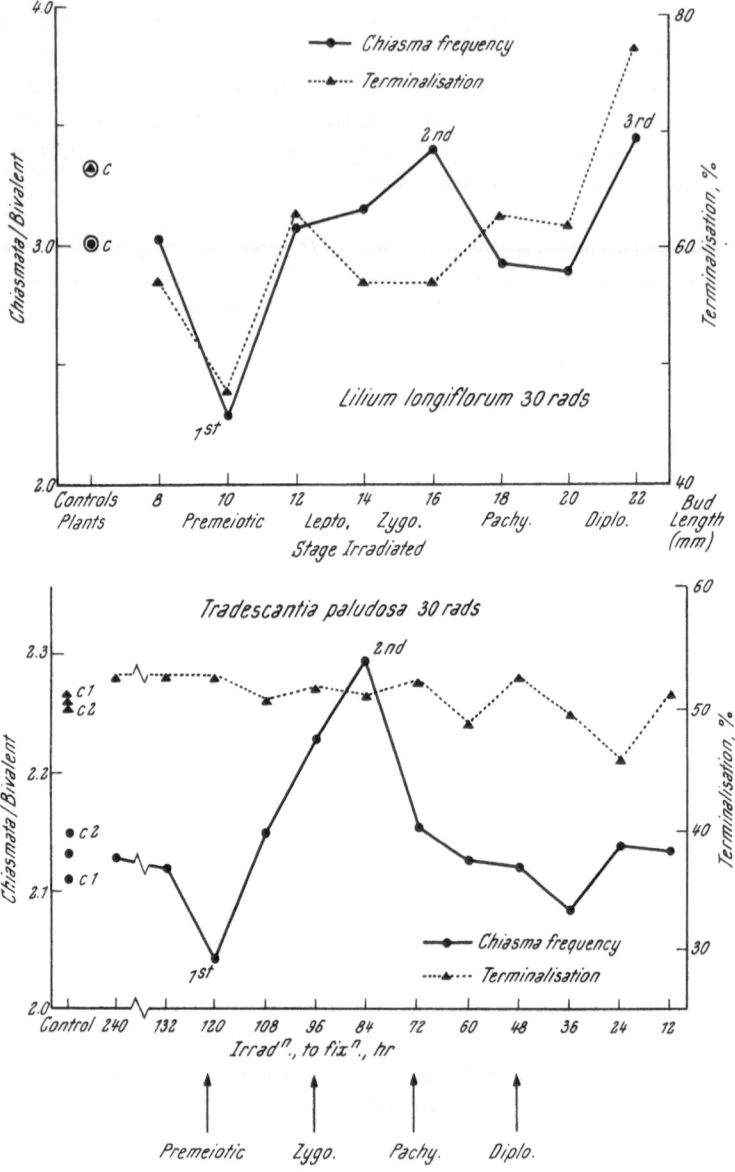

Fig. 183. The influence of radiation on the frequency and terminalisation of chiasmata in *Lilium longiflorum* and *Tradescantia paludosa* (LAWRENCE 1961a and b).

been shown to occur (TAYLOR and MACMASTER 1954). Anthers irradiated during this period show a much reduced chiasma frequency. They are also distinguished by disturbances in chromosome coiling and the presence of univalents. The second and third radio-sensitive periods were early pachytene and late diplotene and irradiation at both these periods produced an increased chiasma frequency.

The response at the third stage Lawrence interpreted as resulting from a failure of terminalisation rather than an actual increase in crossing-over but that due to pachytene irradiation appears to be a genuine effect on crossing-over itself.

Despite considerable differences between *Lilium* and *Tradescantia* with respect to the number and position of chiasmata, the response of chiasma frequency to irradiation appears to be fundamentally similar. It is true that the delimitation of meiotic stages is considerably less accurate in *Tradescantia*. Nevertheless both appear to agree in showing a depression of chiasma frequency following late pre-meiotic irradiation and an elevation after irradiation of late zygotene- early pachytene. In *Tradescantia*, however, plants irradiated during late diplotene do not show the increase in chiasma frequency present in *Lilium*. But significantly in *Tradescantia* there is no increase in chiasma terminalisation*.

Both these studies point to the same general conclusion, namely that while events at DNA replication may influence the cross-over response there is also a post-replication phase when recombination occurs. It is therefore possible to alter the recombination response controlled by these two events quite independently of one another. And chiasmata seem to result directly only from events taking place at the post-replication phase. This is also in line with the idea that chromosome duplication involves not only DNA replication but assortment and individualisation of the products of replication (Peacock 1963).

If, therefore, the breakage-exchange hypothesis should eventually prove to be wrong it will not be because of the evidence so far adduced against it by microbial genetics. Of one thing we can be assured, spontaneous breakage can and does occur and it can and does occur at pachytene-diplotene (see pg. 27). While we cannot specify the conditions which lead to breakage, the phenomenon itself is beyond dispute. And while the torsion theory may leave the precise nature of breakage inadequately explained this should in no way detract one from accepting a theory involving breakage and exchange. Thus in *Paeonia californica* (Walters 1956) chiasma formation is largely confined to the distal portions of the chromosomes. Significantly spontaneous breakage is also distally localised, which suggests that the conditions involved in chiasma formation are also involved in the production of breaks. Again most kinds of spontaneous breakage appear either to have a genetic basis or, at least, to be modified in frequency by genetic factors. The same is true of chiasmata.

The similarity implicit in these comparisons between meiotic crossing-over and the exchange processes initiated by other kinds of chromosome breakage make it worth considering in a little more detail what light induced-breakage throws on the problem of crossing-over.

There have been two main ideas regarding the mode of formation of irradiation-induced chromatid aberrations—the breakage first hypothesis and the exchange hypothesis. According to the first of these, the primary event in the production of the aberration is a chromatid or iso-chromatid break. The free ends at the breakage site may then rejoin or, if breaks have occurred close enough in space and time, ends may join with free ends from other breaks. In the exchange hypothesis, however, the primary event is not a break but some other unspecified disturbance in the chromosome. When two such events occur and are close

---

* See note 5 Appendix.

enough in space and time they may be followed by a stage of exchange initiation and when exchange is incomplete, discontinuities arise which are indistinguishable in appearance from direct breakage. This implies, therefore, that chromatid breakage does not occur in a direct sense but only in association with an exchange process (REVELL 1955, 1959).

Table 78. *Effect of Protein Inhibitors on Chromosome Breakage and Rejoining in Vicia faba Seeds.*
(Data of WOLFF 1959.)
Following treatment seeds were allowed to germinate and two days later, when the first divisions occurred in the root tips, 300 metaphase cells were scored for 2-break aberrations.

| First dose | Interval | | Second dose | Two-hit Aberrations (Yield per 100 cells) | Expected Aberration Yield | |
	Time	Treatment			With Rejoining	Without Rejoining
600 r	75 mins.	Water	300 r	25.9 ± 2.4	19.7	39.3
		Chloramphenicol (300 µg/ml.)		37.3 ± 3.6		
	110 mins.	Water		30.7 ± 4.5	23.4	48.0
		Aureomycin (300 µg/ml.)		51.5 ± 4.1	27.6	55.2
	90 mins.	Water		31.6 ± 3.7	30.8	62.0
		Penicillin (300 µg/ml.)		28.9 ± 3.5	26.5	52.7

Should REVELL's idea prove to be correct then the difficulty of conceiving a mechanism of direct breakage will no longer apply. At the moment, however, it is still easier to think of exchange in terms of physical breakage followed by reunion. Indeed it is virtually impossible to discuss the problem in any other terms as yet.

By fractionating a dose of radiation and then applying specific metabolic inhibitors between the fractionated doses, WOLFF (1960) showed that, in *Vicia* seeds:

(a) The rejoining of breaks is inhibited by all the various inhibitors of cellular respiration and of ATP formation. This means that rejoining is an active metabolic event requiring energy.

(b) Specific inhibitors of protein synthesis also keep breaks open. Thus the antibiotics chloramphenicol and aureomycin, when applied between the two fractionated treatments, both lead to breaks from the first dose remaining open so that they are capable of forming exchanges with breaks induced by the second dose (Table 78). Penicillin which is not a protein inhibitor has no such effect.

Essentially similar results have been obtained in ovular tissue cells and PMC's of *Trillium kamtschaticum* (MATSUURA et al. 1962). MATSUURA and his colleagues also have evidence that mitomycin C produces a marked increase in the yield of two-hit aberrations. This antibiotic is known to specifically inhibit the DNA synthesis of *Escherichia coli* but whether it performs an equivalent role

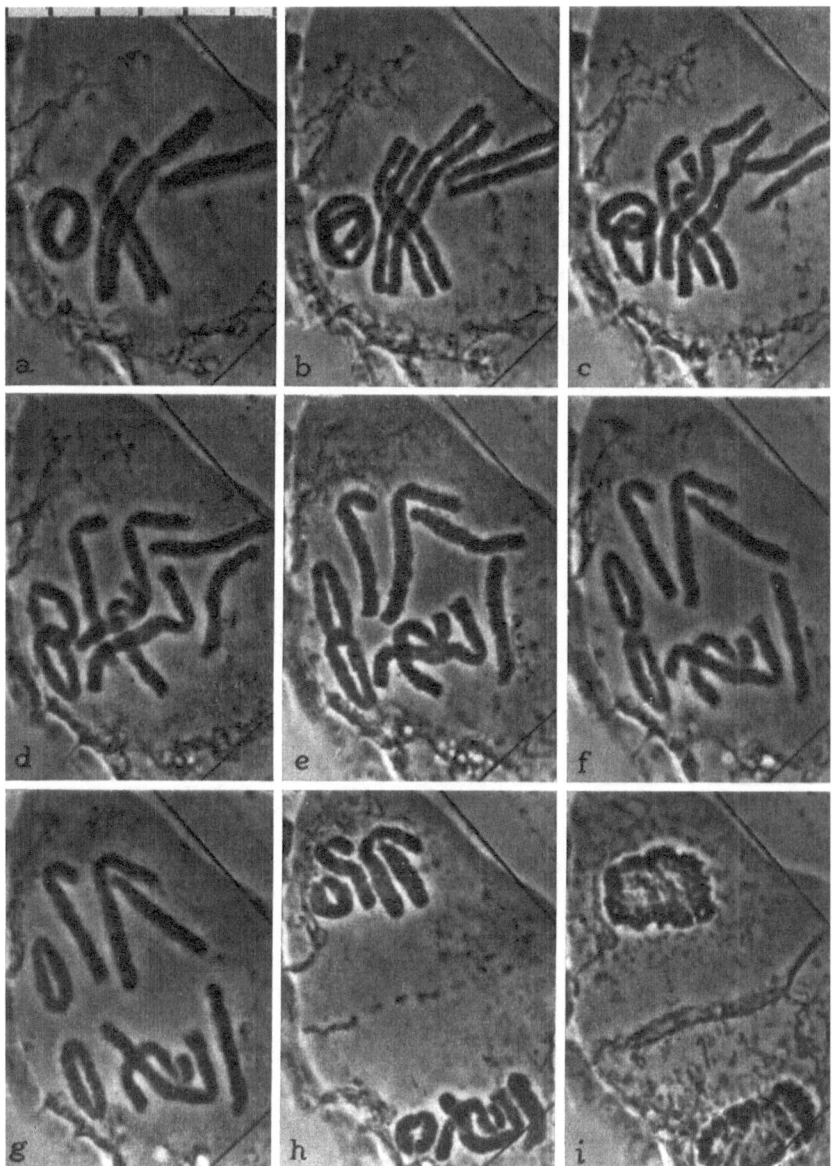

Fig. 184. The behaviour of a spontaneous ring chromosome with chromatid interlocking during anaphase in *Haemanthus katherinae* in a small micronucleus containing only 4 chromosomes. At anaphase both rings stretch (c-e), the lower one breaks (f) and within 2 mins. rejoins (g). Time after a–b (7.5 mins), c (11 mins), d (17 mins), e (26 mins), f (30 mins), g (32 mins), h (56 mins) and i (39 mins). Phase contrast, flattened endosperm preparation (Bajer 1963).

in *Trillium* remains to be established. Taylor, Haut and Tung (1963) have some evidence that DNA replication may be necessary for the reunion of at least some simple categories of chromosome breakage. Fluorouracil deoxyriboside (FUDR) is an analog of thymidine which is a specific inhibitor of the enzyme thymidylate synthetase when used at very low concentrations $(10^{-7} M)$. It therefore blocks the synthesis of thymidylic acid and, when cells are deprived of this, DNA-replication is stopped. Following such treatment the chromosomes of root cells of *Vicia faba* show simple breakage. But the lesions heal if thymidine is supplied to the cell. When X-irradiation is combined with FUDR treatment there is a marked increase in the number of acentric fragments and a decrease in the number of dicentric bridges in comparison with an X-ray control series. TAYLOR *et al.* argue that this demonstrates that reunion requires DNA synthesis but this has been challenged by KIHLMAN (1963). Indeed, BELL and WOLFF (1964) have recently shown that the capacity of FUDR to produce lesions in chromosomes is independent of DNA synthesis in that it can occur in G_2 cells*. In any case exchanges do not occur in the presence of FUDR and crossing-over is an exchange of the interchange type. The evidence, as far as it goes then, does not require DNA replication at exchange. Rather it requires protein synthesis. It would follow, of course, that if breakage occurs in DNA then the rejoining of broken ends must presumably involve the repair of DNA but why should this necessitate DNA synthesis? It is generally assumed that healing or union of broken ends is an interphase event, even when breaks occur or are induced during division as in the case of the breakage-fusion-bridge cycle (see page 94). BAJER (1963) has shown, however, that broken ends produced by mechanical tension at anaphase in interlocked ring chromatids and twisted dicentric chromosomes actually unite during division. In fact, fusion occurs within three minutes of breakage during anaphase (Fig. 184).

A number of authors have expressed the belief that an improved understanding of recombination mechanics depends on improved knowledge of the chemical organisation of the chromosome. Indeed much hope has been pinned on the possibility of reinvestigating crossing-over at the molecular level. But if the lessons of chemistry as applied to other aspects of biology have any meaning it is debatable to what extent the answers one gets in molecular terms will actually solve biological problems.

So long as we are ignorant of the manner in which DNA is organised in chromosomes we shall not be in any position to interpret the details of crossing-over in molecular terms although some speculations have already been made (WHITE-HOUSE 1963). The "near synthesis of the seemingly conflicting evidence on chromosome structure derived from tritium labelling, enzyme digestion, X-ray breakage and electron microscopy" which Prof. D. LEWIS (1963) has held to be now possible remains distant and is likely to continue to be so for some considerable time.

III. Coiling

1. Patterns of Coiling

Among the most conspicuous phenotypic changes associated with nuclear division are the various types of coiling shown by the chromosomes. Three types of coiling are generally recognised and these were first distinguished by DAR-LINGTON (1935). They are:

---

* See note 6 Appendix.

(1) Relational coiling.—This, as the term suggests, refers to the twisting of one strand (chromatid or undivided chromosome) around another: as such it is an external coiling. Relationally coiled strands are plectonemically related, in other words the coils interlock so that the two strands cannot separate laterally unless the coils are undone. The two polynucleotide columns of the double DNA helix are related in this way too. Relational coiling is seen between sister chromatids at mitosis. This is gradually undone during prophase by the rotation or rolling of sister chromatids in the opposite direction and most, if not all of it, is generally resolved by metaphase (see also below).

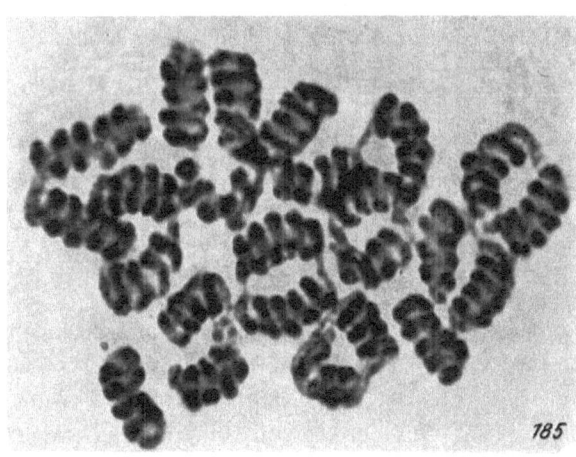

It is shown too by paired homologues at meiotic prophase. There is some disagreement as to when this relational coiling is developed. But relational coils are seen at early diplotene (Fig. 9, page 7), when they may be confused with chiasmata unless the two chromatids of each chromosome can be resolved, and in some cases, at least, they are visible during pachytene also (DARLINGTON 1935, 1936).

Fig. 185. The major internal coils at first metaphase of meiosis in a tetraploid *Tradescantia*. Photograph kindly supplied by Mr. C. G. VOSA.

It will be appreciated that, while all relational coils may have the same ultimate basis, the immediate origin of the relational coils between sister chromatids at mitosis and paired homologues at meiosis is different. The products of reproduction (chromatids) are coiled in the first case and the components of pairing (homologues) in the second. The former could result from a spiral cleavage plane but the latter must involve an actual rotation, for the homologues at leptotene are separate. Relational twisting between sister chromatids at meiosis (Fig. 10, page 7) is, of course, comparable with that at mitosis.

(2) Internal coiling.—This coiling is individual to a chromosome thread and it is generally held to be responsible for the shortening and thickening of chromosomes at mitosis and meiosis, though changes in the actual length of the thread cannot be excluded (see below). Shortening and thickening would result from an increase in diameter of the gyres of an internal coil at the expense of their number. As this results in thickening many have called the process coiling, as indeed we have done; but, as it depends on a decrease in the number of turns in the coil, others describe it as uncoiling (SWANSON 1947). There may be more than one order of internal coiling and in this respect mitotic chromosomes may differ from those at the first division of meiosis (and the second division too in some cases—see below) which have been held to have at least two orders of internal coiling, major (Fig. 185) and minor (DARLINGTON 1935, KUWADA and NAKAMURA 1933, 1934). The existence of a minor spiral was first claimed by FUJII (1926)

for *Tradescantia*. But some workers (e. g., HUSKINS 1941 in *Trillium*) regard the so-called minor spiral as a mere waviness in the chromosome thread. Other workers who have studied *Trillium*, however, have presented reasonable evidence for a minor spiral in this type (COLEMAN and HILIARY 1941, KEEFFE 1948).

(3) Relic coiling—This describes the more-or-less relaxed spirals which are seen in the prophase chromosomes of mitosis and meiosis. As DARLINGTON's apt term suggests, they are held to represent the incompletely resolved internal coils of the previous division, the resolution of which was impeded by the confines of the nuclear membrane. In this respect, therefore, prophase is a continuation of telophase rather than the reversal of it. Thus, in those organisms with a prolonged interphase between the two meiotic divisions, the major coils of the first division are seen as relic coils during the second division. In this event, the second division chromosomes are thinner than those at the first division (see for example Figs. 134–137, pages 151–152). But where the two meiotic divisions are not separated by an interphase the second division chromosomes are the same size as those of the first and the comparable relic coils are not seen until the following mitosis. In diplonts this represents the first cleavage division but in others it is the first division of the haplophase. For the flowering plants, this is the first pollen grain mitosis and the comparable division in the embryo sac.

The spiral organisation of mechanically active chromosomes was seen as long ago as 1880 by BARANETSKY but, of course, its universal occurrence was not recognised at that time. It was a subject of extensive and intensive study for twenty years between 1925 and 1945. But since then it has drawn the attention of fewer investigators. Consequently, little can be added to the reviews, summaries and theories of twenty years ago.

In certain Protozoa the spiral structure is clear without any special prefixation treatment (CLEVELAND 1949). But while there have been some studies on animals (e. g., grasshoppers—COLEMAN 1943; MAKINO and MOMMA 1950) most of the work has been undertaken with plants, notably *Trillium* and *Tradescantia*. Despite extensive investigation (or perhaps because of it) agreement has not been reached. There is little in the above summary which is in dispute but disagreement exists at two levels, observational and interpretive. Let us first consider the points of agreement in relation to the internal coiling of meiotic chromosomes:

(a) It is agreed that the direction of coiling, right hand or left, is not an inherent property of a chromosome region. The same segment in different cells and homologous segments in the same cell may be the same or different in respect of the direction of coiling.

(b) It is agreed that the chromosome is not a unit of coiling for the direction of coiling may not be consistent along its length. Changes in direction may occur at the centromere to give opposite coiling in the two arms of V- or J-shaped chromosomes and they may occur at chiasma to give a reversal of coiling within a chromatid.

(c) It is agreed that sister segments coil in the same direction at the first meiotic division though the terminal movement of chiasmata is expected to bring oppositely coiled chromatids together when crossing-over takes place between homologous segments with opposite coiling. In this connection it should be pointed out that, while the internal (major) coils of sister chromatids at meiosis are individual to each of them, they do not lie separate and parallel at diakinesis-metaphase like the chromatids of mitotic metaphase. They are then jointly coiled in the sense that the spiral of one fits into, and is free to separate from, the spiral of the other. In short, their relationship is paranemic.

(d) It is agreed that the internal coils must develop as a result of internal forces. They cannot be created by external forces and rotation of the ends because, apart from the conceptual difficulties, internal coiling is shown by ring chromosomes which have no ends.

Table 79. *Summary of Observations on Direction of Coiling in Chromosome Arms Across Centromeres and Chiasmata in Species of Tradescantia.*
(Modified from DARLINGTON and VOSA 1963.)

Type and Reference	Association or Stage		Across Centromere		Across Chiasma	
			Discord.	Concord.	Discord.	Concord.
T. reflexa (NEBEL 1932)	2 x	II's	61	83	...	...
T. reflexa (SAX and HUMPHREY 1934)	2 x	M I	84	256	74	91
		A I	38	90	...	...
T. sp. (WATANABE 1960)	4 x	I	165	235	...	...
		II and IV	147	153	...	...
	2 x	II's	63	17	22	18
T. bracteata (DARLINGTON and VOSA 1963)	< 3 x [1]	AA (in TAA)	115	25	34	36
		TA (in TAA)	...	...	41	29
		TA (in TA)	53	17	45	25
	4 x	II, III, IV	47	21	24	31

[1] Sub-triploid in which one of three homologues was represented by a telocentric (T) chromosome while the other two were normal (A), see Fig. 186.

There is also some agreement at a quantitative level. In relation to (a) above, it is agreed that right- and left-hand coiling are equally likely. But in connection with (b) the quantitative data are not consistent. In fact all the three possible relationships have been found with regard to changes in direction at the centromere. Dissimilar coiling across the centromere (discordant) has been described as more, equal, and less frequent than, similar coiling across the centromere (concordant).

A random relationship has been claimed for *Trillium*, but analysis here is complicated by the centric localisation of the chiasmata (see MATSURA 1935, 1937 and WILSON and HUTCHESON 1941). Reversals at the centromere have been described also in *Rhoeo* (SAX 1935) and the data for *Tradescantia* are summarised in Table 79.

The observations on differences in the direction of coiling on opposite sides of a chiasma are more consistent and a random relationship seems to be general (SPARROW, HUSKINS and WILSON 1941). However, in a study on diploid, triploid and tetraploid *Tradescantia* DARLINGTON and VOSA (1963) found that while this was true in the case of a chiasma between equal sized homologues (in bivalents and multivalents) it did not hold when unequal chromosomes were involved. The case they studied was a "sub-triploid" in which one chromosome was telocentric (Table 79, Fig. 186). In the chains of three involving the telocentric (TAA) the coiling relations across the T-A chiasma were 41 discordant: 29 concordant while in the unequal TA bivalents it was 45 discordant: 25 concordant. A discordant relationship was therefore 1.4 times as common in the first case and 1.8 times as common in the second, the over-all excess being 1.59 which is significant (P < 0.01). It will be appreciated, however, that in *Tradescantia* the chiasmata are localised terminally.

Further, DARLINGTON and VOSA "avoided configurations with interstitial chiasmata." Consequently the relations they studied were effectively those between paired arms. The *Trillium* workers, on the other hand, were dealing with interstitial chiasmata and coiling within arms.

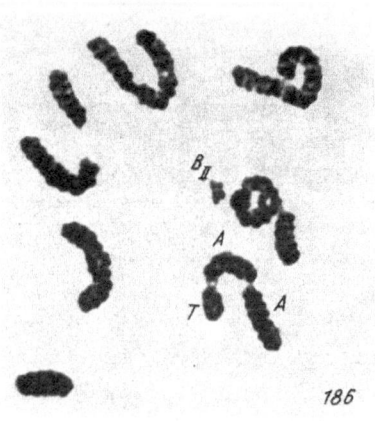

It has been claimed that, except for the interference introduced by crossing-over, the direction of coiling is consistent within a chromosome arm in a particular cell. But intra-arm reversals have been held to occur independently of chiasmata. In *Rhoeo*, where chiasmata are localised pro-terminally, they were found to be rare (SAX 1935). They are uncommon (NEBEL and RUTTLE 1935) or absent (DARLINGTON and VOSA 1963) in *Tradescantia* also, which has a similar distribu-

Fig. 186. A sub-triploid *Tradescantia* with a deficiency for one arm, producing a telocentric (T), and two minute B-chromosomes associated as a bivalent. Photograph kindly supplied by Mr. C. G. VOSA.

tion of chiasmata. But the clearest evidence for the occurrence of intra-arm reversals comes from the studies on asynaptic and desynaptic *Trillium* (HUSKINS and WILSON 1938, WILSON and HUTCHESON 1941, SPARROW, HUSKINS and WILSON 1941). It is claimed that here the number of adventitious reversals of coiling is a function of length or, more specifically, the number of gyres. A similar claim was made by MATSUURA (1937).

If the relic coils of one division are equivalent to the internal coils of the previous one, and the evidence is that they are (see below), then reversals in the direction of relic coiling in one division can be taken as evidence for reversals in the internal coiling of the previous division. WHITE (1940) has described intra-brachial reversals in the relic coiling of the virtually telocentric X-chromosomes of grasshoppers at spermatogonial mitosis. HUSKINS and WILSON (footnote page 287, 1938) aware of this work from a personal communication, concluded that "obviously, these cannot be due to chiasmata." But, if as recent evidence suggests, sister strand exchange may be quite common at mitosis (see page 212) then perhaps such an exchange could effect a reversal of coiling similar to that brought about by chiasmata at meiosis. In fact, on the basis of observations made on the internal mitotic coils themselves, it has been claimed that not only do reversals in direction occur within an arm, but sister chromatids may be coiled in opposite directions over corresponding segments (MANTON and SMILES 1943). Similar observations were made by CLEVELAND (1938, 1949) in *Spirotrichonympha* where the reversal sites in sister chromatids differ. It will be remembered that the chromatids at mitosis do not show the joint paranemic coiling observed at meiosis (see above).

Relic Coiling

If relic coils are the remains of the internal coils of the previous division then, clearly, the pattern of internal coiling in one division should be reflected by that of the relic coiling at the following division. This question has been examined in some detail by the workers on *Trillium*. In this genus the two meiotic divisions are not separated by an interphase. Consequently the major internal meiotic coils appear as relics at prophase of the first pollen grain mitosis. SPARROW, HUSKINS and WILSON (1941) summarise their result as follows—"changes of direction in the relic coil have been analysed for comparison with changes at first meiotic anaphase. The mean percentage of gyres with intra-brachial changes are 17.8% for relic and 18.1% for the major coil. It has been shown previously that the number of gyres at early microspore prophase is only slightly less than at second anaphase. These data provide direct evidence that the coils of microspore prophase are really relics of the major coils of meiosis, as DARLINGTON's name for them implies." It will be appreciated that the intra-brachial changes referred to above include those owing to chiasma formation. UPCOTT (1938) in a study of *Hyacinthus*, on the other hand, found intra-arm reversals in the direction of relic coiling in only 3 of the 95 nucleolar chromosomes she studied and DARLINGTON (1935) found a consistent direction within arms in *Fritillaria*.

In passing we many note that, in the investigation by WHITE mentioned above, right- and left-hand relic coiling in the X-chromosome during spermatogonial mitosis seemed to be equally common. Thus for *Microcentrum* he found 41 right-handed and 30 left-handed X-chromosomes in the same individual, X-chromosomes showing reversals being excluded in this analysis. It would appear therefore that the internal coiling of mitosis like that of meiosis may be in either direction and so the direction of coiling is not an inherent property of the chromosome.

Relational Coiling

When two lengths of wire are maintained in parallel so that there is no twisting of one around the other and the two are coiled around a cylinder, then, if following the removal of the cylinder the coils are straightened, the two lengths of wire will be found to be plectonemically related, interlocked and incapable of separating without twisting. In other words, if the coils produced as described above are "pulled-out," the two strands will be relationally coiled. The relational coils thus produced will be in the same direction as the pulled-out gyres.

Further, if there are n gyres in the original helix, then there will be $2n - 1$ relational twists in the extended state. (Each overlap in the relational twist is here considered as a twist though some may prefer to call it a half-twist.) In making the above numerical comparison it was assumed that a constant direction of coiling was maintained in creating the original gyres.

If, however, reversals are introduced then, prior to any unwinding or cancellation of relational twists in opposite directions (see below), the number of relational twists is reduced by one by each reversal. Thus, where n is the number of gyres in the original coil and r the number of reversals of direction in it, the number of relational twists in the extended state is $2n - 1 - r$. Where there are n gyres, there are $n - 1$ adjacent gyre-pairs and this equals the maximum possible number of reversals. For the relational twisting the corresponding value is given by $2n - 2$.

Earlier (see page 201) we drew attention to the claim that half-chromatid units had been resolved optically in the meiotic chromosomes of some types which include *Trillium*. It is further claimed that while sister chromatids are paranemically related, half-chromatids are plectonemically disposed.

Thus, the model discussed above can be compared with a chromatid at the first metaphase or anaphase of meiosis. The gyres would then be equivalent to the (major) internal spirals and the two components comparable with the half-chromatids. But in view of the connections which have been established between the internal coils of meiosis and the relic coils of the first pollen grain mitosis in a type like *Trillium*, the gyres of the model can be compared also with the relic coils of the post meiotic mitosis. But can the two strands then be regarded as the equivalent of the sister chromatids? The *Trillium* workers believe that they can. Thus SPARROW, HUSKINS and WILSON (1941) state categorically that "since the chromatids which are relationally coiled at microspore prophase are the products of the tertiary split of meiosis, it is obvious that the arrangement ... of the resultant half-chromatids must bear an intimate relationship to the mechanism of relational coiling."

We have already discussed the conceptual difficulties and the conflicting evidence in connection with this problem. But having given the viewpoint of the *Trillium* workers let us consider their observations. SPARROW, HUSKINS and WILSON (1941) found that, in the pollen grains of *T. erectum* and *T. grandiflorum*, the direction of relational twisting was random across the centromere except in the case of the so-called B-chromosome in their nomenclature. The C-, D- and E-chromosomes have median or nearly median centromeres while the centromere of the A-chromosome is virtually terminal. The exceptional B-chromosome is unique in having unequal arms, the long arm exceeding the other by a factor of about three to four. With regard to this chromosome, examples showing reversals across the centromere were fewer than those without reversals, the respective ratios being 7 : 27 for *T. grandiflorum* and 1 : 9 for *T. erectum* (Table 80). Although the numbers are low, concordance appears to predominate (cf. direction of internal coiling across chiasmata in unequal bivalents—see above).

When the frequency of intra-arm reversals in the direction of relational twisting at mitosis in pollen grains and root tips were compared, reversals were found to be twice as common per twist in the former. This difference was accounted for in terms of the role of chiasmata in effecting reversals in the direction of internal coiling, a factor believed to be operative only at meiosis. But the point we made earlier in connection with relic coiling applies equally here.

Reversals in the relational twisting at pollen grain mitosis were compared also with those in the relic coils of the same division and with the (major) internal coils of meiosis. To facilitate comparison all sets of data were expressed in the same terms, namely, the number of reversals observed over the maximum number of reversals possible. The latter was given in terms of a relational coil as described above and the fraction was expressed as a percentage.

As quoted above, the percentage of gyres with intra-brachial changes in direction was 17.1 and 18.1 for the relic and major coils respectively. When expressed as a percentage of the theoretical maximum number of possible changes these values become 11.15 and 11.4 respectively.

Table 80. *Direction of Relational Twists on Opposite Sides of the Centromere at Prophase of Mitosis in the Pollen Grams and Root Tips of Two Species of Trillium.*
(Data of SPARROW, HUSKINS and WILSON 1941.)

Chromosome Type		R · R	R · L or L · R	L · L	R · U or U · R	L · U or U · L	U · U
Microspores							
T. grandiflorum	B	16	7	10	2	4	5
	C	9	9	7	5	8	6
	D	8	11	8	9	1	7
	E	4	22	12	1	1	4
Total		37	49	37	17	14	22
T. erectum	B	5	1	4	35	14	14
	C	6	11	7	20	17	12
	D	9	12	5	17	14	16
	E	9	25	15	8	15	0
Total		29	48	31	80	60	42
Root Tips							
T. grandiflorum	E	7	25	12	4	2	0
T. erectum	E	12	19	3	2	1	0

Comparison with the number of reversals in the relational twisting was complicated by the fact that the percentage of reversals here did not appear to be constant. The data were not tested statistically but it appeared that the percentage of possible changes realised increased with a decrease in the number of relational coils in the arm. Thus, for arms with 2, 3, 4, 5, 6–7 and 8–11 coils the respective percentages were 24, 25, 20, 19, 18 and 15. Since the relational twisting resulting from the "pulling-out" of the gyres in the model considered above is in the same direction as the coiling of the gyres, the relational coiling on the two sides of a reversal of coiling in the gyres will be in opposite directions. Consequently, unwinding resulting in the cancellation of relational twists is expected to proceed from points of reversal. This means that the number of twists will decrease more rapidly than the number of changes in direction of these twists. It is in these reasonable terms that SPARROW, HUSKINS and WILSON

explain the situation they found to obtain. They further argue that if the very
early stages of prophase prior to any cancellation could have been studied, the
percentage of possible changes realised in the relational twisting in the pollen
grain nucleus, which they found to tend from 24% to 15%, would reach the
same value (ca. 11%) as that found for internal and relic coils.

On this basis, the *Trillium* workers regard the relational coil to be the result
of the plectonemic relationship between the half-chromatids in the internal coils
of the previous division or, as we may put it, the consequence of a spiral plane
of cleavage. Half-chromatid units in the mitotic chromosomes of grasshoppers
were described by MAKINO and MOMMA (1950) after special treatments but they
did not comment on their coiling relationship.

Results at variance with those obtained in *Trillium* were described by UPCOTT
(1938) in her study of *Hyacinthus*. This work was concerned with the behaviour
of the nucleolar chromosome. This is mediocentric and the organiser is about
half-way along one of the arms. The number of comparisons available was small
but "so far as they go, however, the results show that relic and relational coiling
may be in either direction in each arm; that either type of coiling may be in
the same or opposite direction in the two arms and that ... there appears to
be no correlation of directions of relic and relation coiling in the same arm ..."
UPCOTT does not make it clear, however, whether the two types of coiling in
a given arm were compared over the same region. The relational coiling referred
to by UPCOTT is what she calls "true relational coiling" in respect of which no
intra-arm reversals were recorded. But what must presumably be called false
relational coiling is also described. UPCOTT (1938) writes that "at the earlier
stages of prophase there are several *apparently* relational coils whose direction
has not been marked (i. e., indicated in the drawings). These I *assume* to be
coils compensating for the relic coiling to which they are opposite in direction."
(The italics and comment in brackets are ours.)

UPCOTT believed that the direction of internal coiling for a given arm was
consistent in a given division as was that of the relational coiling. And contrary
to the *Trillium* workers who regarded the relational twisting in one division as
a reflection of the internal coiling of the previous one, UPCOTT regarded the changes
responsible for their formation "to be concerned at the same time with the
formation of the new internal coils." Further, in view of the lack of consistency
observed between the direction of the relic and relational coiling, she concluded
that "the direction of coiling in one division has an indeterminate relationship
with that of the next." The *Trillium* workers, on the other hand, and as we
have discussed, consider the relational twists and their direction to be determined
by the internal coils and their direction at the previous division. It must be
agreed, however, that the direction of internal coiling in one division has no
relationship with that of the next or, of course, that of the preceding one.

2. The Basis and Mechanism of Chromosome Coiling

Observational differences of the kind considered above are at the root of the
disagreements regarding the basis and mechanism of coiling. The various theories
which have been proposed fall into one or other of two main schools of thought—
the torsional or molecular school and the matrical or pellicular school.

The most comprehensive hypothesis of the molecular school is that presented by Darlington (1935, 1937, 1955) who proposes that "spiralisation is associated with a compensating internal twist (the molecular spiral) due to a rearrangement of the constituent particles, a rearrangement either between molecules or within molecules ... Since the molecular spiral must determine both the major and minor spirals, these must be in the same direction, and opposite to the twist that conditions or determines them." Nebel (1939) suggested a variant of this scheme which allowed for the occurrence of changes of direction within an arm.

The matrical school, on the other hand, proposes that coiling is a consequence of differential length changes (owing to contraction and/or expansion) between

Table 81. *Chromosome and Chromonema Lengths in Trillium erectum.*
(Data of Sparrow, Huskins and Wilson 1941.)

Stage	Length (in Microns)	
	Chromosome	Chromonema
Leptotene	...	920
Zygolene	...	1040
Pachytene	...	640
Diakinesis (early)	86	109
Diakinesis (mid)	125	187
Metaphase-I	99	320
Anaphase-I	93	327
Anaphase-II	76	310

the chromosome thread(s) or chromonema (ta) and an enclosing matrix or pellicle which acts as a confining cylinder. This proposition is favoured by Huskins, Wilson, Smith, Humphrey and Sax amongst others. There is, however, some variation of opinion within this school. Thus, Sax and Humphrey (1934) concluded that the major coils of meiosis in *Tradescantia* result from a contraction of the matrix which forces the longer thread into a helical coil. But Huskins and his co-workers (Huskins 1937, Wilson and Huskins 1939), following a suggestion by Bêlar (1928), have postulated the opposite change for *Trillium*, namely, the elongation of the thread within a matrix of constant length. It is true that Sparrow, Huskins and Wilson (1941) obtained some data which indicated that a slight contraction of the matrix may occur in *Trillium* but that it was not of comparable magnitude to that recorded for *Tradescantia* (Table 81).

Yet a further variant has been proposed by Coleman and Hiliary (1941) who consider the elongation of the thread to be the result of the straightening of the minor spirals. These workers have also postulated that the chromosome ends may initiate the major coiling in *Trillium*. Wilson and Huskins (1939), however, do not accept the existence of a minor spiral (see also above). On the basis of Astbury's (1934) demonstration that keratin fibres may be reversibly stretched by unfolding to double their length, these workers argue that "since the chromonema may likewise double its length during coiling, it is possible that the observed lengthening may be due to a similar molecular reorientation. The pellicle may be an interface at which a change of surface tension might account for stretching."

Thus the matrical school is not averse to an explanation involving molecular orientation and the torsional hypothesis of DARLINGTON does include an appreciation of the role of the chromosome surface (see page 32, DARLINGTON 1935). SWANSON (1957) has said that the logical extension of the molecular theory would be to consider "that the configurations of the nucleic acids and proteins in the chromosome framework govern the coiling of chromosomes, and from studies of the nucleic acid and protein structures it appears that the natural state of these molecules is in the form of helices." The latter part of this contention can hardly be doubted but this is not true of the first part. Thus TAYLOR (1962) has suggested a skeletal chromosome structure which can take various forms and which, in its ladder-form, would be a ribbon which has the property for coiling, i. e., differential contraction of the middle and sides. Such a ribbon will roll into a trough-shaped cylinder and take a helical shape. Such coils, which could occur at random, left or right handed, would be the large coils observed with the light microscope, and therefore not directly related to any coiling of the DNA double helix (TAYLOR 1959).

In summary we can see that while there is some observational agreement concerning chromosome coiling there are many conflicting observations and even greater dissention in the matter of explanation. But there are two main features which any hypothesis on the nature of coiling must be able to accomodate. First, the relationship between the directions of the different types of coils and their direction in a given region in different divisions. Second, the behaviour of ring chromosomes.

In connection with the first point it is clear that if the relational coiling between paired chromosomes at pachytene arises from uniformly concordant internal torsion in each of them, then it cannot "be related, as was earlier supposed (DARLINGTON 1935, 1936) to the internal coiling in which ... partner chromosome arms ... are likely to be discordant" (DARLINGTON and VOSA 1963). However, the simultaneous occurrence of relational coiling between paired regions within and outside relatively inverted segments would require a change or changes in the direction of the supposed underlying torsion within an arm. And as we pointed out above, the relational coiling between pairing chromosomes is rather different from that between chromatids of dividing chromosomes.

Ring chromosomes coil normally at mitosis and they can disjoin normally. But they may also give rise to double-sized dicentric ring chromatids or to two, mono-centric, interlocked ring chromatids of normal dimensions (McCLINTOCK 1938). The skeleton model proposed by TAYLOR (1962) allows rotation, but it is not clear that this would work at the chromosome level. Clearly, if the chromatids of ring chromosomes, like those of two-ended chromosomes, are plectonemically coiled at prophase of mitosis, then there must be some way of compensating for the interlocking which would otherwise result. It will be recalled that UPCOTT's "false relational coiling" was in the opposite direction to the relic coiling.

Compensation would result if the number of right-hand and left-hand relational twists were equal, for there would then be a cancellation of coiling from the points of reversal. This problem may be related to the first one, for if a pair of plectonemically related strands are coiled around an axis without rotation and in the opposite direction to the relational coiling, the internally coiled strands thus produced come to have a paranemic relationship. This can, of course, be completely effective only if the number of turns made in creating the two sets of coils do correspond. It will be appreciated that models of this kind can be misleading for they are made by using an external force to hold and rotate chromosome ends while the chromosome undergoes comparable changes even when it has no ends.

Nebel envisages three orders of coiling in somatic metaphase chromosomes and four in the metaphase chromosomes of meiosis. The first and lowest order of coiling is believed to be very stable as to its dimensions and to be represented by the 30 Å fibrils seen with the electron microscope. On this is superimposed a second order of coiling. This, like the higher orders is expected to vary with degree of microfibrillar packing and thus with stage. But its gyres are of the order of 200–500 Å in diameter. The third and final order in somatic chromosomes is the one occasionally seen with the light microscope. The gyres of this have a diameter of one to two thousand Angstrom units. At meiosis a fourth order of coiling exists—the (major) internal coils—with gyres of about 10,000–30,000 Å.

Each chromatid is held to have many microfibrillae (32 in mouse and man) which show paranemic packing so that separation is possible. At the replication phase these fibrils are considered to be uncoiled (except for the first order coiling) and to be arranged in flat sheets perhaps in bundles of four. The sheet may become a bundle by rolling so that separate groups of bundles are not then detectable. And a closer grouping of the fibrils at the leading edge of the rolling sheet may give the appearance of a core which is sometimes seen at meiosis with the electron microscope. In his diagrams Nebel draws lines at right angles to the wavy lines which represent the first order 30 Å fibrils. The former "are considered to be not the true microfibrillae but rather the cross links of the gene strings." Nebel points out that this scheme is at variance with the *ex cathedra* model proposed by Taylor (1957).

Unfortunately our comprehension of chromosome coiling ends here and so therefore must our discussion of it.

3. Allocycly

The sequence of coiling and uncoiling considered above is that typical of most parts of most chromosomes. It is associated with changes in chromosome length and diameter which it, at least in part, determines. It is associated also with changes in pycnosity and chromaticity. But under certain conditions—environmental, genotypic or cellular—some chromosome segments, whole chromosomes or even whole chromosome complements behave differently in this connection, i. e., they exhibit allocycly. The regions of chromosomes which are most commonly allocyclic are the centromere regions, the telomeres and the nucleolus organising regions while the most common types of chromosomes showing this behaviour are the sex chromosomes, the supernumerary or B-chromosome and the sex-limited chromosomes of certain dipterans.

The commonest type of allocyclic behaviour is that where the segments con. cerned appear over-condensed relative to regions following the standard cycle-

This so-called positive heteropycnosis may be manifest during the resting stage or at any time during division except, of course, meta-anaphase when regions showing the standard behaviour are at their most condensed state. The opposite allocyclic condition, negative heteropycnosis, is considerably less common and it can be shown at any time during division.

Different regions may show different types of allocyclic behaviour in the same cell cycle and the same region may behave differently in different cycles. One of the best examples of the former condition was described by SMITH (1952) in males of the beetle *Tribolium confusum*, while the X-chromosome in the males of short-horned grasshoppers illustrates the second condition. Both examples show that the chromosome is not the unit of behaviour in respect of cycly.

The male of *Tribolium* has eight pairs of autosomes and a neo-XY allosome-pair. At pachytene, highly condensed, nearly spherical blocks can be seen on each side of the autosomal centromeres. The other parts of these chromosomes have the typical, beaded pachytene appearance. The short arms of two of the autosome pairs consist solely of the so-called centric half-blocks. The X-chromosome is similar to these but its short arm includes also a segment of about equal bulk to a centric half-block and to which the nucleolus is attached. This nucleolar-associated region has a condensation cycle indistinguishable from that of the centric half-blocks. This means, of course, that the existence of a centric half-block in the short arm of the X is inferred by comparison with the situation in the other two-armed chromosomes. The Y-chromosome is homologous with the long arm of the X which thus constitutes its pairing segment in the heterogametic sex; the nucleolar arm is differential. The centromere of the Y appears to be strictly terminal so that it contains only one centric half-block. Thus, at pachytene, the centric blocks and the nucleolar segment of the X-chromosome are positively heteropycnotic while all other regions are isopycnotic. But by metaphase all chromosomes are isopycnotic except for the Y-chromosome. Its centric block is still deeply stained but the rest of it is negatively heteropycnotic. SMITH (1952) calculated the length changes shown by the various parts of the sex-chromosomes. He found that the segment of the long arm of the X which was isopycnotic at pachytene behaved in the same way as the comparable regions of the autosomes. Its spiralisation coefficient (pachytene length divided by metaphase length) was 13. But the coefficient of the homologous region of the Y-chromosome, with which it was paired at pachytene, was only about 4. The segments which were already condensed at pachytene did not appear to condense further. But the centric half-block of the Y-chromosome appears at metaphase to be only about half its diameter at pachytene though its length seems to be the same. It, then, may be said to have a spiralisation coefficient of a half.

In tomato also regions adjacent to the centromere are heterochromatic at pachytene (GOTTSCHALK 1951). It further resembles *Tribolium* in that the nucleolar arm in both is wholly heterochromatic. However, at pachytene, the chromatic regions of tomato are longer and comparatively thinner than those of *Tribolium* and a chromomeric structure can be discerned in them. It is perhaps for these reasons that the coefficients obtained for tomato (BROWN 1949) differ from the corresponding ones for *Tribolium*. Thus, in tomato, the chromatic segments adjacent to the centromere must contract further because the length of the chromatic segments at pachytene exceeds that of the whole chromosome at late diakinesis (11.0 ± 0.36 μ versus 4.8 ± 0.13 μ). Brown concluded that this contraction depended on the shortening of what he called the "achromatic interchromomeric regions" within the centromere-adjacent chromatic segments. Now, at mitosis, the nucleolar chromosome is approximately isobrachial if one neglects the length of the extended nucleolar constriction. But at pachytene it is heterobrachial. Its short arm, as pointed out above, consists of a chromatic segment. A similar segment of nearly equal length (4.8 ± 0.21 μ versus 4.6 ± 0.15 μ) is found adjacent to the centromere in the long arm which further includes an isopycnotic segment. This segment is 26.9 ± 1.31 μ at pachytene but, clearly, it can contribute to the metaphase length an amount approximately equal to only the difference between the lengths of the two chromatic regions in their most contracted state.

Segments adjacent to the centromere which show heterochromacy at meiotic, but not mitotic, prophase have been described in many organisms, e. g., *Oenothera*, *Agapanthus*, *Kniphofia*, *Plantago*, *Zea*, *Rhoeo*, *Blaberus* and *Tipula* and such segments have been discussed elsewhere in terms of block reactions and position effects (Darlington 1957, Lima-de-Faria 1956, Lewis and John 1963 a).

The X-chromosome in the males of the short-horned grasshoppers is typically unpaired, the sex-determining mechanism being XX/XO. In these organisms the X-chromosome shows negative heteropycnosis in the early spermatogonial mitoses. The extent of this decreases with successive divisions and in the later mitoses of the male germ-line it behaves isopycnotically. But at telophase of the last pre-meiotic mitosis the X-chromosome remains condensed and it consequently appears as positively heteropycnotic at the first meiotic prophase. At metaphase of this division, however, it again becomes negatively heteropycnotic but by interphase it is once more positively heteropycnotic, a condition it returns to after the completion of the second meiotic division also. This is the general sequence but all parts of the X do not behave in exactly the same way. Thus, when the X is negatively heteropycnotic at first meta-anaphase it is usual for the one or both ends—distal and proximal to the terminal centromere—to show the appearance to a less marked extent (see Figs. 15 and 16, pages 10–11).

White (1940) found that the diameter of the X-chromosome was smaller than that of the autosomes when it was negatively heteropycnotic but wider when it showed positive heteropycnosis. Also the negatively heteropycnotic X was narrower, and the positively heteropycnotic X wider, than the X-chromosome in its isopycnotic condition at comparable stages of division. But the length of the negatively heteropycnotic X in the early spermatogonial divisions was the same as that of the thicker, positively heteropycnotic, univalent-X at diakinesis.

White also examined the early prophase stages of some of the early spermatogonial mitoses and found that the X, like the autosomes, showed relic coiling: evidently the even-more negatively heteropycnotic X of the preceding division had a spiral structure although this could not be observed directly. Further the threads showing relational coiling were thinner in the X than in the autosomes. In the long-horned grasshoppers, however, the X-chromosome shows positive heteropycnosis during prophase of the spermatogonial mitoses and in these the threads showing relic coiling are thicker in the X than the autosomes.

In view of these findings WHITE concluded that the variations in thickness corresponding to the variations in pycnosity were a reflection of the thickness of the thread being coiled and not to a difference in coiling *per se*. He envisaged the somatic metaphase chromosome as a compact helix, i.e., one in which the gyres are in contact and with virtually no space in the core. Among the properties such a helix is the independence of its length from the diameter of the thread composing it (see PÄTAU 1948). This will of course determine the diameter of the helix but its length will be related only to the length of the thread being coiled. More specifically the diameter of the cylinder enclosing the helix will be $(2\,r + r\sqrt{3})$ where (r) is the radius of the thread. This gives a packing factor (diameter of helix divided by diameter of the thread—DARLINGTON and UPCOTT, 1940) of 1.87. The length of the helix will be $\dfrac{2\,l}{\pi\sqrt{3}}$ where (l) is the length of the thread: it is thus independent of the radius of the thread and gives a constant equal to 2.72. The reciprocal of this constant gives the cosine of the angle between the helix and its vertical axis. The number of turns in a solid helix will clearly vary with both the length of the thread under consideration and its diameter. But the number of turns in a solid helix with a length (or height) of L is given by $\dfrac{L}{0.576\,D}$ where (D) is the diameter of the helix. There is thus an inverse relationship between the diameter of a solid helix and the number of turns in it. And WHITE found that "the number of coils per unit length is greater in negatively heteropycnotic chromosomes although it is possible that the relation between chromosome diameter and the number of coils per unit length is not directly proportionate."

In male acridoids it is not uncommon for the centric ends and sometimes the non-centric ends of acrocentric autosomes to show pre-condensation at meiosis in the male. But in many species there is also an autosomal bivalent which shows more extensive pre-condensation. Indeed, more than half its diplotene length may be positively heteropycnotic (Fig. 9, page 7). Unlike the X however this segment does not become heteropycnotic until zygotene and it shows its heteropycnosity only during the meiotic prophase.

A difference between the allocyclic segments of autosomes and allosomes is apparent in *Tribolium* and other beetles also. Thus, while the allocycly of the sex-chromosomes is seen at meiosis only in the male, the centric regions of all the chromosomes behave in the same way at male and female meiosis. A comparable difference between the sexes has been described for spiders too (HACKMAN 1948).

It is clear from these few examples, and many more could be cited, that there are many kinds of allocyclic behaviour and whilst there is a relationship between length, diameter and chromaticity in some cases, it is by no means universal. Further, diameter changes other than those owing to coiling appear to be involved in some cases and this may apply to certain length changes also. Thus, negatively heteropycnotic segments can be revealed by cold treatment.

4. Chiasma Terminalisation

In some species the chiasmata at metaphase are as frequent as they are at diplotene and they occupy the same, or much the same, positions at the two stages. Commonly, however, there is a progressive shift in the distribution of chiasmata between the two stages. This movement along the arms of paired chromosomes from their points of origin to more distal positions was called terminalisation by DARLINGTON. The process may be partial, or complete terminalisation may be achieved so that all chiasmata become terminal by first metaphase.

The degree and progress of terminalisation can thus be expressed in a coefficient which gives the fraction of the chiasmata that are terminal at a given stage. Terminalisation coefficients seem to be higher in species with small chromosomes and in those which have some measure of distal localisation initially.

Three main hypotheses have been advanced to account for terminalisation:

(a) Electrostatic hypothesis—according to this hypothesis two kinds of repellant force are responsible for terminal movement (DARLINGTON and DARK 1932, DARLINGTON 1937). First, a general body repulsion which is distributed evenly along the length of the chromosome. This force is held to operate not only within but between bivalents also and leads, in this latter instance, to the movement of bivalents away from each other and towards the periphery of the nucleus at diakinesis. Of course, as the chromosomes contract they will necessarily appear to be further apart at later stages even if there is no actual movement. This effect would be especially pronounced if the chromosome ends were "attached" to the nuclear membrane. Be this as it may the generalised force is held to have a constant effect in all organisms. Second, a repulsion which is localised between the centromeres. This is held to be the stronger force.

Various observations have been cited in support of the hypothesis and the forces on which it rests. Thus, the circular form assumed by the loops defined by the chiasmata at diplotene/diakinesis and the relative rotation through a right angle of the segments on opposite sides of a chiasma imply a generalised force acting equally on all parts. Further, terminal movement is more extensive in loops than in half-loops and greater, it seems, in centric half-loops than non-centric loops. It should be greater also in small loops. What is more, distributed surface charges are characteristic of ampholytes like nucleoprotein and there is evidence to indicate that chromosomes react electronegatively.

Thus, the greater repulsion within the centric loop (or half-loop) allows it to expand at the expense of the adjacent, more distal, region and non-centric loops extend at the expense of the non-centric half-loops which occur between a chiasma and a chromosome end. Thus, on these bases, there must be a general movement of chiasmata towards the non-centric ends of the chromosomes until the process is complete or an equilibrium is reached.

(b) Coiling hypothesis.—This hypothesis is due largely to SWANSON (1942) who describes it as "terminalisation through despiralisation" (i. e., through spiralisation as we have used the word—see page 224). SWANSON's argument runs along the following lines. During meiosis there is a continuous and progressive decrease in the number of (major) internal coils and a corresponding increase in gyre diameter. This means a change from a long lax thread to a shorter, more rigid one. Under these conditions of increasing rigidity, the chiasmata will tend towards the spatial distribution which will spread the tension developed by coiling over the greatest possible length. Thus, terminalisation is accounted for in terms of the same increasing rigidity which leads to the rotation of chromatids and the resolution of the relational coiling between them at mitosis. On this basis, as on the electrostatic hypothesis, a bowing-out of loops is expected. But, initially at least, either proximal or distal movement of chiasmata could occur depending on the original distribution of chiasmata. But, as SWANSON points out "the only direction in which a constant reduction in the strain imposed by coiled threads could be brought about would be a distal one."

(c) Elastic chromosome repulsions.—ÖSTERGREN (1943) suggested that among the "many factors controlling chiasma terminalisation" were those owing to elastic repulsions. He argues that solids have a definite shape and that efforts to change this shape meet with elastic resistance. Owing to the fact that chiasma formation tends to distort, the angle made by the segments on the side of a chiasma adjacent to a second chiasma (i. e. within a loop) will be greater than that described by segments between a chiasma and an end. Consequently, the potential energy and elastic resistance afforded will be greater also. A terminal movement of chiasmata will decrease the angles within the loop and result in a decrease in the total potential energy associated with the chiasma. And since systems tend to lose potential energy, terminalisation can be predicted.

These, then, are the principal hypotheses that have been advanced to account for the terminal movement of chiasmata. Since they all aim to account for the same phenomenon, the expected behaviour on the basis of any of them is the same in a wide range of circumstances: they differ mainly in the nature of the force they postulate. But, clearly, since various hypotheses have been proposed, bases for preference must exist also.

As we have seen, the electrostatic hypothesis postulates a force of repulsion between centromeres which is greater than the generalised body repulsion. Some, however, have doubted the existence of centric repulsion. In support of its existence DARLINGTON and DARK (1932) cite "the departure from circular form of the attachment (i. e., centric) loop and the exceptional state of tension seen in the chromosomes between the spindle attachment and the first (proximal) chiasma." Others, however, have pointed out, quite correctly, that this departure is not usually seen until after the nuclear membrane has broken down and the centromeres are interacting with the materials of the spindle (SWANSON 1942). But the movement apart of homologues after pachytene usually proceeds from the centromere although the nuclear membrane is intact at this time (see Fig. 7, page 6). This has been observed many times (COOPER 1941, HUGHES-SCHRADER 1943) even in material treated with colchicine where spindles are inhibited (JOHN and LEWIS 1957) and it indicates a force localised in the region of the centromere.

It must be admitted that to designate it "repulsion" is nothing more than to describe the effect and not the nature of the force. Thus, if coiling too proceeds from the centromere, which itself remains uncoiled, then a bending of the chromosomes could occur between the more coiled and the less coiled regions, while the centromere serves as a hinge just as it does in the V- or X-shaped chromosomes which are frequently observed following the colchicine treatment of mitosis (see Östergren 1943). Here too greater contraction appears to be involved. What is more, a gradient of coiling from the centromere would also dispose of the problem of the initial proximal movement of chiasmata which is to be expected under certain circumstances on the basis of Swanson's hypothesis.

It will be recalled, however, that claims have been made for the role of chromosome ends in the initiation of coiling in *Trillium*. But this type does not show terminalisation. Indeed there may be even be a slight movement towards the centromere in *Trillium* (see Swanson 1942). Further, the procession of coiling would not accomodate Mather's (1940) contention that when terminalisation is incomplete only the most distal chiasmata are involved in the movement.

While the centromere is generally regarded as the initial site of homologue separation at early diplotene, other patterns of parting have been described. Thus, terminal initiation of withdrawal has been claimed in *Sorghum* where, it appears, the centric regions are the last to separate (Magoon and Shambulin-gappa 1963), while less defined patterns have been recorded in *Pleurozium* (Vaarama 1954) and *Cypris* (Dietz 1958). Indeed, on the basis of observations on the prophase separation of homologues in tomato, Moens (1964) has offered a re-interpretation of the early meiotic stages. Using the position of the pollen mother cell in relation to the longitudinal gradient of the pollen-sac as an indication of age, he concluded that the centromeres do not separate until the diffuse stage which precedes diplotene. Telomeres also may be late in separating while other regions move apart at an earlier, pre-diffuse stage which Moens calls schizonema. He argues that others may have mis-interpreted this stage as a zygotene with incomplete pairing which was initiated pro-centrically and/or near the ends.

The second force postulated by the electrostatic theory certainly exists though it is not always operative. It is, however, virtually impossible to determine its magnitude and therefore one cannot know how effective the force might be in relation to terminalisation. Sister-chromatids in colchicined "anaphases" do fall apart and come to lie parallel to one another at a distance (Fig. 126, page 132). But the distance is not large, though paired threads may be expected to repel one another more strongly. What is more, in those organisms which have an achiasmate meiosis but in which bivalents are formed, there is a prolonged parallel pairing of homologues which are peeled-off one another from the centromere. On the electrostatic hypothesis one must conclude that the surface charges are reduced or neutralised in such cases as they would be in cases of "stickiness" and somatic pairing. The prolonged parallel pairing in achiasmate types does not offer as much difficulty to the other hypotheses but there is strong presumptive evidence that it obtains in some chiasmate systems as well (see John and Lewis 1957, 1960).

All three hyptheses must offer some explanation for the variation between species in respect of terminalisation coefficients. The electrostatic hypothesis, as we have pointed out, invokes variation in the strength of centric repulsion. Swanson, however, mainly on the basis of the higher coefficients which obtain in types with smaller chromosomes, has related to this to the degree of coiling. Since coiling is universal and terminalisation may be virtually absent, it is necessary to postulate a threshold value so that no terminal movement takes place when the degree of coiling falls below this threshold. MATHER (1940) too has stated that terminalisation is virtually an all-or-nothing reaction.

In support of his hypothesis SWANSON (1942) quotes experimental data obtained by subjecting *Tradescantia* to various temperature shocks (SWANSON 1941). He found that heat treatment normally caused a greater contraction of the chromosomes but in some cells "long chromosomes" resulted from treatment at 40° C for one day. These chromosomes had as many as 32 (major) internal coils compared with about 8 per chromosome in the control material. In some cases, the long chromosomes had a chiasma frequency which though higher (11.66 versus 11.51 per cell), was not significantly different from that of the control. But their terminalisation coefficient was only 0.27 compared with the control value of 0.64. Comparisons of this sort are equivocal, however, unless one can show that the initial distribution of the chiasmata has not been affected also. Comparisons involving cells with different chiasma frequencies are even less cogent. Thus, where a chromosome arm has only one chiasma it can be terminal to give a coefficient of one. But there cannot be more than one terminal chiasma per arm whatever the number of chiasmata in it. This is, of course, obvious but sufficient attention has not been paid to it. Perhaps a better comparison would be one based on the extent to which the maximum coefficient is realised. Thus, in his comparisons of the degree of terminalisation in different inbred lines of rye, REES (1955) expressed terminalisation as the number of terminal chiasmata over the number of paired arms. In the case above studied by SWANSON, the cells with long chromosomes had a higher frequency of univalents than the controls (0.55 versus 0.33 per cell) so that the higher number of chiasmata were shared by fewer bivalents.

We see, then, that some evidence, inferential or observational, can be adduced in support of the various suggestions. All in all we favour a mechanical rather than an electrostatic explanation. It must be admitted, however, that unless some sequential process like the processional coiling mentioned above is invoked, neither of the mechanical hypotheses proposed so far can account for terminalisation in rod bivalents because differential stress can be developed only if there is more than one point of exchange. It is generally agreed that terminalisation is less extensive in rods but it does occur. The difference between the intensity of centric and body repulsions proposed by the electrostatic hypothesis obviates the difficulty. But the extreme departure from the circular form of the centric loop at metaphase and the considerable attenuation which can exist between the centromere and its nearest chiasma provide clear evidence that the force(s) operative in the centric loop is (are) greater at metaphase-I than it (they) is (are) at earlier stages. Yet, as the attenuation itself shows, chiasmata are very reluctant to move at this time. Perhaps terminalisation is impeded by the joint coiling of the associated chromatids, a factor which does not obtain at diplotene. Further, terminalisation appears to be the rule in types with non-localised centric activity (see page 143).

Finally, two points which are related to one another. First, it was at one time believed that terminalisation could result in chiasmata slipping-off the ends of chromosomes. Such a claim was made by MOFFET (1932) for species of *Ane-*

more. But this is not now held. Second, two opinions have been expressed regarding the chromatid relationships which can obtain at a terminal chiasma. On DARLINGTON's view a terminal chiasma represents a terminalised chiasma which formed interstitially; and its maintenance depends on a special "terminal affinity" which develops after the lapse of attraction between homologues. Thus, the complete terminalisation of two reciprocal (2-strand) or complementary (4-strand) chiasmata to the same end would give a "reductional" relationship at the terminal association. In other words, sister segments would be associated on each side of the terminal association so that the end-to-end connection would be between non-sister chromatids. WHITE (1959), on the other hand, following a suggestion by CALLAN (1949), has tentatively suggested that there is always an "equational" relationship at a terminal chiasma; non-sister segments are associated on both sides of the terminal association so that sister chromatids are terminally connected across the chiasma.

On this view, terminal association, is the result of a delay in the reproduction of the telomere or else a delay in the individualisation and separation of the products of its reproduction. In this event the "fusion" of 2- and 4-strand double chiasmata would not result in a terminal association but a cancellation. WHITE figures a single cell which he interprets in this way. In connection with "accidents" of this kind he points out that:

(i) Terminalisation is usually greater in types having only one chiasma per arm, and

(ii) In types with more than one chiasma per arm, terminalisation should be absent, negligible or confined to either the proximal or the distal chiasma.

IV. Chromosome Movement and Segregation

So far we have discussed only those mechanical changes which affect the morphology of the chromosomes. Here we want briefly to indicate the bodily movements they undergo and which affect their distribution.

1. Random Distribution

The random behaviour of the chromosomes at anaphase of meiosis depend on the movements which the chromosomes undergo on the spindle prior to metaphase. These are of three kinds. First, there is the orientation of the chromosomes relative to their partners and the poles. Second, the pre-metaphase stretching and the subsequent re-approach of co-oriented centromeres. Third, the congression movement which determines their distribution on to the metaphase plate. Only when these movements are complete do the meiotic chromosomes undergo anaphase separation. The hypotheses that have been advanced to account for these movements are legion. They embrace almost every conceivable force including traction fibres, electrostatic forces, tactoid and sol-gel transformations and jet-propulsion. But as chromosome movements are the subject of a separate monograph in this series (DIETZ and WENT 1966) we will not discuss them. Some of them are undoubtedly influenced by chromosome size and valency.

This is particularly well seen in groups with pronounced hollow spindles (Fig. 187). Here univalent chromosomes and small chromosomes not uncommonly lie within the centre of the spindle (Fig. 162).

2. Systems of Non-random Distribution

Extensive studies in maize and *Drosophila* have clearly demonstrated that gene segregation behaves in accordance with the principle established by MENDEL. And this, in turn, depends on the fact that the orientation of a bivalent or half-bivalent on the metaphase plate is at random relative to that of the other members of the complement and the poles. In both genera, however, cases are known where a specific chromosome passes to a particular pole and segregation is accordingly preferential.

The detection of non-random segregation is, of course, a difficult task. But since it leads to distorted genetic ratios one can use such ratios to screen possible instances of it. However, a disturbance of gametic ratios may depend not so much on directed chromosome movement as on the failure of gametes carrying particular chromosomes to be formed or else to function normally. Thus, ovule abortion and megaspore competition, to mention but two factors, must be excluded before any aberrant genetic ratio can be ascribed with confidence to preferential chromosome movement.

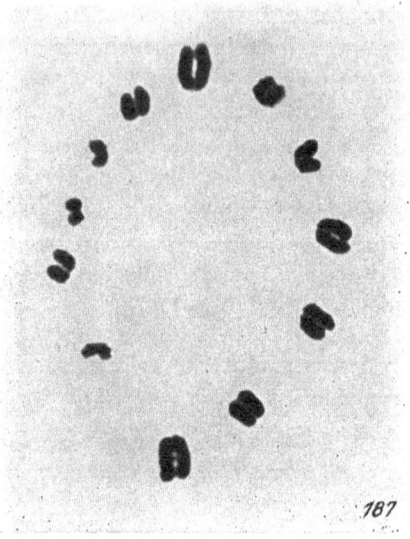

Fig. 187. Polar view of second metaphase in *Locusta migratoria* with the chromosomes arranged on the periphery of a hollow spindle (orcein, ca. × 1500).

Non-random movement itself can be detected cytologically only if the poles and the chromosomes are suitably marked. Thus the poles can be distinguished on the female side in plants for here an ordered linear tetrad of megaspores is formed and only one of these is functional. Even so, the first naturally occurring example of non-random recovery of meiotic products to be described was in the male of *D. obscura* (MORGAN, BRIDGES and STURTEVANT 1925). In this species it was found that males hemizygous for an X-chromosome carrying the "sex ratio" gene (*sr*) produced predominantly X-bearing sperm so that practically all the progeny were female. Meiosis proceeds perfectly normally, however, in females heterozygous or homozygous for sex-ratio X. Subsequently this system was also found in a number of related species including *D. affinis, athabasca, aztecta, pseudoobscura* and *paramelanica*.

In 1936 STURTEVANT and DOBZHANSKY made a cytological study of meiosis in sex-ratio males of *D. pseudoobscura* (Fig. 188). Autosome behaviour was normal but at first division the Y was precociously condensed and well separate from the X. The X itself was usually, though possibly not always, quadripartite evidently having duplicated not once but twice. At first metaphase the Y-chromosome does not move but remains as a condensed body lagging on the spindle. It is excluded from both products of the first division and ultimately forms a small micronucleus and is eliminated. The quadripartite X, on the other hand, divides at both first and second meiotic divisions*. The net result of this sequence is that all four sperm carry an X-chromosome. In *D. pseudoobscura* this anomalous system does show some irregularity since occasionally some of the sperm are Y-bearing, some are XY and others lack both sex-chromosomes (nullo-XY).

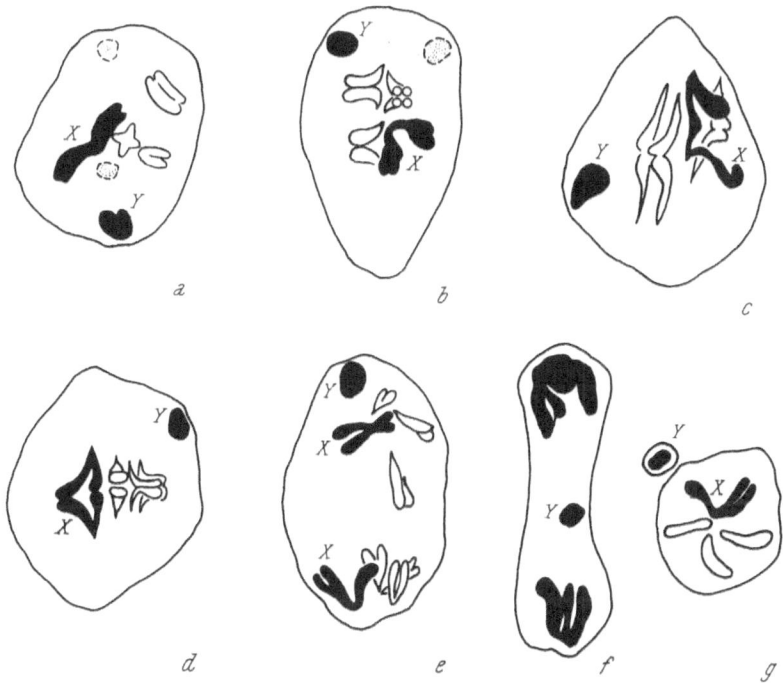

Fig. 188. Behaviour of the sex-chromosomes (shown solid) in sex-ratio *Drosophila*. *a*—Diakinesis; *b–d*—Meta-phase-I; *e*—anaphase-I; *f*—telophase-I; *g*—metaphase-II (STURTEVANT and DOBZHANSKY 1936).

These two latter classes are expected to produce XXY females and XO males. And, in fact, XO males have been found in the progeny of sex ratio males of this species. One final point, in *D. affinis* the sex-ratio system does not operate in the presence of Y-autosome translocations.

More recently SANDLER, HIRAIZUMI and SANDLER (1959) have discovered a quite different system of preferential segregation in a natural population of *D. melanogaster*. Here chromosome-2 carrying the locus "segregation-distorter" (*SD*) is passed from heterozygous male parents to functional sperm with a frequency well in excess of the expected 50%. They termed this behaviour segregation-distortion and showed that it was characterised by a number of interesting features:

(a) It occurs only in heterozygous *SD*-males. It is not operative in male homozygotes and not at all in females, emphasising yet further the distinctive nature of the female meiotic system (see page 24).

_____

* See note 7 Appendix.

(b) For *SD* to be effective the *SD* region must pair with the homologous *SD*+ region. If pairing is disturbed by introducing a marked inversion such as curly (*In [2 LR] Cy*), which interferes with the synapsis of the *SD* region, segregation is normal. On the other hand *In [2 L] Cy*-homozygotes, in which synapsis is normal near *SD*, show a typical *SD*-effect (SANDLER and HIRAIZUMI 1960).

(c) The *SD*-locus is probably located in the centric heterochromatin of the right arm of chromosome-2.

The precise cytological basis for this system is not known. SANDLER *et al.* (1959) have suggested a model but there is no observational evidence to support it and it would be surprising if it proved to be true*.

Under normal circumstances, non-homologous chromosomes assort independently at meiosis. But under abnormal or special conditions exceptions to this principle occur in *Drosophila*. Some of the best known instances of this are found in *D. melanogaster* for here viable zygotes can be obtained after non-disjunction of the sex-chromosomes or the small IVth autosome. The significant observation is that in the presence of extra chromosomes or in the presence of translocations and inversions, non-random assortment takes place. Thus, in experiments using translocations and inversion where specific chromosomes in diploid females were made to behave as univalents, regular preferential segregation could be demonstrated between Y and II (OKSALA 1958), Y and IV (GRELL 1957, 1959b), X and II (FORBES 1960), *Dup.* (X) and IV (GRELL and GRELL 1960) and X and IV (GRELL 1959a). That the non-random distribution of chromosomes in these cases is the consequence of directed segregation can be established by the equality of reciprocal classes. Other phenomena, such as preferential inclusion of certain chromosomes in the egg nucleus or differential mortality of zygotes would not be expected to produce equal reciprocal classes.

All the above mentioned authors suggested that preferential segregation depends on non-homologous pairing leading to non-random assortment. But since the behaviour of the chromosomes during meiosis in the female of *Drosophila* is not amenable to cytological analysis this has not been demonstrated. It is suggested, however, that non-homologous pairing occurs when two or more potentially univalent chromosomes are present at meiosis. Of course, univalents can be obtained not only by decreasing the homologous pairing affinity by means of structural changes but also by introducing extra chromosomes. If non-homologous pairing does determine non-random assortment one might, therefore, expect segregation in triploid females to give some evidence of it. And this, in fact, has been demonstrated (BEADLE 1934, 1935; SANDLER and NOVITSKI 1956; FROST 1961).

RAMEL (1962) has recently carried out an extensive study of non-disjunction in the long second chromosomes of *D. melanogaster*. He found that:

(a) An extra Y-chromosome in the female increased the amount of non-disjunction by about ten-fold.

(b) A significant increase in non-disjunction also occurred with heterozygosity for *InDC × F* in chromosome-3, and

---

* See note 8 Appendix.

(c) According to their effect on non-disjunction the different genotypes studied could be arranged in the following order:

$$\widehat{XX} > \widehat{XX}Y = XXY > XX\widehat{Y} > \text{X-inversions.}$$

If we turn to the factors at the level of the chromosomes themselves it could be supposed that heterochromatin plays an important role in determining non-homologous pairing since, as we have seen (page 155) heterochromatin exhibits less specific pairing behaviour than euchromatin and an unpaired centromere should be even more reactive for non-specific pairing.

OKSALA (1958), on the other hand, has suggested an alternative hypothesis to account for the interchromosomal effects on segregation based on the mechanics of meiotic pairing. This hypothesis rests on the bouquet polarisation assumed by the chromosomes during early prophase-I (see page 6). At this stage the chromosome ends remain at that region of the nuclear membrane adjacent to the centriolar system while the centromeres are raised toward the top of the bouquet. These centromeric regions are heterochromatic and frequently form a chromocentre; consequently there is the possibility of pairing competition between homologous and non-homologous chromosomes at the centromere regions. On the other hand completely heterochromatic chromosomes, like the Y-chromosome, remain at the base of the bouquet.

OKSALA argued that if one of the autosomes in an XXY female contains long inversions then the reverse pairing loop may prevent the centromere of this autosome from rising to the top of the bouquet, so that it takes no part in the pairing competition. But this enhances the opportunity for non-homologous pairing between the centric heterochromatin of this autosome and the Y-chromosome. In the same manner a non-homologous association should be possible between two autosomes each of which contains an inversion.

Ingenious as this mechanism is it has no foundation in observation. Indeed there are clear indications that modified pairing patterns of the type required by OKSALA are not pre-requisites for the occurrence of non-homologous pairing. Thus in RAMEL's study the highest frequency of non-homologous pairing occurred in $\widehat{XX}$ females. In this case the *Cy*-inversion should cause the 2nd chromosome to remain at the base of the bouquet. But, as RAMEL points out, there is no reason why the centromere of the attached-X should not rise to the top of the bouquet with the rest of the complement. This means that the close contact of the H-regions necessary for non-homologous pairing could not occur in this case.

Pairing and subsequent chiasma formation between homologues is followed by their segregation except, of course, when abnormalities arise from other causes. On various occasions however, it has been claimed that crossing-over is not always a pre-requisite for regular disjunction. This is clear in types like male *Drosophila* which have an achiasmate meiosis. But the claim has also been made for types where crossing-over is the rule. And the usual example in this connection is the female of *Drosophila* where it has been claimed that chiasma formation is not necessary for the normal disjunction of the X-chromosome (STURTEVANT and BEADLE 1936, COOPER 1945). As in so many genetical experiments where unverified assumptions are made regarding the position and frequency of chiasma formation there is an inevitable element of subjective interpretation in these claims. This is a situation we have already met in connection with presumed sister strand exchange (see page 101).

Chromosomes which pair may fail to form chiasmata but if their proximity can persist until first metaphase a non-random segregation could still ensue. But it would appear that non-random segregation may extend even to non-homologues. Further the evidence shows that in these cases non-random segregation is between chromosomes which failed to cross-over although they were involved in specific pairing i. e. the absence of crossing-over is not a consequence of non-homologous association.

To account for this R. H. GRELL (1962a and b) has suggested that specific pairing leading to crossing-over between homologues (exchange pairing and exchange) is followed by "distributive pairing." Of course, chromosomes which

Table 82. *Expected and Observed Distribution of Univalents at first Anaphase in 43 Egg Mother Cells of Lumbrillicus lineatus* (2n = 3x = 39). (Data of CHRISTENSEN 1960.)

Distribution	$\frac{19}{20}$	$\frac{18}{21}$	$\frac{17}{22}$	$\frac{16}{23}$	$\frac{15}{24}$	$\frac{14}{25}$	$\frac{13}{26}$	P
Observed	19	10	7	3	2	1	1	0.025 — 0.01
Expected	10.812	9.782	8.001	14.405				

have formed chiasmata remain associated and their mode of segregation is fixed within very narrow limits. But those which have not undergone exchange may now associate and while homology is relevant in this respect, non-homologous chromosomes (heterologues) may be involved. Thus non-random segregation is a consequence of prior "distributive pairing." We cannot discuss the hypothesis in detail but the following points are relevant.

There is indisputable cytological evidence that certain chromosomes which never form chiasmata and which may be far apart at diplotene may, nevertheless, subsequently pair, co-orient and disjoin regularly. The clearest example is that of the m-chromosomes in the Heteroptera where the other autosomes show normal chiasmata (see page 191). But we have seen the same phenomenon in certain non-chiasmate sex-chromosomes systems (see page 183). Further an approximately equal numerical separation of chromosomes may be effected more often than expected in types where there are no chiasmata. Thus, at meiosis in the egg mother cells of the parthenogenetic earthworm *Lumbrillicus lineatus* (2n = 3x = 39) only univalents are found. But two nuclei result from the polar movement of undivided univalents at first anaphase. CHRISTENSEN (1960) found that the distribution of the univalents did not conform with the expectation on the basis of random movement (Table 82). The deviation is due mainly to the excess of 19-20 distributions though an approximately equal separation at this stage does not appear to be crucial in relation to the non-reduction shown by this parthenogenetic triploid.

Let us note, however, that the behaviour of the chromosomes in *Lumbrillicus* differs from that of the m-chromosomes in the Heteroptera. In the former there is no evidence for post-pachytene pairing. Indeed, except for a concentration in the region of the nucleolus, the univalents are "evenly spread throughout the nucleus" and a typical metaphase plate has never been observed. Further there is nothing to suggest qualitative specificity in the segregation.

The behaviour of chromosomes in haploids suggests that unexpected, apparently non-specific—even non-chiasmate—associations can obtain at late prophase (see page 106). In the basic meiotic system, associations other than those afforded by chiasmata are expected to lapse at this time. Perhaps, therefore, those in haploids represent a different category of association which is not established until late prophase and which could be compared with GRELLS distributive pairing. Alternatively, this distributive pairing may be compared with the secondary association which has been observed at metaphase and earlier (see page 162) but one which leads to the co-orientation of the members when univalents are involved. Furthermore, non-specific associations of heterochromatin and somatic pairing are both pronounced in Drosophila. It has been shown, however, that like homologous pairing, distributive pairing also is competitive and preferential but the "specificity" must reside elsewhere than the centromere and all of it cannot reside in the proximal heterochromatin (GRELL, E. H. 1963).

But, whatever the details of the situation in female Drosophila, distributive pairing between non-crossover chromosomes cannot be regarded as a standard part of the meiotic system. Rosa canina, for example, represents the extreme opposite for in this type all the univalents move to the same pole at the first meiotic division of the megaspore mother cell.

Finally, it seems to us that GRELL's so-called hypothesis is little more than a restatement of the observation that chromosomes which have not undergone recombination—even non-homologous chromosomes—may, nevertheless, segregate from each other more often than random behaviour leads one to expect. Particularly interesting is the observation that, unlike other chiasmate associations, attached-X chromosomes which have undergone internal recombination do participate in "distributive pairing" (GRELL 1963). The single centromere of this configuration is thus relevant. In our view, therefore, the observations made by the GRELLS are more interesting than the hypothesis which, as might be expected, was based solely on the results of breeding experiments and not on cytological observation. Moreover the hypothesis leans heavily on the false claims of PRITCHARD and PONTECORVO regarding the mechanism of recombination (see page 217). In view of the variety of meiotic systems compiled by natural selection it is premature to postulate a particular cytological basis for a phenomenon which could have a number of explanations*.

There is one further case in Drosophila worth mentioning. There are indications that where half-bivalents are composed of two chromatids of dissimilar length, the shorter of the two is recovered more frequently in female gametes because it is more frequently included in the egg nucleus. Thus NOVITSKI (1951) found that when a female carries two X-chromosomes of different length and crossing-over occurs between them to produce asymmetrical half-bivalents, the shorter of the two members of such half-bivalents is recovered about twice as often as the longer one. There thus appears to be non-random disjunction at the

---

* See note 9 Appendix.

second meiotic division when two structurally dissimilar chromatids compete for inclusion in the functional egg nucleus. NOVITSKI (quoted in KIKUDOME 1959) is of the opinion that this non-random orientation of heteromorphic half-bivalents is established at first-anaphase and is then maintained through the second division.

Table 83. *Knob Segregation at Megasporogenesis in Zea mays.*
(Data of LONGLEY 1945.)
Expectation for random segregation of knob chromosomes is 50% and $P \geq 0.05$ when $\chi^2_{(1)} < 3.84$.

Chromosome No.	Chromosome Arm Type	Total no. Chromosomes Observed	Total Knob Chromosomes Observed		Chi-Square $(\chi^2_{(1)})$.
			No.	%	
1	S	136	79	58.1	3.56 (Not Sig.)
	L	108	68	63.0	7.26
2	S	59	35	59.3	2.05 (Not Sig.)
	L	93	55	59.1	3.11 (Not Sig.)
3	S	113	72	63.7	8.50
	L	169	116	68.6	23.49
4	L	106	87	82.1	43.62
5	L	77	56	72.7	15.91
6 [1]	L	212	125	59	6.81
	L	212	138	65.1	19.32
8	L	90	57	63.3	6.40
9	L	207	148	71.5	38.27
10	L	142	106	74.6	34.51

[1] The two no. 6 chromosomes studied differed morphologically. One had 2, the other 3 knobs and one of the knobs common to the two differed in size, the larger knob being present in the chromosome with 3 knobs.

Similarly ZIMMERING (1955) reported that in *Drosophila* females heterozygous for an interchange (V 4) between chromosomes 2 and 3, which produces a long interstitial segment, complementary female gametes are not obtained with the expected equal frequencies. More particularly, complementary cross-over chromatids are not recovered with equal frequency whereas non-crossover chromatids are. This is due quite simply to the fact that, following crossing-over in the interstitial segment, asymmetric half-bivalents are formed and at second division these undergo non-random disjunction so that there is a preference for the short chromatid to pass to the egg nucleus.

It is generally accepted that the four products of male meiosis are all functional and there is little support from a study of spermatogenesis to suggest otherwise*. Thus even unbalanced male gametes appear to survive and fertilise. Despite this, experiments in *D. melanogaster* using the interchange *T (1,4) B^s* suggest that the type of behaviour shown in the *V 4* female may also operate in the male. If two unequal homologues are present in a male the shorter may appear preferentially in the progeny at a frequency so high that mortality can be excluded as the cause (Novitski and Sandler 1957). Novitski and Sandler

Table 84. *Frequency of transmission to first Back-cross Generation of Knobbed Chromosomes Other than K 10 from Heterozygotes which Are Segregating for K 10/k 10 also or which Are Homozygous k 10/k 10.*
(Data of Longley 1945.)
Expectation for Random Segregation of Knob Chromosomes is 50% and $P \geq 0.05$ when $\chi^2_{(1)} < 3.84$.

Segregating System	Total no. Chromosomes Observed	Total Knobbed Chromosomes Observed		Chi-square $(\chi^2_{[1]})$
		No.	%	
1. K 10/k 10 parent:				
a) Offspring from K 10 megaspore	918	616	67.10	107.4
b) Offspring from k 10 megaspore	664	420	63.25	46.65
c) Offspring from K 10 and k 10 pollen	115	62	53.91	0.70 (Not Sig.)
2. k 10/k 10 parent:				
a) Offspring from k 10 megaspore	195	122	62.56	12.31

also point out that in Snell's translocation in the mouse the frequency of sperm producing viable offspring is about 60%. Yet, as Ford, Carter and Hamerton (1955) have shown, rings-of-four occur in only 8–30% of the spermatocytes: the remaining cells show only bivalents. This should produce no more than 40 or at the most 50% good sperm.

In maize each of the ten chromosomes may possess one or more heterochromatic sections referred to as knobs though many strains have few or none of these. By crossing different strains it is possible to produce progeny heterozygous for a particular knob. As early as 1942 Rhoades had demonstrated that in maize plants heterozygous for a knobbed tenth chromosome (*K 10*) this chromosome underwent preferential segregation during female meiosis. This work was extended by Longley (1945) who showed that *K 10* induces the preferential segregation of other chromosomes of the complement too, provided that one homologue is knobbed and the other knobless. Longley's technique was to backcross F$_1$ plants heterozygous for a particular knob combination to a strain without knobs and then to determine the number of knobbed chromosomes in the backcross progeny. Significant deviations in the mean number of knobs

---

* See note 10 Appendix.

carried by the backcross population from the mean of the two parents then indicates non-random segregation on the female side in the F$_1$ plants.

LONGLEY's results are summarised in Tables 83 and 84. It is important, however, to emphasise that heterozygous knobs on chromosomes other than chromosome-10 do not in themselves determine preferential segregation in the absence of *K 10* (KIKUDOME 1959, contra LONGLEY 1945).

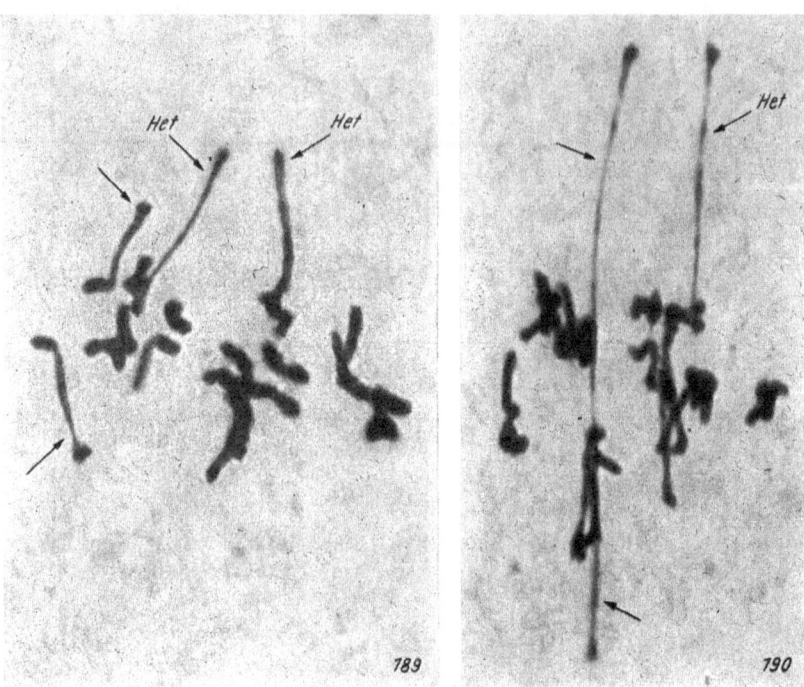

Figs. 189 and 190. Second meta-anaphase in maize to show heteromorphism and homomorphism for neocentric activity (carmine-phase contrast, ca. ×1500).

This effect in maize depends on four principle factors (RHOADES 1952, 1955):

(a) Maize knobs show neo-centric activity in the presence of *K 10* (see page 146 and Figs. 189 and 190).

(b) Crossing-over in the knobbed arm produces heteromorphic half bivalents having one knobbed and one knobless chromatid. At first anaphase in those cells where the knob shows neo-centric activity the knobbed chromatid will be nearer to the pole than its knobless homologue. Presumably, it retains this positional advantage throughout interphase and through the second division so that:

(c) At second anaphase the knobbed chromatids will pass preferentially to the outer products of the linear tetrad, and

(d) There is a physiological gradient in the tetrad so that only the megaspore at the chalazal end is functional. Consequently this will tend to carry a knobbed chromatid.

A comparison of the systems of preferential segregation in *Drosophila* and maize indicates that the processes producing such segregation vary considerably in different organisms. Thus, whereas in *Drosophila* segregation from hetero-morphic systems always favours the shorter of two homologues the reverse is true in maize. Here it is the longer, knobbed chromosome arms which show preferential movement. In both species, however, the production of heteromorphic dyads by crossing-over is a pre-requisite for preferential segregation.

There are indications that preferential segregation may be a more widespread phenomenon. Thus some of the best analysed instances of segregation bias come from the study of the transmission of B-chromosomes (see page 201). Again in the *LD/O* monosomic studied by FRANKEL (see page 72) functional gametes carrying a chromatid from crossing-over in *LD* constitute about $\frac{1}{4}$ on the male

Table 85. *Transmission of Cross-over and Non-crossover Chromatids in LD/O Mono-somics of Triticum aestivum.*
(Data of FRANKEL 1949.)

Cross	42-chromosome Progeny			Ratio		P
	Non C. O.	C. O		Non C. O.	C. O	
	LD/N	SD/N	N/N			
LD/O ♀ × N/N ♂	9	1	9	0.9 / ca 1 : 1	1	> 0.01
N/N ♀ × LD/O ♂	13	2	2	3.3 / ca 3 : 1	1	

side but $\frac{1}{2}$ on the female. A cross-over chromatid thus has a greater chance of reaching the functional megaspore cell than one carrying the duplication. In other words there is a preferential inclusion of cross-over chromatids in the ovules of the monosomic (Table 85). Finally MICHIE (1953) and WALLACE (1953) have demonstrated that, in crosses of different inbred strains of different species of mice, unlinked markers of similar ancestry assort together at meiosis. They have termed such non-random behaviour quasi-linkage and have explained it on the assumption that centromeres of similar ancestry tend to travel to the same pole. This, in turn, will lead to markers situated near the centromeres segregating as though they were linked. Unfortunately this assumed relationship of the loci showing quasi-linkage to specific centromeres remains to be substan-tiated. The situation is complicated by the plain fact that the mapping of mouse chromosomes is still at a fairly crude level; indeed it has not always been recognised that mouse chromosomes are all acrocentric (see WALLACE 1958). While one cannot doubt the non-random nature of segregation in this case there is con-siderable doubt that this is in fact dependent on non-random orientation of homologous centromeres. Indeed this may explain the fact that, in some cases, results interpreted as due to centromere affinity are in blatant contradiction to interference data (PARSONS 1959).

Despite this, WALLACE (1960a and b) has pointed to possible cases of so-called centromere affinity in tomatoes and in cotton. In the latter case the evidence in support of affinity is based on the fact that F_2 and F_3 progeny of F_1 hybrids between *Gossypium arboreum* × *G. herbaceum* and *G. hirsutum* × *G. barbadense*, like those of many equivalent hybrids, show a preponderance of parental-like phenotypes. Unfortunately these hybrids do not combine species that are chromosomally distinguishable so that a direct test for affinity is impossible in this case. Recently, however, PHILLIPS (1964) has carried out a cytological analysis of F_1 hybrids between *G. sturtii* (C-genome) × *aridium* (D-genome) and *G. sturtii* (C-genome) × *lobatum* (D-genome). In these species-pairs the chromosomes are distinguishable in size and these differences are maintained in the hybrids. This analysis failed to show any evidence for centromeric or chromosomal affinity: the members of the C and D genomes orient at random relative to one another even though PHILLIPS' hybrids involved species much more distantly related than are *arboreum*, *herbaceum* and *barbadense*. Indeed no quasi-linkage could be demonstrated genetically in backcrosses of the F_1 hybrids *G. thurberi* (D-genome) × *armourianum* (D-genome) to *G. hirsutum*.

V. The Control of Meiosis

Because of the limited interests and capacities of most investigators, partly for convenience and, perhaps too, as a consequence of the scientist's appetite for pigeon-holing, it has been customary to distinguish heredity from development DARWIN's attitude that heredity is only a kind of development is, however, a more meaningful one so long as we make and maintain the all-important distinction between the phenotype and the genotype. And a part of development, division, is only a kind of heredity—cell heredity.

A distinction, similar in kind to that underlying heredity and development, has been made between the "mechanically active" and the "resting" nucleus. And this distinction has created, or been created by, a difference in attitude and approach. On the one hand, as we have already seen, students of the dividing cell have attempted to analyse the causes and control of mitosis and meiosis on a purely mechanical and physical level. Those whose interest lies with the mechanically inactive nucleus, on the other hand, have, not unnaturally, adopted a physiological attitude, and chemical causes and control have been their concern.

It is now clear, however, that while physical forces undoubtedly play an immediate part in determining changes during division, the only kind of underlying control which can exist is a chemical one. Development is currently considered in terms of the specification of proteins by the genetic DNA through the mediation of various species of RNA. To-date this approach, while rewarding, has not taken us far from the molecule and to the supra-molecular, microscopic or sub-microscopic organelle. But, clearly, it is the right approach. And we must conclude that although the changes which determine, constitute or are caused by division are transient, cyclical and, to some extent, reversible, they nevertheless represent morphogenetic change.

If this is accepted we must conclude that division and its nature are determined and controlled by mechanisms similar in kind to those which operate in the resting, metabolically-active nucleus. Before dealing with the few investigations which have adopted this approach to the problem of cell division and its control, let us consider the various other lines of evidence which have a bearing on the causal analysis of cell division.

1. The Analysis of Meiotic Variation

The first step in elucidating the meiotic mechanism must be to observe and describe, in as much detail as possible, the mechanical sequence one is attepting to analyse. Of course, a complete description of any phenomenon constitutes a complete explanation of it. But to describe *what* happens at one level does not explain *how* it happens at another. The study of any sequence does, however, provide a basis for inference and suggests causal relationships. An obvious and valid one, for example, is that pairing is a necessary prerequisite for chiasma formation.

The meiotic sequence, though more variable than mitosis, is much the same in a very wide range of organisms especially in the "higher" plants and animals. But, even so, considerable variation does exist and extreme departures from the standard sequence is the normal course of events in many species. It is, then, not one, but many meiotic sequences we have to explain. And, just as a study of one sequence suggests causal sequences, so does a comparison of different sequences—though correlations are not always indicative of a direct causal connection.

The variations on the meiotic theme can be classified as to their kind and with regard to the situation in which they occur. Variations may be observed:

a) Between Species

These variations must be regarded as adaptational and selected mutually with the breeding system (DARLINGTON 1963). They must, of course, depend on genotypic differences for only then can they respond to selection. In most cases, however, this argument is presumptive, for genotypic control of the differences can be demonstrated only if crossing is successful.

b) Between Sexes within Species

In general these too must be the subject of mutual selection and depend on genotypic differences. However, there are situations in which males and females are homogametic at the chromosome level so that, in this respect, all the offspring are alike. This condition is observed in such types as *Sciara*, *Miastor* and members of the *Orthocladiinae* in all of which there is differential chromosome elimination in the soma (see WHITE 1954, LEWIS and JOHN 1963a). In the first two, this also is differential between the sexes, but in the orthoclads the soma of males and females has the same constitution.

c) Between Individuals within "Sexes" and Species

These differences may depend on genotypic or environmental differences or both. They may arise "spontaneously" or owing to the manipulations of the investigator at the genetic or environmental level.

d) Between Cells within Individuals

In this category come any differences between divisions on the male and female sides in hermaphrodite individuals. It includes also the variations observed between mother cells of the same type. They may be small and quantitative, like those commonly found in relation to chiasma formation and distribution.

These are unavoidable. But they may be large and qualitative. These are avoidable. Meristic variation of this kind may be symptomatic of genotypic unbalance
or it may reflect environmental upsets. But, although they are avoidable to some
extent, they may not be avoided. Indeed, they may occur as "controlled anomalies" characteristic of the species concerned.

For the most part, intra-individual differences, cannot depend on genotypic
differences. And, where external environmental effects can be ruled out, they,
like any other aspect of differentiation, must be attributed to the differential
action of the genotype or genotypically-determined products. It is true that
genotypic changes do occur during development. Indeed they may even be

Table 86. *Variation between "Generations" (≡ Individuals) of the same Initial Nuclear
Constitution in Miastor.*

All Types initially 12 S + 36 E	Females			Male
	1. Zygotic (from 3 × 4)	2. Parthenogenetic (from 1 or 2)	3. Sexual (from 2)	4. Sexual (from 2)
Somatic elimination	None	36 E eliminated (Soma 12 S)		36 E + 6 S eliminated (Soma 6 S)
Meiosis	Single equational division		Differential first division	
Functional products	Eggs 12 S + 36 E		Eggs 6 S + 36 E	Sperm 6 S

extensive, involving parts of chromosomes (*Ascaris*), whole chromosomes (*Sciara*),
and many of them (*Miastor*), or whole chromosome sets (polyteny, endopoly-
ploidy). But the available evidence suggests that the developmental pattern
determined by the "parental" nucleus has an over-riding influence so that the
genotype of the individual cell or nucleus is less important.

In this category too must be included differences between certain "individuals".
For example, the zygotes of *Miastor* develop into female offspring which give
rise parthenogenetically to female offspring which are both parthenogenetic and
paedogenetic. Although these have the same nuclear genotype as their mothers,
they show somatic elimination of the so-called E-chromosomes set while their
mothers do not. These female larvae can give rise to further generations of
females like themselves in which the oocytes divide by a single equational division
thus avoiding reduction. But eventually, under unspecified conditions, they give
rise, again by parthenogenesis, to sexual larvae which may be male or female.
The course of meiosis differs in sexual males and females and both, of course,
differ in this respect from the parthenogenetic females (Table 86).

This variation between "generations" in a species showing an alternation of
generations, as the zoologist understands the term, is related to the difficulty
of defining an individual in a type with clonal reproduction. And the difficulty
extends to some extent to the habitual inbreeder. What are developmentally
two or more individuals can be regarded as one genetically.

e) Between Nuclei within Cells

The basic unit of organisation, of maintenance and reproduction in plants and animals is the uninucleate cell. But there are cells, tissues and individuals in which more, and often many more, than one nucleus share a common cytoplasm. This condition occurs naturally and it can be induced exceptionally. It may be a temporary condition or a permanent one.

As a rule, nuclei in a common cytoplasm are similar and often synchronised in their behaviour, at least locally. But concordance is not invariable and discordant situations are known both as a rule in nature and as exceptions in experiment.

f) Between Chromosomes within Nuclei

No event in either the resting or dividing nucleus is instantaneous. But, clearly, if orderly reproduction is to be achieved, a measure of synchrony must exist between the participants. Some chromosomes are, however, in a special position and special provisions are made for them. They include the sex chromosomes, the supernumerary or B-chromosomes and those enigmatic additions to the germ lines of types, like those mentioned above, which show somatic elimination. The behaviour of chromosomes of this type though, in a sense, anomalous is typical of the type concerned. It can, therefore, be compared with the "controlled anomalies" seen between cells in types like *Loxa* and *Scutigera* (see page 268). But the differential behaviour of the chromosomes within the nucleus may, like that between cells, be symptomatic of environmental or genotypic disharmony.

g) Between Parts of Chromosomes

In some aspects of its mechanical behaviour the chromosome behaves as a unit. This is true, for example, with regard to its mitotic disjunction in most cases, though such an effect is simply a reflection of its organic continuity. Parts of chromosomes show their greatest autonomy in the metabolic nucleus but differential behaviour is seen also in relation to such things as pairing and coiling. Much of what was said above in relation to e) and f) applies equally here and it makes one realise that as in metabolism, so in division, the cell, the whole nucleus or even the chromosome is not a unit of activity and response in all things.

These then are the situations in which comparisons of meiotic (or, for that matter, mitotic) behaviour can be made. For convenience we will consider some of the relevant investigations in three groups:

(a) Cytological and chemical descriptions of meiotic sequences.

(b) The consequences of genetic or environmental change on the cytological features of meiosis.

(c) The consequences of genetic and environmental change on the chemical features of meiosis.

Clearly, we cannot attempt a comprehensive consideration of these topics here. Consequently, as we have done throughout this monograph we will select the investigations which seem to us to be the most significant. The cataloguing of information, while tedious, can be informative. But, like many people, while welcoming classified information in the works of others, we are disinclined to compile it ourselves. With the reader in mind, however, we will attempt to overcome our reluctance and present, in tabular form, information we do not discuss in detail in the text. And, as much of the relevant cytological information has been presented and discussed earlier in this review, in this final discussion we will be mainly concerned with chemical events and with natural (normal or abnormal) and induced meiotic blocks.

2. The Characterisation of Meiosis

At one time it was, indeed in some circles it still is, fashionable to talk about the potential immortality of *Amoeba* and to dwell on the advent of mortality with the separation of soma and germ lines. But, in a sense, all cells die for they have a limited capacity for growth and, like individuals and species, they have a limited existence in time.

All cells thus die but they may differ in the manner of their death. In the first place, the molecular and, more immediately, the supra-molecular organisation which exhibits the properties we associate with, and by which we recognise and define, life may breakdown. This is cell death as usually understood—death in which the cells, like Saul and Jonathan, were not divided.

But the nuclear/cytoplasmic relationship which constitutes a cell is a temporary one in another sense and it is brought to an end by division. Thus, by dividing, mitotically or meiotically, a cell terminates its own life.

There is a third way of bringing the life of a cell to an end. It can enter into a suicide pact with another cell so that the lives of both come to an end with their fusion.

Paradoxically, suicidal death by fission or fusion is the only way of creating a new living cell, a new nuclear/cytoplasmic relationship. And, unless there are cells which undergo autonomous "cytoplasmectomy," it is the only way also by which parts of the original cell or cells can realise their potential immortality and by which the limits on growth can be extended. Eventually, therefore, growth becomes dependent on division. But the dependence is mutual. Without growth, and sufficient growth, between divisions, cells would become smaller as they divided—so small as to be incapable of survival. In fact this happens in half the mitotic products in certain diatoms.

Of course there are mitotic systems in which a partial divorce is effected, as in coenobia and endopolyploid cells, so that nuclear division or even chromosome reproduction on its own can replace cell division to some extent. But the general correlation between growth and division is clear and it has led to the postulation of a causal connection. Thus as early as 1908 HERTWIG had suggested that as the cytoplasm increased in size, the "Kernplasmarelation" approached the "Wirkungssphäre" of the nucleus and this led to a precipitation of division. It is now clear from studies of mitotic cycles that the relations are not so simple. And there is some evidence from meiotic systems too.

In their study of flowering plants, cytologists searching for buds in meiosis frequently use the length of the bud or anther as a guide of stage. The relations between bud length and various other aspects of development have been studied in detail in *Lilium longiflorum* var. "Floridii" by ERICKSON (1948). He found

that the increase in bud length up to the time of bud opening conformed to
BLACKMAN's compound interest equation:

$$L_t = L_0 e^{rt} \text{ or } \log L_t = \log L_0 + rt$$

where e is the base of natural logarithms,
    r is the growth rate (constant)
    t is the time in days
    L_0 is the initial bud length, and
    L_t is the expected bud length after time t.

Table 87. *Correlations between Bud Length and Division Stage in Lilium longiflorum Var.*
"Floridii" (Data of ERICKSON 1948).

Bud Length (mm.)	Division Stage
3.5 ⎫ 8.0 ⎭	Two peaks (0.08 and 0.15) of pre-meiotic archesporial mitoses in the anther
10.0	No mitotic activity in the anther
13.6	Leptotene (PMC)
22.5	Diakinesis/Metaphase-I (PMC)
24.0	Microspore tetrad
57.0	First pollen grain mitosis
96.0	Meiosis in the megaspore mother cell
128.0	Post-meiotic mitosis in embryo-sac

The growth rate (r) was found to vary somewhat under greenhouse conditions
owing, mainly, to fluctuations in the amount of illumination. ERICKSON also
measured changes in anther length, fresh weight and dry weight and he made
limited observations on the meiotic stages and post meiotic division. His conclusion was that "the logarithm of bud length is undoubtedly a more precise
measure of the age of the bud than any index which could be derived from
chronological data." This fact has been exploited in a number of investigations
(see pages 218 and 265). But one index takes precedence over length and that
is the stage of division. In this respect the meiotic cells are highly synchronised
though pollen mother cells near the apex of the another are about 8–12 hours
ahead of those at the base. This difference, however, is small in relation to the
life of the mother cell and even to the duration of meiosis and for cytological
purposes one can overcome the problems it introduces by consistently selecting
a particular region of the anther for study. Pollen grains are less well synchronised
and, although there are waves of pre-meiotic mitosis, synchrony of division before
meiosis is not good. Results like those summarised in Table 87 have been confirmed and extended by others (TAYLOR and McMASTER 1954, STERN 1960); but
these workers used a different variety, viz. *L. longiflorum* var. "Croft".

Despite the fact that bud length is a very good indication of age, even when
plants are grown under greenhouse conditions with temperature variations of
about 10–20° F and within a wide range of growth rates, length and stage of the
meiotic cycle can be put out of phase. Thus STERN (1960) found that in plants

of *L. longiflorum* var. "Croft" grown at the comparatively high temperature of 85–95° F the first pollen grain mitosis occurred at a bud length of 54–55 mm while under normal growing conditions it takes place in buds 62–63 mm long.

A consideration of the relationship between growth and division (including preparation for it) which are, in a sense, alternatives for the cell at any one time, are important not only in relation to the onset of division but its nature too. More specifically DARLINGTON's precocity theory (1931) proposes that meiosis differs essentially from mitosis in that the division begins earlier in relation to nuclear development. In keeping with this hypothesis is the common observation that pollen mother cells entering division later than their neighbours in the same

Table 88. *Volume Relations (µ³) in the Three Types of Spermatocyte in Arvelius albopunctatus at the "Confused" Stage which Occurs between Diplotene and Diakinesis.* (Data of SCHRADER and LEUCHTENBERGER 1950.) The Six Lobes of the Testis Are Numbered from the Side of Exit of the Sperm Duct (see Fig. 192). The Diameter of the Lobes and the Number of Cells they Contain (in Brackets, column 1) Are Roughly Inversely Proportional to the Size of the Sperm Mother Cells.

Testis Lobe	Nucleus	Nucleolus	Cytoplasm	Cell
3 and 5 (ca. 350 – 500)	1596 ± 108	31	8,278 ± 801	9,874 ± 967
1, 2 and 6	408 ± 43	6.7	2,545 ± 245	2,953 ± 290
4 (> 15,000)	230 ± 32	...	1,146 ± 126	1,376 ± 150

pollen sac are often larger and tend towards a more mitotic division. This tendency may be slight and be revealed only in a somewhat lower chiasma frequency as in rye (REES and NAYLOR 1960) or it may be revealed by more extreme symptoms like asynapsis. Such conditions have been observed in *Allium* (DARLINGTON and HAQUE 1955), *Chrysanthemum* (DOWRICK 1953), *Scilla* (REES 1952) and we have often found it in plants with bulbs or corms. HENDERSON (1962) has shown that in *Schistocerca gregaria* a heat shock of 40° C leads to a reduction in chiasma frequency and to asynapsis, events which he has explained in terms of reduced meiotic precocity. Indeed it also appears possible by heat treatment to obtain premature meiotic induction. This leads to the production of smaller meiotic cells in which the nuclei contain much smaller chromosomes which again show modified chiasma properties.

There are, of course, various ways of measuring age and hence precocity. Clock-time is not a reliable index but, as intimated above, size is a better one. Perhaps, therefore, quantitative measurements of dry weight or protein content are even better. In fact some information is available at this level. In many pentatomid bugs there is a condition called polymegaly which is in some respects comparable to the phenotypic heterospory of plants (BOWEN 1922, SCHRADER and LEUCHTENBERGER 1950). This condition is characterised by the presence, in well defined lobes of the testis, of sperm mother cells of different sizes. In all known cases, two lobes have spermatocytes which are larger than usual while the remaining lobes have mother cells of normal size or else they include one lobe in which the cells are distinctly smaller. The three types are found in *Arvelius albopunctatus* where the differences are very pronounced (Table 88). These differences reside not only in the cells as a whole but in the nucleus, the nucleolus and the cytoplasm separately. Spermatogonial nuclei are all the same size and

Table 89. *The Estimations of Various Components (in Arbitary Units) in Large (a), Medium-sized (b) and Small (c) Sperm Mother Cells in Arvelius albopunctatus.*
(Compiled from SCHRADER and LEUCHTENBERGER 1950.)

The determinations were made photometrically following feulgen (DNA) and Azure A (RNA) staining for the nucleic acids. Protein was determined using Millon's reagent and a Fast Green method. This, however, was different from the one employed by ANSLEY (see p. 268). Following fixation in San Felice, the material was stained in a 0.1% aqueous solution of Fast Green at pH 1.2 for 1 hr. followed by differentiation in absolute alcohol for 1 Day.

Component		Nuclei	Nucleoli	Cytoplasm
1. DNA	a	6.0 ± 0.24	...	...
	b	5.6 ± 0.53	...	...
	c	5.3 ± 0.42	...	...
2. RNA	a	...	0.92 ± 0.09	93.5 ± 6.0
	b	...	0.38 ± 0.013	30.2 ± 3.2
3. Protein	a	26.8 ± 1.58	1.26 ± 0.08	124 ± 12.4
(Millon)	b	6.6 ± 0.57	0.426 ± 0.04	33 ± 5.8
4. Protein	a	...	11.6 ± 1.1	790 ± 64
(Fast Green)	b	...	5.0 ± 0.5	226 ± 19.2

differences do not begin to appear until after leptotene. The three cell types reach their maximum size during the confused or diffuse stage which occurs between diplotene and diakinesis and it is at this time too that the size differentials are at their greatest. The nucleoli also increase rapidly in size after pairing and reach their maximum dimensions in the confused stage. Meiosis follows a normal course in all three cell types and the chromosomes at metaphase do not reflect the size differences between the nuclei. When they are first formed, all spermatid nuclei also are the same size though the cytoplasmic differences are still evident. There is however a differential growth of spermatid nuclei so that the size differences are later re-established. Even so all sperm-head nuclei appear to be the same size.

SCHRADER and LEUCHTENBERGER (1950) did not follow the individual patterns of synthesis but the three types of sperm mother cell were compared at the confused stage. No differences in the amounts of DNA were found. But the amounts of protein and RNA varied in a corresponding way to size and the correspondence in some cases was close (Table 89). There was also parallel variation in the golgi apparatus, the mitochondria and the size of the spindles.

With regard to the precocity theory there are three things to say about this study in *Arvelius*. First the nuclear size differences, at least, did not develop until *after* entry into division. Second since the temporal sequence was not investigated there can be little measure of the degree of concordance between the various components. But third, and most important, is the fact that the situation in *Arvelius* is a controlled anomaly. The findings are thus no more surprising than the corresponding ones which doubtless obtain between micro- and mega-spores in a type like *Selaginella* or between sperm and egg mother cells in animals. The anomalies within anthers in *Allium*, *Scilla* etc., on the other hand, are not controlled; they are symptoms of upsets in the genotype or the environment.

It is clear, then that here, as always, we must distinguish between those conditions which have been subjected to selection and those which are spontaneous (see page 117). This means, for example, that comparisons between related apomicts and sexual species with regard to megaspore mother cell size and type of division are of limited meaning whether the findings are quoted for (BERGMAN 1941) or against (GUSTAFSON 1945) the precocity theory.

3. Provision for Division

The commitment to divide leads to changes in the cell some time before there is any visible change in the morphology of the nucleus. The nature of the provision a cell has to make in connection with division has been extensively studied in mitotic cycles (MAZIA 1963). Mitosis and meiosis have much in common and many of the provisions must be common to both. Thus the studies on mitosis are relevant to an understanding of meiosis but here we can consider only some of the results concerned specifically with the meiotic sequence.

a) DNA, RNA and Protein Synthesis in the Normal Meiotic Cycle

In view of its localisation within the chromosomes, attention has understandably been concentrated on the synthesis of DNA. But the time and rate of both RNA and protein synthesis also have been studied through chemical analysis, photometric measurement and autoradiography. The general findings in relation to normal meiotic sequences are these:

(i) DNA is synthesised prior to the beginning of the meiotic prophase or is completed soon after it (see Table 71, page 202).

(ii) Chromosome histone is synthesised concurrently with DNA.

(iii) When these are synthesised, the synthesis of nuclear RNA is minimal or even non-existent.

(iv) RNA and residual protein continue to be synthesised after DNA and histone production are complete and their production may continue well into the prophase of meiosis.

Results such as these have been obtained in *Lilium* (TAYLOR and MCMASTER 1954, TAYLOR 1959), *Tulbaghia* (TAYLOR 1958) and *Tradescantia* (DE 1961). TAYLOR and MCMASTER's results with regard to DNA synthesis in pollen mother cells, pollen grains and tapetal nuclei of *Lilium* are summarised in Table 90. The results obtained for DNA, RNA and protein synthesis in the pollen mother cells of *Tulbaghia* are summarised in Table 91. Under the conditions of the experiment meiosis took about 6 days in this species and a similar interval included pollen grain interphase and the first mitosis.

Table 90. *DNA Synthesis in Anthers of Lilium longiflorum Var. Croft.*
(Data of Taylor and McMaster 1954.)

Normal, solid and italic type indicate approximate 2 C, 4 C and 1 C values respectively. Thus transition from normal to solid and from italic to normal indicate DNA synthesis. The reverse transitions indicate division.

Bud Length (mm)	P. M. C./P. G.	Tapetum	Stage of Pollen Micro./grain
7.0	46.8 ± 2.7	...	
9.0	31.0 ± 0.6	...	Pre-meiosis
9.5	29.6 ± 1.7	...	
10.0	31,9 ± 0.9	32.8 ± 1.2	
12.0	46.5 ± 1.4	32.0 ± 1,8	Pre-leptotene
12.5	63.0 ± 1.4	32.0 ± 1.3	Leptotene
14.0	...	32.0 ± 1.1	
15.0	63.2 ± 1.5	...	Zygotene
16.0	...	66.2 ± 1.6	
17.0	...	62.9 ± 3.5	
19.0	...	31.3 ± 0.7	
20.5	62.9 ± 2.0	59.0 ± 1.4	Pachytene/diplotene
24.0	32.8 ± 0.8	(56.68) — (a)	Interkinesis
28.0	15.9 ± 0.4	(61.3) — (b)	
39.5	15.2 ± 0.6	...	
45.0	17.0 ± 0.5	...	
48.0	18.2 ± 0.4	...	uninucleate
48.5	17.3 ± 0.4	...	microspore
52.0	(23.8 ± 1.1)	...	
52.5	(20.2 ± 0.9)	...	
55.0	29.6 ± 0.7	...	
61.0	15.0 ± 0.4	...	generative
73.0	(25.8 ± 1.0)	...	nucleus

(a) Eight 4 C and 2 C nuclei, and

(b) One 8 C, two 2 C and seventeen 4 C nuclei (nuclear fusion rather than synthesis appears to be involved here)

There is ample evidence from studies on mitotic cells that DNA synthesis occurs at different times in iso- and heteropycnotic segments whether the latter are naturally positive or induced to be negatively heteropycnotic (see review by Taylor 1962). Comparable evidence is available for the meiotic cycle. In the grasshopper *Melanoplus differentialis* (2 n = 22 + XX/XO) the unpaired X-chromosome of the male is positively heteropycnotic during the meiotic prophase and the previous interphase (see page 4). Certain autosomal segments exhibit allocycly also but these are small compared with the X which forms a block at prophase 2.5 — 3.0 µ in diameter.

Using the autoradiographic method following injections of tritiated thymidine into the abdominal cavity of males, LIMA-DE-FARIA (1959, 1961) studied DNA synthesis in the X and autosomes. Five-to-seven days after injection, sperm mother cells were found at leptotene-pachytene. These were of four kinds with respect to the label, namely:

Table 91. *Rates (Grain Counts/40 μ²) of DNA, RNA and Phosphoprotein Synthesis in the Pollen Mother Cells and Pollen Grains of Tulbaghia.*
(Data of TAYLOR 1958.)

Account must be taken of the duration of the stages (see Text) and the length of exposure to isotope. In fact, organic P^{32}-phosphate was supplied for 30 hours to inflorescences cut 2 inches below the buds.

Stage	Phosphoprotein	DNA	RNA
Mid-pre-meiotic interphase	4.8	0.4	8.3
Late-pre-meiotic interphase	11.5	30.1	2.4
Leptotene	5.2	1.9	8.9
Zygotene	4.8	−0.4	4.0
Late pachytene	5.9	0.0	9.9
Diakinesis	10.7	−1.2	10.5
Interkinesis	7.0	−0.2	3.5
Early tetrad	8.1	15.8	0.4
Late tetrad	9.0	17.0	4.7
Mid P. G. interphase	4.5	−0.1	5.9
Late P. G. interphase	3.9	−1.3	9.4
Prophase PGM 1	4.6	−0.5	9.0
Telophase PGM 1	4.1	0.4	1.4
1−2 day pollen	6.8	1.9	22.8
2−3 day pollen	2.9	13.0	3.2
4 day pollen	1.1	0.0	0.9
5 day pollen	2.1	0.2	−0.3

The Values Are:-

1. Protein—grains after 30. mins. treatment with hot trichloracetic acid (TCA)

2. DNA—difference in grain count between material treated with 1 N/HCl for 8 mins. and TCA treated material, i. e., (DNA + P) − P = DNA

3. RNA—difference in grain count between material treated withcold TCA for 5 mins. and HCl treated material, i. e., (P + DNA + RNA) − (DNA + P) = RNA.

(a) No label in the nucleus
(b) X-labelled, autosomes unlabelled
(c) X and autosomes labelled, and
(d) Autosomes labelled, X unlabelled.

Later stages were presumably not labelled. In the absence of technical faults, these results show that the hetero- and iso-pycnotic segments of a given cell are not coincident in their time of DNA synthesis. These four categories of cells are expected of course whatever the order of synthesis (Fig. 191). The basis of the difference can be distinguished only if the age of the cells can be determined.

The stage reached by a cell at a given time after treatment gives an indication of this. But in view of the long duration of the prophase stages of meiosis, especially pachytene, this is not a very accurate measure of age. Comparisons of the relative frequency of the four cell types at different times after treatment provides an alternative method. In grasshoppers, however, cysts containing cells which are synchronised in respect of DNA synthesis and division stage (i. e. age as used

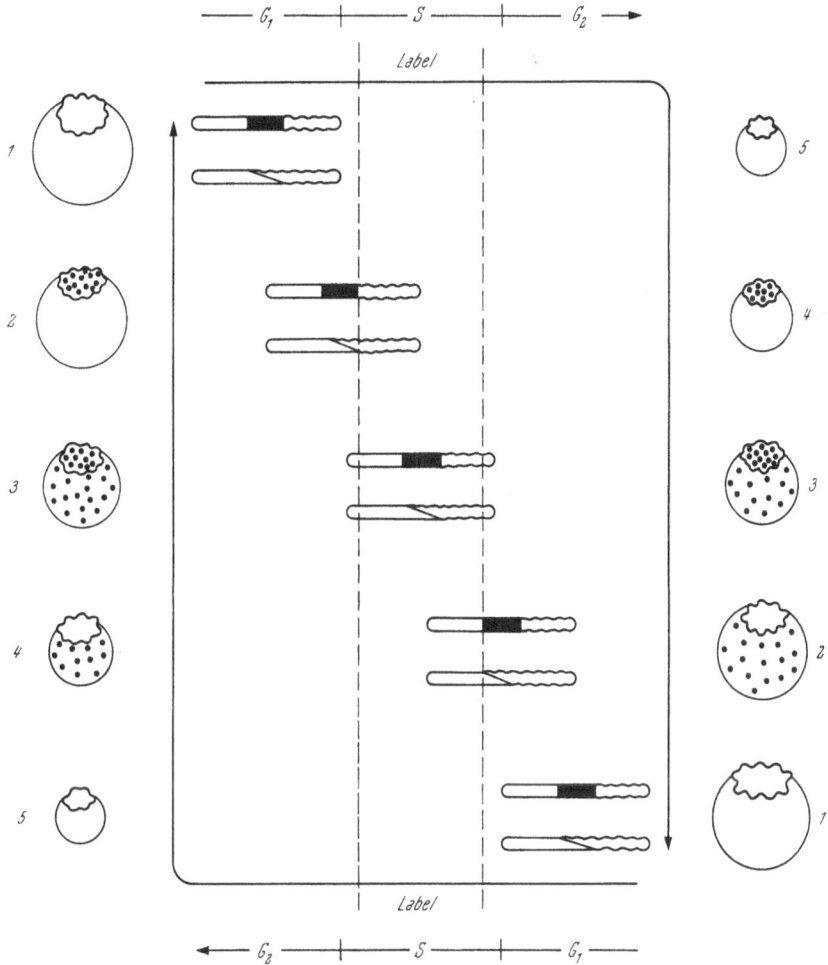

Fig. 191. Diagram illustrating the passage of a chromosome through a period of DNA synthesis (S) during part of which a labelled precursor (e. g., tritiated thymidine) is available. Arrows represent time for the cells. Two main situations are considered: — (i) where heteropycnotic regions (wavy outline) synthesis DNA before the iso-pycnotic (smooth outline) regions (right) and (ii) the opposite condition (left). Two "chromosomes" are drawn to illustrate the conditions where:—(a) the time of synthesis for heteropycnotic and iso-pycnotic segments does not overlap (the interval between the end of one and the beginning of the other is represented by a solid segment). (b) where they do overlap (as seems to be the case in all the studied examples—see text). The period of overlap is illustrated by a diagonal line. To meet the actual situation in *Melanoplus* it should be at a more acute angle to extend over 70% of the chromosome. Even in (a), cells with label in both hetero- and iso-pycnotic regions will occur if label is available for a length of time in excess of that between the two periods of synthesis (as illustrated). If there were a non-synthetic period within S and a short pulse label could be given during it then non-labelled nuclei should appear between 3 and 4. In the case of *Melanoplus*, label was available for about one hour and half of it was taken up in the first 15 mins. Some measure of the duration of the synthetic phases and their degree of overlap could be obtained from the relative frequency of the five kinds of labelled nuclei. In the figure an equal period of synthesis is illustrated by the equal lengths of the two kinds of segment. The cells are numbered according to their order of appearance at a given stage after treatment (i. e. time interval after treatment). Their size illustrates their age. But where the stage at which the cells are examined is a long one, cells of different age can appear at the same stage at a given time (see text). Thus unless some measure of age can be obtained, (i) and (ii) above cannot be distinguished.

here) move along the testicular follicle from the free (distal) end, which contains spermatogonial stages, to the proximal end which contains spermatids and sperm. And it was on this basis that LIMA-DE-FARIA determined the age of the cells observed. The cysts containing cells with label in the heteropycnotic X only were closer to the proximal end and, therefore, older than the others thus showing that this material synthesided DNA later than the isopycnotic autosomes. The period of over-lap is, however, considerable representing about 70% of the S period (cf. Fig. 191). In both the rate of synthesis falls off more abruptly than it builds up. But the X-chromosome, unlike the autosomes, shows a plateau of highest rate over about a quarter of its S period.

Some suggestion of asynchrony was found within and between the autosomes of *Melanoplus* as well. Thus, during the early stages of synthesis (i. e. in those cells where the X was unlabelled) the grains did not appear at random along the length of the chromosomes.

While DNA synthesis is a necessary if not sufficient provision for meiosis, it, itself, must be provided for. The synchronised system within the lily anther has been exploited to study this problem too, though most attention has been concentrated on the microspore interphase and provision for the first pollen grain mitosis (FOSTER and STERN 1959, STERN 1960). This interphase is characterised by short pulses of activity in respect of DNA-ase and thymidine phosphorylase activity which are separated by a short-lived peak of deoxyriboside production. So far only the deoxyriboside pool has been studied in relation to the meiotic cycle. The marked periodicity in the appearance and disappearance of this pool is seen only in relation to the periods of DNA synthesis in the pollen mother cells and their derivatives; it is not found in connection with tapetal DNA synthesis although the amount of DNA synthesised in this tissue after the onset of meiosis is higher than that in the sporogenous cells. Presumably the absence of accumulation in the tapetum is due to the limiting rate of production of deoxyribosides. The only alternative is that the interval between the production of deoxyribosides and their utilisation is short and not well synchronised between cells. But, in view of the synchronous synthesis of DNA, this latter possibility does not seem likely.

Peak levels of deoxyribosidic materials are high (ca. 25–50 times the base values) and a particular peak is found in buds which differ no more than about 1–2 mm. in length. Two millimetres represents about 8 hours growth and this is less than 2% of the life time of the uninucleate pollen grain. Thus the pulses in the presence of deoxyribosides are of the same order of duration as those found with respect to phosphorylase activity and they are, if anything, shorter than those found in regard to DNA-ase.

The sharpness of the periodicity is emphasised by the occurrence of two separate, but close, peaks of production and consumption prior to the first pollen grain mitosis. The interval between these two peaks is about 15–20 hours. The first precedes DNA synthesis but the other may follow it; critical timing is difficult at this stage because, as mentioned above, the synchrony of the pollen grains is limited. If both pulses precede DNA synthesis they may be associated with discontinuous DNA synthesis similar to that found in relation to isopycnotic and heteropycnotic segments (see page 203). But if the second pulse in the twin peak occurs after DNA synthesis it may well be that deoxyriboside production can be associated with synthesis other than that of DNA. Indeed, conjugated deoxyribosidic compounds have been described in animal tissues whose structure is suggestive of a role in lipid metabolism (Sugino 1957, Schneider and Rotherham 1958).

The precise nature of the deoxyribosidic pool has been only partly characterised. If it is chromatographed with a butanol/water solvent, the major part of it becomes distributed in a pattern corresponding to the DNA nucleosides. This suggests that phosphorylated derivatives do not constitute a major fraction of the pool for these do not migrate in a butanol/water solvent. A further indication that the mobile.components of the pool are largely deoxyribosides comes from the correspondance of their Rf values when individual spots are eluted and re-run with a solvent of isobutyric/ammonia.

Evidence for the existence of other compounds in the pool is seen in the observation that the growth-promoting effect of the pool on the test organism *Lactobacillus acidophilus* is enhanced by treatment with crude phosphatase. In this study and, to a lesser extent, in the untreated fractions, a difference was demonstrated between the pre-meiotic and pre-mitotic pools. Thus:

a) When the amount of extracted deoxyribosidic material was assayed microbiologically without pre-treatment with phosphatase, it was found that:

(i) Methanol and trichloracetic acid (TCA) extracted equal amounts during the period of the meiotic pool, but

(ii) In buds of 50 mm. or more, 90% of the pool was extractable with methanol.

b) When the assay was made after phosphatase treatment:

(i) Phosphatase-activated compounds were present only in the TCA extracts of the meiotic pool, but

(ii) In the mitotic interval they were extractable with acidified methanol.

The compounds activated by the crude phosphatase are not mononucleotides for these are as effective as deoxyribosides in growth promotion. They may include nucleotide pyrophosphates, polynucleosides or nucleoside complexes. In the initial work (Foster and Stern 1959) it was thought that a large fraction of the pool was thus constituted. But a re-examination using fresh rather than stored material showed no more than 10–15% of the pool to be of this type.

Assuming, as seems reasonable, that phosphorylation is a necessary prerequisite for polymerisation, the existence of a largely non-phosphorylated precursor pool so close in space and time to the sites and period of DNA synthesis provokes two suggestions (compare with the absence of deoxyribosidic pool in the tapetum—see above):

1) A pool of phosphorylated derivatives does not accumulate because they are incorporated into polynucleotides as quickly as they are formed. This would require the coincident activity of phosphorylating and polymerising systems, or

2) There is a distinct interval of phosphorylation which is even more brief than the pulse of deoxyriboside accumulation so that, in consequence, it escapes attention when estimates are made on pooled anthers. This implies a sequential relationship between deoxyribosidic production, phosphorylation and polymerisation.

Clearly, studies on individual anthers can resolve the issue and, in fact, occasional anthers are found with comparatively high levels of phosphorylated derivatives which favours the second possibility.

b) Origin of the Deoxyribosidic Pool

So far we have used the rather non-commital term "sporogenous region" but let us be a little more specific. A "core" containing about 80–90% of the microspores plus their surrounding substance can be removed without significant damage and only slight contamination by the tapetum. This core can then be analysed separately from the "jacket" layer for deoxyriboside content.

It turns out that about half the pool is outside the microspore core. Deoxyribosidic material could not be found in microspores which had been washed in sucrose to remove surrounding material. The possibility of leaching cannot be excluded here but it certainly looks as though the deoxyribosides originate outside the cells which are, presumably, going to utilise them.

Pre-meiotic DNA-synthesis in the pollen mother cells occurs while the total DNA of the anther is increasing. Indeed, it occurs prior to the two periods of DNA doubling which occur in the tapetal cells at about the zygotene-pachytene stage of the pollen mother cells. Presumably, therefore, deoxyribosidic accumulation prior to meiotic DNA synthesis represents *de novo* synthesis (STERN 1960a and b).

DNA synthesis in the pollen, on the other hand, occurs during a period when the total DNA of the anther is falling considerably. This fall accompanies, but cannot be completely accounted for in terms of the breakdown of the tapetum which begins soon after meiosis. On this ground and the long-established, though never demonstrated, opinion of the nutritive role of the tapetum, the degradation of tapetal DNA would appear to be a first choice as a source of the deoxyribosides. In considering this possibility three points must be borne in mind:

(i) The fall in total anther DNA is gradual while the pulses of deoxyriboside production are sharp.

(ii) The amount of DNA degraded is far in excess of the deoxyribosidic compounds accumulated.

(iii) Auto-radiographic experiments in which tapetal DNA was labelled with P^{32} did not show any transfer from the tapetum to the DNA of the sporogenous tissue. Likewise labelling tapetal DNA with H^3—cytidine does not lead to any meiotic labelling of the pollen mother cells (TAKATS 1959).

The pulse-like phases of enzyme activity and in the appearance of their products, under conditions in which the enzymes themselves are being assayed, clearly lead to an inference of enzyme induction.

4. Controlled Meiotic Anomalies

We have already discussed the work of SCHRADER and LEUCHTENBERGER (1950) on the chemical characterisation of polymegaly in pentatomid bugs (see page 259). It will be recalled that the different dimensions attained by nuclei, nucleoli and cytoplasm in the small, normal and large cell lobes of the testis of *Arvelius* are the result of different degrees of synthetic activity. These influence the RNA and protein content of the various lobes but do not affect the DNA value. Some of the other controlled meiotic anomalies have been similarly characterised chemically and these too have something to tell us about the causal aspects of meiotic behaviour.

ANSLEY (1954) compared the amounts of DNA and basic protein (believed to be histone) at different stages in the micro- and macro-testis of *Scutigera* using feulgen and fast-green staining together with the Million reaction and in conjunction with photometry. His findings were as follows.

In the macrotestis, which has a normal meiosis, both DNA and histone were produced during the first prophase of meiosis. The synthesis of both continued during zygotene, but diplotene nuclei showed the 4 C DNA value. The production of both components was in step so that the two fractions showed parallel variation. This constant relationship will be described as 1 : 1. It will be appreciated, of course, that both components are measured in arbitrary units so that the ratio has no meaning in terms of absolute quantities. Actually the arbitrary values obtained for DNA and histone coincided very closely but as ANSLEY points out this is due to the similarity of the extinctions of the colour reactions for the dyes employed. The only meaningful comparisons therefore are those between the ratios and those between the values for a given compound. The occurrence of DNA synthesis during the meiotic prophase is unusual but not, apparently, unique (see page 202) and presumably, the same applies to the histone fraction. However, photometry would appear to be a less accurate method of assessment than uptake of labelled precursors.

The microtestis has an abnormal meiosis; the first division of it is non-chiasmate and equational, the second is reductional for all segments. The mitotic-like first division shows reduced precocity in that the 4 C DNA value is attained prior to the onset of the mechanically active phase. The early microspermato-cytes have twice as much histone also as those in the macrotestis so that here too the DNA : histone ratio is still 1 : 1 because the values of both are double (4 : 4 versus 2 : 2). But in secondary microspermatocytes the corresponding ratio is 1 : 2. Now, the DNA value is halved by the equational first division so that the ratio in comparison with the early microspermatocytes is 2 : 4. There must, therefore, be a further production of histone in the microspermatocytes and analysis showed that this did not occur during the first division itself. The reduction in the amount of histone during the second meiotic division in the microtestis does not follow the simple halving which might be expected. Secondary microspermatocytes have 1.75 ± 0.6 units of histone at the early stages and 1.50 ± 0.06 at metaphase. The microspermatids, however, have only 0.44 ± 0.03 so, just as additional histone must be produced earlier, some histone must now be lost. Of course, conversion rather than loss may be involved, and there was evidence to indicate that micro- and macro-spermatocyte nuclei did not differ in their amounts of total protein (see also below).

Thus, the two kinds of spermatocyte differ with regard to the time of DNA and histone synthesis and in the DNA : histone ratio during division. It is worth noting, however, that the differences between the micro- and macro-testis are not precipitated by the advent of meiosis but are presaged at much earlier stages. Thus volume differences are apparent during the first three spermatogonial generations and morphological differences can be observed from the fourth cell generation onwards. Thus, the late spermatogonial nuclei in the microtestis are not visibly different from the primary microspermatocytes. It is interesting to note in this connection that the later microspermatogonia, unlike the corresponding macrospermatogonia seem to rest in the G_2 phase so that most of them have the 4 C DNA value. A similar difference obtains in respect of the histone fraction so that the DNA : histone ratio in both is 1 : 1.

ANSLEY (1957) carried out a similar study in *Loxa*. Here the testis consists of seven compartments of lobes. BOWEN (1922) has arbitrarily numbered these lobes consecutively, considering the one nearest the side at which the spermduct enters as the first. In *Loxa*, as in *Arvelius*, there is a constant difference in the size of the cells found in the different lobes (polymegaly). The harlequin lobe (no. 5) is characterised by very small spermatocytes. Lobes 4 and 6 contain unusually large spermatocytes while the rest of the lobes carry spermatocytes of normal dimensions (Fig. 192).

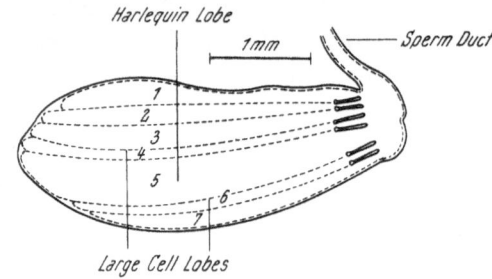

The abnormal meiosis in the harlequin lobe of this species (see page 63) resembles that in the microtestis of *Scutigera* in that:

(i) The sperm mother cells are smaller,

(ii) They are asynaptic, and

Fig. 192. The structure of the testis in *Loxa picticornis* (SCHRADER 1960).

(iii) The first division is equational.

The two types agree also at the chemical level in that:

(iv) The asynaptic cells come to have a DNA : histone ratio which departs from 1 : 1 in favour of excess histone. In both, the excess is of the same order, ca. 2 : 3; and this is reached by "zygotene" in the harlequin lobe, and

(v) A 1 : 1 DNA : histone ratio is maintained in normal cells and both are synthesised during the early prophase of meiosis. In the harlequin lobe, peak values are reached during "zygotene."

The two types differ in that:

(i) Aside from the asynapsis, stages equivalent to those of normal meiotic prophase can be observed in the harlequin lobe, but

(ii) By early diplotene 12 univalent autosomes are present, so that, if pairing has occurred earlier, it has been followed almost immediately by desynapsis. Then during the confused stage, there follows an unprecedented series of events. In every cyst of the harlequin lobe there is an irregular fusion of spermatocytes as well as some amitotic divisions. This leads to the production of first metaphase cells that contain as few as 2 and as many as 200 chromosomes (2n = 14).

(iii) The spermatocytes in the harlequin lobe do not show the reduced chemical precocity found in the microtestis of *Scutigera*. They start meiosis with the 2 C-DNA value and a 1 : 1 DNA : histone ratio. Thus, their behaviour in respect of the time of DNA synthesis is similar to that of cells in the normal lobes.

In *Loxa* also, differences are apparent at the spermatogonial stage. Thus, while the DNA : histone ratio is 1 : 1 at interphase in small spermatogonial nuclei, excess histone is synthesised in the harlequin lobe so that in large spermatogonia the ratio is about 2 : 3 or 4. Presumably therefore this extra histone, like that at meiosis in both types, is "lost."

Comparisons between normal cells and the abnormal cells of the two types are suggestive (Table 92). The abnormal cells of *Loxa* and *Scutigera* agree in the DNA : histone ratio they achieve but differ in the time at which they achieve it. The time of synthesis in *Loxa* is the same as that for normal cells and prophase appears normal. It may be therefore, that the univalent condition in *Loxa* is better regarded as desynapsis while genuine failure of zygotene pairing (asynapsis) may obtain in *Scutigera*. The relevance of this argument in relation to DAR-LINGTON's precocity theory is clear.

Table 92. *Comparison of Normal and Abnormal Meiosis in Loxa and Scutigera (see text).*

| Meiosis | Normal Loxa and Scutigera | Abnormal | |
		Loxa	Scutigera
Precocity of division	Extreme precocity: DNA and histone synthesised during leptotene/zygotene	As for normal	Reduced precocity: synthesis completed prior to leptotene
DNA: histone	1 : 1	ca. 2 : 3	ca. 2 : 3
Prophase	Normal	Meiotic-like	Mitotic-like

Before leaving these investigations we must consider a question we deferred earlier regarding histone synthesis, conversion and loss. In the alkaline Fast Green method (ALFERT 1953, ALFERT and GESCHWIND 1953) fixation in 10% neutral formalin is obligatory after which the DNA is extracted (with trichloracetic acid) and the material is stained in fast green at pH 8. The method is supposed to be specific for proteins of the histone type. It would appear that DNA, histones, acidic proteins and, perhaps, other materials including RNA, polysaccharides and phospholipids are intimately related in the living chromosome (POLISTER 1952). The main components do not stain, however, unless the chromosome is hydrolysed or partially digested. Indeed the histones will not stain until the DNA is removed. ALFERT describes the action of fast green as one involving the titration of the bonds rendered available by the disruption of the DNA-histone association. On this basis ANSLEY suggests that the DNA : histone ratio may always be 1 : 1, so that parallel increases (concordance) in the two do represent synthesis. But an apparent increase (discordance) in histone is due to an increase in the number, or a change in the kinds of bonds between DNA and histone. Apparent loss can be explained, *mutatis mutandis*, in the same way. In view of the heterogeneous nature of "histones" changes in the kinds of histone may be involved and ANSLEY mentions the possibility that the atypical cells in *Loxa* and *Scutigera* may possess histones of a much higher iso-electric point than those in normal cells. Indeed, a small amount of abnormal histone could result in a large quantitative change in the ALFERT-GESCHWIND reaction. Lest we have over-stressed the alternatives to increased synthesis we should point out that similar results were obtained in the case of *Scutigera* using the Million reaction.

A further study of this problem has been made in *Loxa* (ANSLEY 1958). The results suggested that the asynaptic cell-lines approached somatic cells not only in dye-binding capacity but also in the amino-acid content of the histones. It would appear that only about 30% of the α-amino groups and under 10% of the imidazole groups of histidine are expected to bind with the dye, the groups mainly responsible for the alkaline fast green reactions being the guanidine groups of arginine and the amino groups of lysine. Now, acetylation at room temperature principally affects the latter and is thus expected to inhibit staining

Table 93. *Mean Relative Values of Alkaline Fast Green Staining For Histones in Loxa flavicolis Before and after Acetylation.*
(Data of ANSLEY 1958.)
Germ line cells are mixtures of spermatocytes and spermatogonia.

Cell Type and Number	Untreated	Acetylated	Difference	% Difference
Somatic (80 and 70)	2.55 ± 0.04	1.32 ± 0.03	1.23 ± 0.05	48.2
Asynaptic (81 and 62)	2.70 ± 0.04	1.42 ± 0.04	1.26 ± 0.06	47.4
Synaptic (67 and 52)	1.79 ± 0.04	0.73 ± 0.03	1.06 ± 0.05	59.2

owing to lysine. ANSLEY compared the alkaline fast green reaction of asynaptic cell-lines, synaptic cell-lines and somatic cells both before and after treatment with acetic anhydride for two hours (Table 93).

The principal point to emerge was that the asynaptic line and the somatic cells resemble each other much more closely in:

(i) the amount of histone they contain, and

(ii) the effect of acetylation which effects an 11% greater reduction in synaptic cells.

This, on the basis given above, suggests that the histones of synaptic-line cells contain 11% more lysine and a correspondingly lower amount of arginine.

It should be stressed that the asynaptic and synaptic estimates were not made on sperm mother cells alone but on a mixture of spermatocytes and spermatogonia. It would appear that the cell lines rather than the meiotic cells alone are differentiated.

A common feature of the controlled anomalies discussed above is the synthesis of normal amounts of DNA. NICKLAS (1961) has also established that the amount of DNA in the nuclei of the harlequin and the normal lobes in *Mecistorhinus* and in *Neodine* are identical. By contrast a mechanically-standard meiotic sequence may obtain in cycles whose only departure from normality is the synthesis of abnormal amounts of DNA. Thus in several species of the dipteran genus, *Tipula*, oocytes and certain oogonial cells contain not only chromosomes and a nucleolus but a large feulgen-positive body which is organised in contact with the two X-chromosomes of the female. This body suddenly appears during interphase in the two secondary oogonia produced by the mitosis of a nucleus which does not possess such a body. Each of these two cells undergoes further division and after three mitotic generations each has contributed equally to a population of sixteen cells. The body persists during these divisions and it can be seen lying in the spindle between the separating anaphase groups. It is, however, incorporated into only one of the daughter nuclei and a similar body is not organised in the sister nucleus which did not receive it. Thus, of the 16 cells eventually produced two contain such a body and fourteen do not. These together with one of the body-containing cells develop into nurse cells while the remaining body-containing cell forms the oocyte (BAUER 1932, 1952, BAYREUTHER 1955).

At the last pre-meiotic anaphase the body varies in shape but spherical ones have a diameter of about 1.5 μ. It enlarges during oocyte formation, reaches a maximum diameter of ca. 6.0 μ at pachytene, becomes vacuolated during the prophase and disintegrates fairly rapidly at diplotene, i. e. on the third to fourth day of pupal life.

By injecting tritiated thymidine into suitable larvae and then making auto-radiographs of feulgen-squashed ovaries, LIMA-DE-FARIA (1962) was able to show that DNA is synthesised in the body both in oogonial cells and in the meiotic cycle. In the latter, at least, the feulgen-positive body and the chromosomes incorporate thymidine at different times, although the periods of synthesis overlap. At pachytene about 60% of the total nuclear DNA is contained in the body. Due to dilution the dispersion of this material cannot be followed at later stages. (STICH 1962) claims that in *Cyclops* too the oocytes contain about twice as much, or more, DNA than expected on the basis of the Boivin-Mirsky rule. But the first and second division meiotic products have the "expected" 2 C and 1 C DNA values respectively. According to STICH's account of the fifth cleavage, DNA-containing granules, which arise from the terminal hetero-chromatic segments of the chromosomes, leave the chromosomes at prophase, accumulate at the equator of the spindle and are excluded from the daughter nuclei. But S. BEERMAN (1959) claims that parts of the heterochromatin itself are shed (compare with *Ascaris*). Further, the BEERMANNs (personal communi-cation) have re-investigated the situation microspectrophotometrically and con-cluded that the germ line nuclei prior to reduction contain about three times as much DNA as the somatic nuclei which have undergone diminution. "The amount lost during diminution corresponds to the difference between the two. Thus, there is no necessity to involve extra DNA production." In *Cyclops furcifer* it would appear that heterochromatic material is shed not only from the ends but from interstitial segments near the kinetic regions. This poses interesting questions in relation to chromosome structure and organisation.

5. Additional Division Requirements

In the preceding sections we have seen something of the provisions cells make for mitosis and meiosis. So far we have concentrated on the materials of the chromosomes and more especially on nucleic acid. But clearly there are other preparations to be made. Among those which have received some attention are the synthesis of spindle precursors and the energy provisions for division. Most of this work has been concerned with mitotic rather than meiotic cycles but the two are not likely to differ significantly in respect of the processes concerned.

(a) Sulphydryl and disulphide levels in anthers.

Sulphydryl (SH) compounds are widely involved in metabolism and they, together with disulphide (-S-S-) bonds, have been implicated in spindle organisation (see Brachet 1957, Mazia 1963). Since a large fraction of the cell protein is tied-up in this structure it is clear that its synthesis must precede its organisation by some considerable time. Evidence in favour of this conclusion comes from the study of mitotic spindles (Went 1959a and b, Taylor, E.W. 1959).

So far as the meiotic cycle goes, Stern (1956, 1958) has studied the levels of soluble and protein-associated sulphydryl in whole anthers of *Lilium longiflorum* var. Croft and *Trillium erectum*. His investigations also determined the variations in disulphide levels over the same periods. The two species showed essential agreement but there were interseting differences.

In *Trillium* there is a gradual rise in soluble-SH (the major component of which is probably glutathione) from September to December. The peak concentration is reached early in zygotene. The pachytene level is somewhat lower and more or less constant. Owing to the comparatively long duration of this stage the soluble-SH level forms a plateau during it. Diplotene/diakinesis heralds a sharp drop which continues to the end of meiosis. Interphase sees a decrease in the rate of decline which increases again during the first pollen grain mitosis.

Protein-SH levels do not alter much until diakinesis when there is a sharp drop. However, this is followed by a sudden rise which continues during the remainder of meiosis so that the level at the end of this division is about the same as that at the beginning. Post-meiotic interphase again sees a decline which is continued during PGM-I.

Disulphide levels, like protein-SH, remain fairly constant until diplotene. But while sulphydryl levels then fall, the disulphide level rises sharply (the levels are, however, very variable) before falling.

Thus, the premeiotic rise in soluble-SH is associated with constant levels of protein-SH and disulphide. This argues in favour of *de novo* synthesis rather than reduction of endogenous sources of oxidised glutathione. Further, the diakinetic increase in -S-S- does not account entirely for the drop in -SH. But dehydrogenation rather than disappearance of thiol compounds appears to be the principal change. Furthermore, the sharp and opposite changes in level of soluble-SH as opposed to protein-SH during the post-diakinetic stages of meiosis suggest a causal connection consistent in general terms with the ideas of Rapkine which have been discussed in detail by many workers recently (see Brachet 1957, Mazia 1963, Stern 1959).

The behaviour of sulphydryl and disulphide groups in *Lilium* are in essential agreement. The major difference is that while the soluble-SH level falls slightly during the post-meiotic interphase of *Trillium*, it rises to about the pre-meiotic level in *Lilium*. Total soluble -SH and -SS shows a similar increase and protein

-SH levels remain constant compared with a fall in *Trillium*. At first sight these may appear to be a major differences. But as STERN points out, they can be explained on the basis of the habits of the two species.

Meiosis and pollen grain mitosis in *Trillium* occur when the plants are vegetatively dormant and often covered with snow. During this period the anthers increase by about 50% in fresh weight and protein nitrogen and this must be effected at the expense of food reserves in the corms. Thus, the anthers of *Trillium* are in some respects, comparable to yolk-laden eggs in that provision for pollen grain mitosis, or at least of sulphydryl compounds, is made even before meiosis has occurred. In *Lilium*, on the other hand, meiosis and pollen grain mitosis take place not only while the anthers are growing (and growing much more than those of *Trillium*), but while the whole plant is vegetatively active. In keeping with the idea that the lily has less thought for the morrow than *Trillium* is the fact that the pre-meiotic -SH level of the latter is four times as great when expressed on the basis of protein nitrogen. The differential is even greater when expressed in terms of fresh weight for while the protein nitrogen constitutes about 1% of the fresh anther weight in *Trillium*, the corresponding value for *Lilium* is about 0.3%.

We see, therefore, that apart from the pre-division build-up, the major changes in -SH compounds occur during periods of chromosome contraction and spindle-movement. But any interpretation of the role of thiol groups in these connections must take into account the wide range of other metabolic activities which are influenced by thiol compounds. Significantly, however, as we shall see below, the thiol levels at various stages do not correspond with those of triose phosphate dehydrogenase activity. We may also add that while ideas on spindle structure and organisation appeared to be crystallising a few years ago with the work of MAZIA on sea-urchin eggs, the situation now is again in a state of flux.

(b) Energy requirements of division.

The study of this question has a long history in which the names of RAPKINE, WARBURG, SWANN and BULLOUGH are prominent. The situation has been reviewed recently by BRACHET (1957), MAZIA (1963) and BULLOUGH (1952) among others.

After a number of conflicting reports and rather inaccurate studies, it now appears certain that rhythmical variations in respiratory rate, as measured by oxygen uptake, do occur in relation to mitotic cycles, division itself being accompanied by a decreased rate. In animal eggs, the fluctuations are not large (ca. 5%) which accounts partly for the disagreements observed in the earlier reports. Fluctuations have also been observed in developing animal eggs which show no growth between divisions and in which the interphases are protracted. Thus, division events themselves appear to be the determining factors.

Similar variations have been found in respect of whole anthers in *Lilium* and *Trillium* (ERICKSON 1947, STERN and KIRK 1948) over a period covering meiosis and pollen grain mitosis. The pre-meiotic and post-meiotic resting stages are long in these plants and this, as we have seen, facilitates accurate pinpointing on a time scale. Growth does occur during these intervals, however, but the fluctuations in oxygen uptake do seem to reflect the division cycle though cause and effect are not easily distinguished.

Cells vary in their response to treatment with inhibitors of energy metabolism. But with few exceptions, like the yolk-laden eggs of frogs, entry into mitosis (but not the continuation and completion of existing mitoses) is arrested by inhibitors of glycolysis.

NASATIR and STERN (1959) determined the changes in the activities of two glycolytic enzymes, aldolase and triose phosphate dehydrogenase. Two high peaks of activity were found a week before and four days after the first pollen grain mitosis respectively. Division itself occurred in a trough of activity. If, however, aldolase, which was limiting at all stages, was added to the pollen grain suspension, a minor peak of D-glyceraldehyde-3-phosphate dehydrogenase activity was obtained during a short period which embraced pollen grain mitosis. Aldolase itself did not show such a peak. Notice that although the activity of many glycolytic enzymes is influenced by thiol groups, the fluctuations shown by the latter follow a nearly opposite course to the activities of the enzymes described above. But the actual meiotic sequence has not been studied in this direction.

6. The Developmental Control of Meiosis

Earlier we mentioned differences between cells within individuals in respect of meiosis which could not depend on genetic differences between those cells. We want now to consider some aspects of this problem as it arises in individuals which are more-or-less balanced and cultured under "normal" environmental conditions. In other words, variations which can be considered as symptomatic of the normal developmental pattern of the types concerned. For the most part we will confine ourselves to variation within "tissues", that between sexes having been discussed earlier (see page 25).

a) The Suppression of Meiosis and the Failure of its Products in Normal Individuals

Development in most multicellular organisms involves the differentiation of three kinds of cell in respect of division—those which are not required to divide again, those which will do so only under exceptional circumstances (wound healing, regeneration etc.) and those which have every expectation of division, mitotic or meiotic. Once the main organs have been differentiated it is usually possible to point to the cells which have a mitotic future in development and those which have a meiotic expectation in sexual heredity. But in many plants, potential sporogenous cells do not always fulfil their early promise. And, on both male and female sides, hopes may be dampened either before or after meiosis. Some examples are cited below most of which were gleaned from MAHESHWARI (1950) which should be consulted for detailed references.

Embryology, even more than cytology, has, however, accumulated a long record of exceptional behaviour. And it is not always easy to determine from the published accounts whether a particular behaviour is typical, common or exceptional for a species. Be this as it may there is no reason for believing that either genotypic or environmental anomalies are involved in determining the events described below. They are determined by what is the normal course of development for the type concerned.

(i) Male Side

a) Pre-meiotic—In some members of the *Gentianaceae*, e. g. *Swertia perennis*, in which the tapetum is poorly developed, some of the potential pollen mother cells fail to undergo meiosis and, presumably, play a tapetum-like role in anther development. Thus, a situation obtains which is similar to that in the capsules of liverworts where elaters are differentiated from the intermingled spore mother cells.

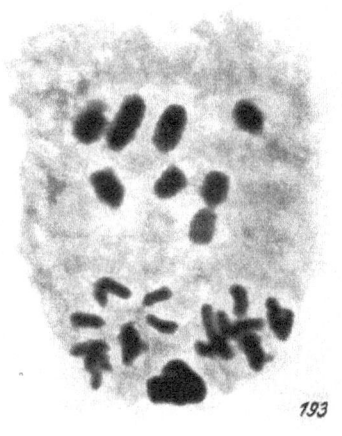

In *Holoptelea* and *Ophiopogon* some potential sporogenous cells of the anther do not even reach the mother cell stage. The extreme situation of this kind is seen in the development of the megasporangium in types like *Selaginella*. Here the micro- and megasporangia are indistinguishable during the early stages of development and each contains numerous potential spore mother cells. But while they all function in the microsporangium with the consequent production of many small spores, only one (or in some cases, two) of them is effective in the megasporangium so that four (or eight) large megaspores are produced.

Fig. 193. Pollen grain mitosis in *Eleocharis palustris microcarpa* (n = 8). The upper nucleus is the functional one. The three other meiotic products are abortive. Note chromosome size difference between functional and non-functional nuclei (carmine, ca. ×1250).

An even earlier differentiation within the potential sporogenous tissue is observed in the anthers of *Zostera*. Here some cells divide longitudinally to give the elongate pollen mother cells. But in others of the same generation the last mitosis divides the cell transversely. The products of this division do not undergo meiosis and are eventually crushed by the functional cells.

There is no basis for believing that these situations are in any way comparable to those found in animals like the orthoclads (BAUER and BEERMANN 1952) where a differential mitosis defines normal from aberrant sperm mother cells.

b) Post-meiotic—Typically, in plants and animals, all four meiotic products are functional on the male side except of course in cases of segregational sterility. This "sterility" may be normal as, for example, in types like *Sciara* and *Miastor* which have controlled chromosome elimination. There are, however, situations in plants, where post-meiotic failure is observed which cannot be attributed to segregation. Thus, throughout the Cyperaceae only one pollen grain is produced per mother cell. The mother cells form a single layer in the anther loculus and they are wedge shaped. Three of the meiotic products migrate to the thin end of the wedge and degenerate (Fig. 193). A comparable situation obtains in some members of the Epacridaceae.

In *Kigelia*, degeneration takes place later. Meiosis is completed and four products are formed but some spores fail to develop though others in the same tetrad proceed to maturity.

(ii) Female Side

a) Pre-meiotic—There are many examples in which more than one megaspore mother cell is differentiated in the ovule. In some cases many of them divide

Linear Tetrad	Example
Micropylar ⟷ Chalazal end end	
1.	 Most common type
2.	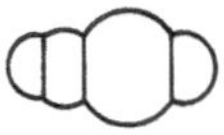 *Aristoletia racemosa*
3.	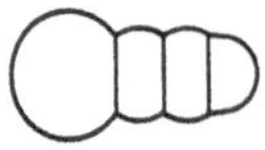 *Rosa, Elytranthe, Langsdorffia, Leucopogon, Balanophora,* a few composites, e. g. *Taraxacum laevigatum*
4.	 Occasionally in *Rosa*
5.	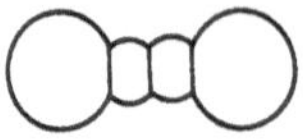 *Cucurbita reflexum*, not unusual in the *Compositae*. Temporary condition with RENNER effect operating
6.	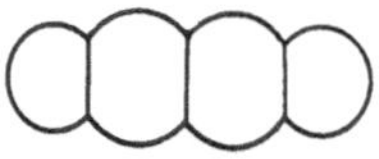 All or any—*Gloriosa, Ostrya, Poa, Casuarina*
7.	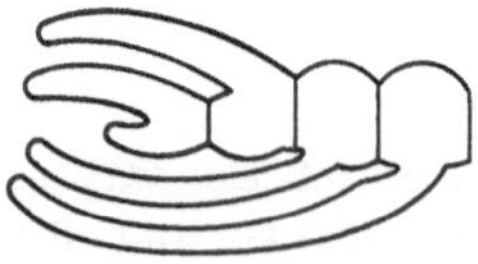 *Rosularia, Sedum, Laurus, Galium, Potentilla, Putoria*

Fig. 194. Atypical behaviour in the megaspore tetrads of plants. The micropylar end of the linear tetrad is to the left.

meiotically. For example in *Urginea indica, Ruppia* and *Butomus*, two adjacent mother cells may divide to give a row of eight spores; while in the Rubiaceae and Compositae there are several mother cells all of which may undergo meiosis.

In other cases, however, as sometimes occurs in the Onagraceae, only one, usually the central, cell of a group of half a dozen is functional though one or two others may reach the mother cell stage.

b) Post-meiotic—In plants and animals, only one meiotic product is typically functional on the female side. But Fig. 194 illustrates some of the variation found in plants.

c) Intra-meiotic—Here we are concerned not with a breakdown resulting in inviability but with a failure, partial or complete, in one of the first division products.

In most plants the functional megaspore is differentiated at the chalazal end of the linear "tetrad". But, not uncommonly, the micropylar cell and the chalazal cell diverge in development at the dyad stage. Thus, while the second division proceeds normally in the lower (chalazal) cell, the upper cell may

(i) Divide normally and synchronously with the micropylar cell of the dyad,

(ii) Divide normally but somewhat later than the lower cell,

(iii) Divide normally but without subsequent wall formation, or

(iv) Undergo an abortive division.

Breakdown at interkinesis is revealing because it shows that all the provisions for the completion of meiosis cannot always be made before the division begins. Perhaps, those with and those without an interkinetic resting stage differ in this respect.

b) The Synchronisation of Meiosis

The clear-cut, meristic, life-or-death variation between cells within the potentially-sporogenous tissue of the anther which was considered above, and especially that determined before meiosis, can be regarded as an extreme symptom of the asynchrony which occurs in respect of meiosis even within a pollen-sac.

In the description given earlier we stressed the high degree of synchrony which is generally found within an anther or even between the anthers of a bud in types like *Lilium*. This uniformity which is found also in respect of mitosis in tissues like the young endosperm and which can obtain both in the presence and absence of intervening cell walls, is often regarded as an indication of a general uniformity in the distribution of materials necessary for, or which trigger, meiosis. It is true that a high measure of synchrony is the rule in anthers, but distinctive patterns of asynchrony are found. Even in lilies synchrony is not perfect and cells at the apex of the anther are about 8–12 hours in advance of those at the base with a gradual gradient in the intervening segment.

In rye the asynchrony is greater and it has been studied in detail (REES and NAYLOR 1960, REES 1962). The anthers were cut transversely into four segments of equal length and numbered I–IV from the base. The filament is inserted at the junctions of segments I and II while the vascular strand passes upwards through segments II and III and terminates shortly after its entry into segment IV.

Consistent timing differences were found between these segments. Segment II was the most advanced, segments I and III were similar and segment IV contained cells which had progressed least. These segments are not clear-cut of course and asynchrony exists within them.

A study was made of the relationship between timing differences and two other chromosome characters, one meristic (chromosome breakage, see page 27), the other metrical (chiasma frequency), both of which are known to be under the control of the genotype. When different anthers were compared it was found that pollen mother cells reaching metaphase-I early had higher chiasma frequencies than those which reached it later. This finding is consistent with the precocity theory. A similar relationship between sections was not ruled out but there was clear evidence for interaction between segments.

With respect to breakage it was found that:

(i) Earlier-dividing basal segments of the anther had less breakage than later-dividing apical segments, but

(ii) Within segments the earlier-dividing cells showed more breakage.

The patterns of behaviour are not, then, as simple as might have been supposed. And REES interprets them in terms of supplies reaching the cells via the vascular strand of the filament. Sources of supply are obviously important but it is worth remembering that the gradient in lily anthers seems to run from the apex to the base while radio-active material traverses the anther in the opposite direction.

From the experimental point of view the value of these "chromosome characters" is considerable for they allow an analysis of the phenotypic variation which arises within cells with time. They are easily, though painstakingly assessable and, in the case of metrical characters, each cell can be represented numerically with considerable accuracy.

We must stress that the intra-anther variations in rye, like the conditions described previously, are characteristic of normal development in the types concerned. They show that what is, in one sense and to some extent, a uniform tissue is nevertheless heterogenous at other levels. The comparable condition is known of course in mitotic tissues, e. g., mouse epidermis (GELFANT 1963).

There are many examples where abnormal differentiation arises within the pollen-sac, some of which were mentioned earlier. Relevant to this aspect of meiotic control would be a discussion of the various patterns of meiosis found within ovules especially those of apomictic species. They have been described in detail by GUSTAFSSON (1946, 1947, 1948) and, as there is a monograph on apomicts in the present series (NARBEL-HOFFSTETTER 1964), we will not consider them here.

c) The Reversible Arrest of Meiosis in Female Animals

Most plants are either haplontic or haplodiplontic and with rare and occasional exceptions (e. g., ascospore fusion in yeast) the immediate products of meiosis do not function as gametes. Thus, the division preceding gamete formation in most plants is mitosis. Most animals, on the other hand, like some algae, are diplontic and the gametes are the immediate products of meiosis.

The eggs of animals are differentiated for the most part under the influence of the diploid, maternal nucleus; their development is pre-meiotic. And considerable variation exists regarding the time of entry of the sperm relative to egg development (Table 94). In the present connection the interest of this phenomenon lies in the fact that female meiosis maybe temporarily arrested and may proceed no further until the sperm has entered the egg.

Some of this variation may be determined by the mating behaviour. For example, in sea-urchins the eggs are not accessible to the sperm until female meiosis is complete. Given the opportunity, however, they will penetrate earlier but development does not then follow. Nevertheless in many cases it is clear that the completion of female meiosis must await sperm-entry. And the interest of this phenomenon in the present discussion is that the temporary block occurs at different stages of meiosis in different types though some intra-specific and intra-individual variation exists in this respect.

In the sea-urchin the sperm moves towards the egg nucleus aster-first. Sooner or later this gives an amphiaster which is the fore-runner of the asters of the cleavage spindle. Here, then, it would appear that further development cannot proceed until the sperm enters the egg, because it is responsible for the organisation of the first cleavage spindle. This role may be of significance in some pseudogamous apomictic animals where the sperm does not make a nuclear contribution

Table 94. *Sperm entry in Relation to Female Meiosis in Sexually Reproducing Animals.* (Compiled fron WILSON 1925.)

Type and Example	Stage of Sperm Entry
1. **Brachycoelium**-type. Also in *Otomesostoma* and *Saccocirrus*	Sperm enters immature oocyte
2. **Ascaris**-type. In various platodes, molluscs and crustaceans	Sperm enters mature oocyte prior to first division or else at time of breakdown of the nuclear membrane of the egg mother cell nucleus
3. **Cerebratulus**-type. Also in *Chaetopterus* and *Dentalium*	Female meiosis temporarily blocked at first metaphase awaiting sperm entry. Division proceeds immediately following entry
4. **Ophyotrocha**-type	First division may advance to first anaphase prior to entry
5. **Rana**-type. Also in many chordates	Meiosis proceeds to, but is blocked at, second metaphase
6. **Siredon**-type. Also present in bats	Meiosis may proceed to second anaphase but can go no further without sperm entry
7. **Sea-urchin**-type. Also in certain coelenterates	Meiosis is completed prior to sperm entry. Under natural conditions the sperm does not have earlier access to the egg

to the embryo (see, however, below). But if, as appears to be the case, the centriole or part of it, is a self-duplicating genetic element, it is by no means obvious why the maternally derived one should die in the egg.

What part sperm entry plays in overcoming temporary blocks which occur during meiosis itself is even less clear though the cytoplasmic changes accompanying sperm entry have been widely studied.

In passing, let us also note yet another instance of the changes which can be wrought by natural selection so that events which are positively correlated in one species can become unrelated or even negatively correlated in another. Above we have seen various examples in which sperm entry allows meiosis to proceed. But in *Cognettia glandulosa*, an enchytraeid worm, the entry of a sperm suppresses the second division! This, however, is a parthenogenetic species and a pseudogamous one (CHRISTENSEN 1961).

7. Meiotic Variation Controlled by Genetic Differences

The genetic control of meiosis can be considered in three dimensions. To begin with the effects of structure must be distinguished from those of genotype. Initially, this distinction was required and made in relation to the meiotic sequence of structural hybrids because structure (in the sense used here) and genotype can have similar consequences on the sequence. For example, to take the most obvious situation, failure of pairing in hybrids may be due to non-homology (i. e. structure as used in this connection) or to genotypic unbalance. Again, as we have seen (see page 29) bridges and fragments may be the secondary consequences of a primary structural change or they may be the result of a primary chromosome mutation in unbalanced genotypes.

We will not discuss here the means of distinguishing between the effects of structural hybridity and those of an unbalanced genotype since we have considered the matter in some detail elsewhere (LEWIS and JOHN 1963a). We may note, however, that different methods of distinguishing these effects may give conflicting results. For example, abnormalities in diploid hybrids, especially those involving univalence, are generally attributed to non-homology (structure) if normal pairing is shown by the corresponding allotetraploid. In fact this is regarded as a test. It does, however, neglect any differential effects of dose. Now, in the male hybrids produced on crossing the two moth species, *Pergesa curtula* ($2n = 58$) and *P. pigra* ($2n = 46$), there are few if any bivalents at first metaphase (FEDERLEY 1931). But in triploid male hybrids of the AAB type, homologous chromosomes form bivalents while the set represented once remains unpaired. From this result it would appear that the failure of metaphase pairing in the first generation diploid hybrids was the consequence of non-homology and not of genotypic unbalance. But, on the other hand, the diploid female hybrid shows almost complete pairing thus indicating a genotypic basis for the incapacity of the males. A comparable difficulty arises in the case of hybrids between species of the plant genus *Geum* (GAJEWSKI 1953).

Although the distinction between structural and genotypic control was made initially in relation to hybrids, the concept can be extended to embrace structural and numerical homozygotes. In fact, the aim is to distinguish those effects owing to the *arrangement* of the structural and genetic components of the chromosomes which influence its behaviour from those which are a consequence of the combined *activity* of the genetic material. Position effects thus intrude on the distinction as they do in the parallel one between gene and chromosome mutation. It is true that numerical chromosome changes may affect meiosis in a manner comparable to the effects of structural rearrangements. But unequal numerical changes like polysomy are expected to result in genotypic unbalance.

Clearly, except in the case of mechanically unfit chromosomes like acentrics and dicentrics, structural features of this kind considered above do not enter into a discussion of mitosis for there the mechanical unit is the chromosome and not a bivalent or higher association as at meiosis.

The third dimension is, however, common to both mitosis and meiosis. It is not easily defined succinctly but its consequences may take the form of a block-reaction or gradient effect at one extreme and an inter-chromosal interaction on the other.

Table 95.   *Genotypic Control of Meiotic Processes.*

Process	Species	Reference
1. Pairing of homologues (Asynapsis) a) No (or little) pairing	*Rumex acetosa ♂* *Hevea brasiliensis* *Hyoscyamus niger* *Alopecurus myosuorides* *Lycopersicon esculentum*	Yᴀᴍᴀᴍᴏᴛᴏ 1934 Rᴀᴍᴀᴇʀ 1935 Vᴀᴀʀᴀᴍᴀ 1950 Jᴏʜɴѕѕᴏɴ 1944 Sᴏᴏѕᴛ 1951
b) Variable pachytene pairing	*Zea mays* *Triticum aestivum* *Picea albies*	Bᴇᴀᴅʟᴇ 1930, 1933 Mɪʟʟᴇʀ 1963 Lɪ, Pᴀᴏ and Lɪ 1945 Aɴᴅᴇʀѕᴏɴ 1947
2. Prevention of pairing between homoeologues	*Triticum aestivum*	Sᴇᴀʀѕ and Oᴋᴏᴍᴀᴛᴀ 1958 Rɪʟᴇʏ and Cʜᴀᴘᴍᴀɴ 1958 Rɪʟᴇʏ 1960
3. Chiasma formation a) Occurrence (Desynapsis)	*Allium amplectans* *Secale cereale* *Ulmus glabra* *Triticum aestivum*	Lᴇᴠᴀɴ 1940 Pʀᴀᴋᴋᴇɴ 1943 Eᴋʟᴜɴᴅʜ-Eʜʀᴇɴʙᴇʀɢ 1949 Sᴇᴀʀѕ 1954 Kᴇᴍᴘᴀɴɴᴀ and Rɪʟᴇʏ 1962
b) Frequency	*Allium fistulosum × cepa* *Agrostis hybrids* *Secale cereale*	Mᴀᴇᴅᴀ 1942 Jᴏɴᴇѕ 1956 Rᴇᴇѕ 1955, 1957 Rᴇᴇѕ and Tʜᴏᴍᴘѕᴏɴ 1956
c) Localisation	*Allium fistulosum × cepa* *Secale cereale*	Lᴇᴠᴀɴ 1936, 1941 Mᴀᴇᴅᴀ 1942 Eᴍѕᴡᴇʟʟᴇʀ and Jᴏɴᴇѕ 1945 Rᴇᴇѕ 1955
d) Terminalisation	*Secale cereale*	Rᴇᴇѕ 1955
4. Chromosome coiling	*Matthiola incana* *Hordeum vulgare* *Lolium perenne* *Zea mays*	Lᴇѕʟᴇʏ and Fʀᴏѕᴛ 1927 Mᴏʜ and Nɪʟᴀɴ 1954 Tʜᴏᴍᴀѕ 1936 Rʜᴏᴀᴅᴇѕ 1956
5. Spindle formation	*Drosophila simulans* *Lolium perenne × Festuca pratensis* *Zea mays* *Clarkia exilis*	Wᴀʟᴅ 1936 Dᴀʀʟɪɴɢᴛᴏɴ and Tʜᴏᴍᴀѕ 1937 Cʟᴀʀᴋ 1940 Vᴀѕᴇᴋ 1962
6. Orientation and disjunction	*Zea mays*	Bᴇᴀᴅʟᴇ 1932 Rʜᴏᴀᴅᴇѕ and Vɪʟᴋᴏᴍᴇʀѕᴏɴ 1947 Rʜᴏᴀᴅᴇѕ 1952

Process	Species	Reference
6. Orientation and disjunction	*Secale cereale* *Drosophila melanogaster*	THOMPSON 1956 GOWEN 1933 LEWIS and GENCARELLA 1952 SPIELER 1963
7. Chromosome movement	*Drosophila melanogaster*	LINDSLEY and NOVITSKI 1958 PTASHNE 1960
8. Cleavage of cytoplasm	*Zea mays*	BEADLE 1932

The only primary effect on meiosis of a viable structural change is that on pairing—a mechanical aspect of chromosome behaviour in which the whole chromosome is not the unit of action. But, following crossing-over, secondary anomalies, including further structural changes, can arise. These have already been discussed (see page 67) and only the other two dimensions of genetic control will be considered here.

a) Genotypic Control of Meiosis

Development is controlled by heredity. Consequently, the examples of developmental determination of variation in the meiotic sequence we described earlier (see page 255) are, in a sense, genotypically controlled. But here we are concerned with effects owing to genotypic differences rather than those resulting from the differential activity of the genotype in different cells and tissues (see page 254).

Genotypic control has been demonstrated for many chromosome "characters" and by various means. The principal types of investigation involve a comparative study of:

(i) Mutant individuals and related forms of the same species.—Many of the earlier investigations in maize were of this kind and, understandably, major gene effects were demonstrated by this means. Inevitably, however, the modifying effects of the background genotype made themselves felt. In this type of material, the phenotypic change is large for such a character-difference is required for the recognition of the mutant form. And its meiotic pattern usually lies outside the normal range of variation for the species.

(ii) Inbred lines produced by inbreeding a normally outbreeding species.— Most of the studies of this type were undertaken with rye and, again understandably, and in contrast to the first type of investigation, multi-genic effects were revealed which were within the range of normality.

(iii) Hybrids of three kinds—those between the mutant and normal forms mentioned in (i), those between the inbred lines mentioned in (ii) and finally species-hybrids.

(iv) Types unbalanced by numerical change involving A- or B-chromosomes. —The effects of B-chromosomes have been studied in *Sorghum* (DARLINGTON and

Thomas 1941) and *Pyrgomorpha* (Lewis and John 1959). But the most intensive studies are those in relation to A-chromosomes and these have been performed especially in wheats (Riley and Chapman 1958, Sears and Okamoto 1958, Riley 1960, Riley, Kimber and Chapman 1961, Riley and Kempanna 1963).

Some of the effects demonstrated by these various means are summarised in Table 95. In compiling this table we have made no attempt to be exhaustive or even comprehensive but rather exemplary. There is, of course, a difficulty which attends a classification and interpretation of effects such as these. During meiosis, failure of pairing, for example, must needs be followed by a failure of chiasma formation; it is, we may say, epistatic. Where, therefore, there is no pairing, we cannot say whether there exists some property which would affect chiasma formation *per se*. Similarly, in the presence of univalents, especially in large numbers, there is often an extreme elongation or general failure of the spindle but this is not a necessary consequence. Where, therefore, we find both spontaneous univalents and spindle anomalies are we to attribute the latter to the former or are we to conclude that two distinct failures are involved? Pleiotropy is not easily resolved.

It is not our intention to discuss genotypic control in detail for it has been recently reviewed by Rees (1961). For our present purpose all we need note is that, owing to the unbalance of the genotype, meiosis may fail or may depart from the standard sequence in different ways and the first symptom of anomaly may occur at different stages and hence at different times.

In some hybrids, for example, meiosis may not even begin although mother cells are differentiated, or it may fail immediately after pachytene although zygotene pairing is normal (e. g. *Pavo* × *Numida*, Poll 1920) or immediately after a normal first metaphase (e. g., *Cairina* × *Anas*, Crew and Koller 1936) or sister products of first division may not show their usual synchrony at second division (*Lilium* hybrids, Brock 1954) and so on (see Stebbins 1958, White 1954, Lewis and John 1963a and also Table 57, page 149).

Many of the investigations which have demonstrated genotypic control, unlike those in rye, were not expressly designed for that purpose. But the demonstration of genotypic control by whatever method, intentional of otherwise, justifies the conclusion that genotypic differences are involved in determining the meiotic variation which is found between species, even when this cannot be shown experimentally. Thus, every investigation of meiosis is in effect a study of genotypic control.

In dealing with the chemical characterisation and environmental modification and analysis of division, we have had frequent occasion to point out that by far the majority of the investigations have been on the mitotic cycle. But genotypic control stands out as a conspicuous exception for much more is known about the effect of the genotype on meiosis than mitosis. The reasons for this are clear and have been pointed out many times. First, meiosis is a more complicated system which, presumably, requires a more intricate system of control and more genes to control it. Consequently it is likely to go wrong in more ways, more often and to be more sensitive to genotypic change. It means also, of course, that it should be more sensitive to environmental change but amenability of material has determined the greater attention to mitosis in this case. Second, mutations leading to mitotic abnormalities, unless their effect is confined to terminal mitoses (see below), would lead to death in development and for this reason they would escape detection. An abnormal meiosis, on the other hand, leads to death only in heredity and its causes are, therefore, more easily recognised.

The post-meiotic mitoses in higher plants are in a special category for they too are concerned more with sexual heredity than with development. And mutations are known in maize (e. g., the well-known polymitotic mutant) which affect these divisions but not the diploid mitoses. Mutations such as these are interesting in showing that the products of loci which affect the course of division are not always effective; they have their most pronounced effect only under certain cell conditions. The same conclusion is reached, of course, from the occurrence of more than one meiotic pathway within the individual. And it holds also for a gene in *Drosophila* which affects the imaginal cells of the larva but not those involved in the development of the larva itself (STAIGER and GLOOR 1952).

b) Position Effects at Meiosis

We come now to that third something, as HORACE would have it, which lies somewhere between structure and genotype. Position effects are well known and some of them can be seen cytologically. There are, for example, indications of such effects in relation to neo-centric activity at meiosis. The genotypic component in this effect is clear from the studies on rye and maize, so also is the developmental aspect of its control for it is seen only at meiosis (see page 251). But the genetic features are seen at two main levels. First, neo-centric activity is found only in plants of a particular genotype. Second, where the activity is present, it is exhibited only, or more often, by particular chromosomes. Now, localised and highly specific activation/inhibition of genic activity is an indispensable condition for controlled development—including the control of division (see page 288)—and one must conclude that this type of regulation characterises the self control which the genotype exercises in respect of its metabolic work. But while development necessitates a co-ordinated disenchainment of functional units, division requires a high degree of synchrony and similarity of mechanical behaviour. Thus, differential response to genotypically-controlled cellular conditions is not expected between chromosomes or parts of chromosomes within a nucleus. Indeed, this is one of the symptoms which can frequently be used to distinguish between the effects of structural hybridity and genotypic unbalance. But differential responses do obtain. And in the present connection, the most enigmatic problem of all is that the ability of a segment to respond may be determined not only by its nature but by its position as well.

This third aspect also is seen in the case of neo-centric activity, for it would appear that the controlling elements do not act directly on the regions that respond but through the centromeres of the chromosomes of which the elements are a part. Thus, a region of potential response does not show neo-centric activity if it becomes separated from a centromere as it may do by being included in an acentric fragment following crossing-over in a heterozygous paracentric inversion (RHOADES 1952).

Comparable conditions obtain in the case of coiling in the Y-chromosome of *Tribolium* (see page 235) and the disjunction of the paternally-derived X-chromosome in *Sciara* (see Darlington 1957 and Lewis and John 1963a for a detailed discussion). One example, however, is of particular significance in view of the distinctions we drew above regarding metabolic control and mechanical behaviour. It is this. In the XX females of mammals one X-chromosome becomes positively heteropycnotic during early development. The effect is random between X-chromosomes so that the paternally-derived X is affected as often as that derived from the mother. Once effected, the condition persists and the heteropycnotic X forms the sex-chromatin (Barr 1959) or so-called Barr-body of the resting nucleus. Cell-lines, however, are consistent; cells with a pycnotic paternal-X give rise to cells in which the same X is pycnotic and the same applies, *mutatis mutandis*, to those cells in which the maternally-derived X-chromosome is pycnotic. Lyon (1961, 1962) has argued that pycnosity is associated with inhibition of activity so that females are phenotypically mosaic. Here then we appear to have block-control which operates not only in relation to cytological appearance but metabolic activity as well. Indeed, the iso- and hetero-pycnotic X-chromosomes differ also in their time of DNA synthesis (Taylor 1963).

The mechanism involved in barring one X-chromosome, or at least parts of it, from making a major contribution to the metabolism of the cell and the nature of the memory involved in determining the consistency within cell-lineages are intriguing but unresolved issues. But, in connection with our present discussion, the most provocative observation is that autosomal segments translocated to an X-chromosome are induced to behave in an X-like manner both in respect of appearance and activity (Ohno and Cattanach 1962) though the effect appears to be limited in respect of both appearance and activity (Russell et al. 1962)*.

But as we mentioned earlier, special provision is often made for special chromosomes and we now see that this may apply as much to control of metabolism as it does to mechanical behaviour. It is worth noting also that while the sex-chromosomes, or parts of them, are frequently heteropycnotic at meiosis in the heterogametic sex, they are isopycnotic in the homogametic sex. Thus, in mammals, the sex differential in the soma is the reverse of that in the gamete mother cells.

Even more intricate situations obtain in the monogenic strains of *Sciara*, a genus in which the egg carries one, and the sperm transmits two, X-chromosomes. Sexual differentiation here involves many components:

(i) An X-borne determinant which is never carried by males and which, consequently, never exists in the homozygous condition in those females possessing it. The influence of this determinant extends even into those offspring which, owing to normal segregation at meiosis in the female, no longer have it.

(ii) In the individuals produced from females lacking this X-borne determinant, a heterochromatic segment received from the male parent and derived ultimately from the maternal grandmother, causes the chromosomes including it to be

---

* See note 11 Appendix.

eliminated at the 7–8th cleavage divisions. This segment is normally a constituent of the X-chromosome so that both paternally-derived X-chromosomes
are usually eliminated from such individuals which, consequently, are male.
But if translocated to an autosome, this paternally-derived H-segment will cause
its elimination. The effect is chromosome-limited. The homologous segment
and chromosome derived from the female parent is not affected at this time. But
at meiosis the H-segment causes the maternally-derived chromosome possessing
it to undergo non-disjunction at second division so that it is transmitted in
duplicate. Again the maternal X is normally involved but the autosomes can be
implicated by translocation (CROUSE 1961).

(iii) In those offspring produced from females containing the above mentioned
X-borne determinant—even those which do not have the determinant itself—a
factor it initiates decrees that only one of the paternally-derived X-chromosomes
shall be eliminated from the soma. The "choice" exercised here can be compared
with that in the case of Barr bodies to some extent. But in *Sciara*, the two
X-chromosomes from the male have the same history for at one stage they
represent sister chromatids.

(iv) The type of elimination determines the sex of the gonad and the sex of
the gonad determines the meiotic pattern which is normal in females, anomalous
in males.

We discussed the situation in greater detail in an earlier book (LEWIS
and JOHN 1963a) and, unlike one reviewer of it (DAVIDSON 1964), we would
never begrudge space for *Sciara*. Indeed we have not discussed it at greater
length here only because we expect it to figure prominently in the contribution
made to this series of monographs on modes of mitosis and meiosis (LEWIS and
JOHN in preparation).

We trust that this short discussion of the genetic control of chromosome
behaviour has been sufficient to show that the chromosomes do exercise self-
control in respect of their mechanical as well as their metabolic activities and that
the means of communication they adopt in transmitting information to themselves
is often intricate and indirect. At one extreme, the messages are transmitted or,
at least, acted upon only locally. But the system of communication can transcend
the confines of the cell and may even cross the boundary between sexual generations which is bridged only by the fusion of the gametes.

8. The Effects of the Environment on Meiosis

The term environment as used here is intended to cover all those agents,
other than the genotype, which affect the phenotype. And, since the genotype
and the environment are the only components in the determination of the phenotype, every observation of meiosis is as much a study of environmental effects
as it is of genetic effects. But here too we must be concerned with the differential
effects of the environment.

To study the effects of various physical and chemical agents on the characteristic processes of living organisms is a popular sport among biologists. And
the dividing cell has been subjected to such treatment as much as any biological
system. Meiosis has again received less attention. This has its fortunate and
unfortunate aspects.

It is unfortunate in that information which could be of value to an understanding of meiosis in causal terms is not available. It is fortunate in that many of the investigations of this type (if those on mitosis are anything to go by) have been undertaken with no more sense of inquiry than to determine the effect of X on Y. Now, where environmental variations of the kind likely to be experienced by the organism are concerned, these studies are of value because they relate to, amongst other things, the homeostatic mechanisms of the organism and, often, to the environmental determination of the meiotic and other aspects of the genetic system. But otherwise the results, for the most part, represent answers to questions which have not yet been asked and, for this reason their meaning is obscure. The useful studies are those motivated by a desire to test hypotheses. And these are few.

There are many studies on the effects of radiation on meiosis some of which we have already considered (see page 218) but we will make no attempt to review them there. Indeed, most of them are not relevant to our present purpose for they have been concerned with the effects of radiation on chromosome structure rather than on the division sequence itself. There have also been some studies on the effects of temperature and various chemical agents (see for example pages 241 and 37). But we will confine ourselves to a discussion of one investigation which to date seems to us to represent the most fruitful line of approach.

The integration of metabolism impedes an attack on specific steps and processes. But results of value are likely to be obtained only by determining the effects on division of agents whose immediate influences on metabolism can be specified with some degree of precision.

Controlled phenotypic differences between cells in space and time must be explicable in terms of the differential activity of gene products. This, in turn, may reflect genetic differences and depend on mutation, or else only differences at one or other of the various steps by which the genetic material mediates in the control of the peripheral phenotype. The first is expected to operate between individuals, the second, within individuals and, with time, even within cells. Because of the second possibility, genetically different cells need not be phenotypically different any more than genetically identical cells or individuals need be the same. Thus, in dealing with development and especially with periodic changes in a given cell of the germ line, mutation can be ignored. This means that the variations in, for example, the enzyme activities described earlier (see page 275) must be explained in terms of the mechanism by which genes regulate protein synthesis or by factors which affect the action of the proteins once formed. With regard to the former, a number of steps are now agreed upon and they are outlined in Fig. 195.

Until comparatively recently it was usual even for geneticists to explain periodic changes in terms of post-protein-production events (e. g., unmasking of inactive protein, inhibition of one activity leading to an accumulation of certain materials or their diversion to other pathways etc.), while changes in initial gene action were thought to initiate differentiation. But, particularly since the work of JACOB and MONOD (1961), there has been an increasing and, perhaps, exaggerated tendency to think in terms of gene activation rather than the activation of gene products.

Now, if an activity at a given stage (i. e. time) is a reflection of new synthesis, then agents which interfere with protein synthesis at any of the steps illustrated in Fig. 195 should have the common effect of suppressing that activity if they are applied prior to its normal time of appearance. STERN and HOTTA (1963) have studied this question by applying protein "inhibitors" to whole anthers of *Lilium* and *Trillium*, *in vitro*. In the absence of these inhibitors cultured anthers complete meiosis and the first pollen grain mitosis in a normal way and the normal periodicity in the levels of various enzyme activities etc. is maintained.

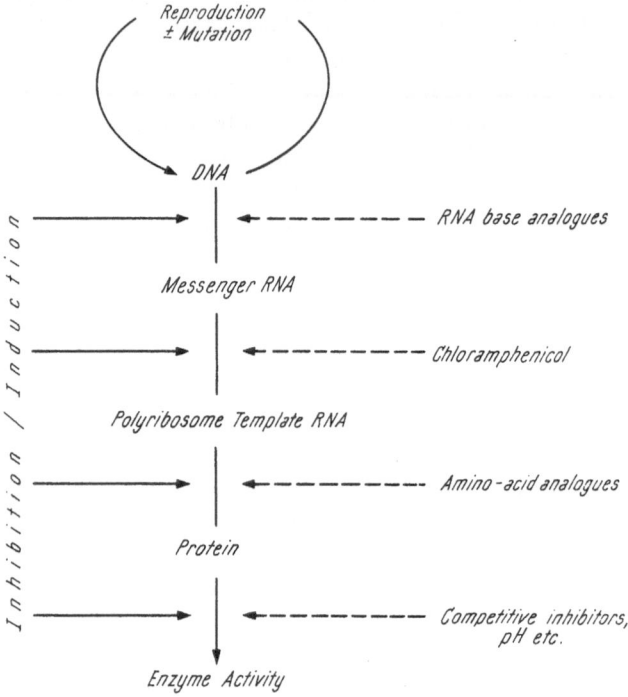

Fig. 195. The genetic mediation of enzyme production.

Three kinds of "inhibitors" were employed:

(i) RNA analogues (Azaguanine). These are expected to lead to mistranslations of the genetic code from the alphabet of DNA to that of messenger RNA.

(ii) Amino-acid analogues (5-methyl tryptophan, ethionine). These affect the translation from ribosomal template RNA to the language of proteins.

(iii) Chloramphenicol. It would appear that this acts at an intermediate step and interferes with the transmission of messenger RNA to the ribosomes.

By isotopic labelling it was found that 3–4 days elapsed between the time of application of these compounds and their arrival in the sporogenous cells. Although this is a long period it is comparatively short in relation to the length of the life-cycles of the cells concerned.

With regard to azaguanine treatment it was found that:

(i) Application after DNA synthesis but before about mid-leptotene delayed first anaphase. But, although the delay was often a matter of weeks, both meiosis and PGM 1 were eventually completed in most cases.

(ii) Mid-prophase of meiosis was comparatively insensitive so far as delay was concerned.

(iii) Application during tetrad formation was inhibitory and no reversal was observed even after 30 days, and

(iv) Treatment of young pollen grains delayed the first mitosis but it did eventually occur.

Significantly, administration of C^{14}-labelled lysine and cytosine to control anthers indicated two marked periods of RNA and protein synthesis which correspond more-or-less with the sensitive periods (i) and (iii) above. The same stages were sensitive to the other inhibitors also, but there were these differences:

a) In the case of chloramphenicol, meiosis may be completed eventually following treatment at stage (i) but first metaphase may not occur until about 90 days compared with 18 for controls and 30 for azaguanine treated cells.

b) In the case of amino-acid analogues, treatment at stage (i) led to the arrest of first division and this was not reversed even after 90 days. Control anthers complete meiosis in about 24 days and PGM 1 in about 40 days after culture.

It was further shown that the sharp peak in thymidine phosphorylase activity (which occurs between peaks of DNA-ase activity and pre-mitotic DNA synthesis in control anthers—see page 265) was completely inhibited by the above substances provided that they were applied at least a day before the normal time of appearance of the peak. Clearly, therefore, there are periodic changes in sensitivity which appear to be related to periods of synthesis.

The completion of division following treatment does not necessarily mean a completely normal sequence and absence of effect. In fact, there is a third sensitive period between the other two. It appears to be a complex one involving a number of distinct though narrow sub-intervals but these have not, as yet, been defined exactly. Treatment during this period allows cells to proceed through division but with various attendant anomalies.

Chromosome breakage was never observed but uncoiling, over-condensation (to about a quarter normal size in some cases), the appearance of "achromatic gaps" and failure of chromosomes to aggregate into single interphase nuclei were observed. With regard to the last feature, some cells with a complete haploid set of five chromosomes were found in which the chromosomes were distributed between 1–5 nuclei. But it appeared that even where each chromosome formed a separate micro-nucleus, mitosis was possible.

Although various morphological chromosome anomalies were found there was no correlation between morphology and behaviour: some chromosomes of abnormal appearance behaved normally while some of normal appearance behaved abnormally. Among the commonest behavioural anomalies were:

a) Unequal but polar segregation at anaphase-I to give, in extreme cases, an 8–2 distribution of half-bivalents followed by normal second division.

b) No polar movement at first division with cytokinesis delimiting various numbers of chromosome groups each of which proceeded to a normal second division. Since these two anomalies (a and b) tended to occur in different anthers it would appear that different synthetic processes are involved.

c) Unequal second division was not common but second telophase was often delayed and micronuclear formation was frequent, and

d) Various cleavage anomalies were found and wall formation also was subject to error.

Studies such as these are in an early stage but they represent an approach which is likely to contribute more to an appreciation of the control of meiosis than the *ad hoc* testing of anti-metabolites.

VI. The Regulation of Meiosis

In the preceding sections we have seen something of the course of meiosis in a variety of organisms and something of the events leading to meiosis in some of them. We have seen that a standard system—that generalised sequence which the elementary student learns along with the generalised cell—obtains with but minor variations in by far the majority of them. We have also seen that numerical and structural changes in the chromosome can alter the pairing arrangements of chromosomes and, thereby, they may change the mechanical units of orientation and disjunction. Finally we have seen too that the most bizarre sequences characterise some organisms, especially certain male animals. And, although the matter is not always amenable to direct experimental analysis, we must conclude that differences in the nucleus are responsible for determining the variation in their own meiotic behaviour (Table 96).

But variations on the meiotic theme are not exclusively specific or sexual. Nor are they exclusive symptoms of genetic differences for they may obtain within individuals. And here, in this recapitulation, we want to reflect on what the investigations described earlier tell us about the meiotic sequence and its regulation.

Chromosome mutations as such do not have a pronounced effect on the essential steps of meiosis. To this extent, their influence on meiosis is comparable to that which they have on exophenotypic properties (LEWIS and JOHN 1963a and b). Such changes have, of course, a meaning for the organism; they can also be exploited by the practical plant breeder in a variety of directions. But, unless they lead to genotypic unbalance, they tell us little about the regulation of meiosis itself. Thus, studies on meiosis in polyploids, for example, tell us that even homologous segments can associate only in pairs at zygotene. A study of structural hybrids, on the other hand, shows that even non-homologous segments can become related owing to the torsion set-up by the specific pairing of adjacent homologous regions. Again, observations on multiple and multivalent associations tell us something of the conditions necessary for co-orientation. And so on.

But if we ask—what determines a cell to enter meiosis; by what means is the separation of paired homologues delayed until after orientation and congression; what determines the lapse of attraction which is held to herald the onset of first anaphase—in fact if we put any question relating the sequence of events in time, then meiosis in chromosome mutants does not tell us anything that cannot be inferred, rightly or wrongly, from a study of meiosis in structurally homozygous diploids. We must, therefore, look elsewhere for information relevant to such questions.

It is likely to be found in investigations comparable to those which have met with such initial success in elucidating the way in which the genetic material conveys its information and controls its own information transfer. Division, however, is likely to elude analysis even longer than the more conventional aspects of morphogenesis because it involves everything which is involved in differentiation and something more.   An appreciation of message translation from one chemical language to another is not conceptually difficult.   But minds brought up in classical cytology are likely to boggle at the translation from the chemical to the mechanical level.   By what means, for example, do the genes which are known to affect the formation of chiasmata or the degree of their terminalisation exert their influence?

Table 96.   *Patterns of Meiosis. Some of the "Anomalous" Sequences are Normal for the Type Concerned, others Characterise Abnormal Cells or Individuals.   But they all Reflect the Genotype.*

Event	Type and Sequence									
	1	2	3	4	5	6	7	8	9	10
1. Specific pairing	+	+	+	—	+	+	±	±	+	—
2. Chiasma formation	+	—	—	—	+	+	—	—	?	—
3. Opening-out of bivs.	+	+	—	...	+	+	—	—	—	—
4. Orientation and congression-I	+	+	+	(+)	+	+	+	+	—	—
5. Separation-I	+	+	+	(+)	—	+	+	—	—	—
6. Interkinesis	+	+	+	+	+	+	+	+	—	—
7. Second division	+	+	+	+	±	—	—	—	—	—

Sequence: 1 Normal Meiosis; 2 *Phryne* type. Prolonged parallel pairing but with post-pachytene opening of centric loop. Found also in *Thaumalea* and the Mycetophilidae (see JOHN and LEWIS 1957); 3 Male *Drosophila* type. Non chiasmate meiosis associated with prolonged parallel pairing. Here and in *Pentethria, Scatopse,* some mantids and *Tipula caesia* (see Figs. 35—40) centric repulsion is not evident until metaphase-I; 4 Extreme expression of the asynaptic (*as*) gene in maize; 5 The *Narcissus* variety "Geranium" is a structural/numerical hybrid between *N. poeticus* (2 n = 14) and *N. tazetta* (2 n = 20). Apart from the anomalies owing to chromosome hybridity, the lapse of attraction between homologous chromatids which normally heralds the onset of first anaphase, fails to occur. Restitution ensues. Chiasmate associations may survive the second division also to give double non-reduction (see Figs. 134—137); 6 DOWRICK (1953) studied meiosis in the anthers of an autotetraploid clone of *Chrysanthemum atratum*. Some cells were larger than normal and had a slightly reduced-chiasma frequency (thus giving two measures of reduced precocity). First division orientation and disjunction was normal but the centromeres divide at meta-anaphase-I giving a third symptom of precocity. There is no second division in these cells. Precocious centromere division is sometimes observed in asynaptic maize also where it is accompanied by the absence of interkinesis and second prophase. However, a second division spindle invariably forms (MILLER 1963); 7 Male hybrids between *Drosophila pseudoobscura* ♀ × *D. persimilis* ♂. Non-functional binucleate "spermatids" are produced; 8 Males from the reciprocal cross of 7. Failure of pairing is usually more extensive in this case. There is no cytokinesis and non-functional, multi-nucleate "spermatids" are produced by supernumerary mitoses; 9 Male hybrids between *Pavo cristatus* and *Numida meleagris*. Also in some hybrids in *Triturus* and *Peromyscus* (see STEBBINS 1958); 10 Individuals homozygous for the recessive gene "ameiosis" in maize. Also in some animal hybrids (see Table 55) and abnormal cysts of *Melanoplus* (see pg. 62).

It may well be, of course, that since the neo-classical, information-transfer concepts of developmental genetics have not and, alone, are not likely to elucidate supra-molecular organisation, a study of division may lead to a new attitude towards differentiation just as the reverse influence is now operative. Indeed, a study of division reveals, in principal at least, everything that can be gleaned from a consideration of morphogenesis generally. Thus, the basic premise of genetics—that the phenotype is a product of interaction—is amply illustrated. Second, the genes (or gene products) affecting division are operative only under certain cellular conditions. In other words, the cytoplasm plays a discriminating role in respect of these genes as it does with regard to others. This is seen especially well in hermaphrodites with significantly different meiotic sequences on the male and female sides and in those males showing spermatocyte polymorphism.

Table 97. *Bivalent Formation in 14 Maize Plants Homozygous for the Recessive Gene asynaptic (as).*
(Data of MILLER 1963.)

Plant No.	1	2	3	4	5	6	7	8	9	10	11	12	13	14
Mean II/cell	0.00	0.002	0.15	2.00	4.75	7.03	7.27	8.86	8.88	9.00	9.17	9.76	9.95	9.95
Range	0 only	0—1	0—3	0—7	0—10	2—10	4—10	6—10	3—10	4—10	8—10	8—10	8—10	9—10
No. cells	1000	500	200	50	200	200	100	130	200	50	200	200	450	100

Third, what is superficially at least, the same mechanical sequence and what, presumably, must be determined by similar physico-chemical cellular conditions, can be determined by genotypically different complexes. This is best seen in the breakdown of meiosis in hybrids between even closely related species, especially in animals. This, again, shows us that the phenotype can conceal as well as reveal the genotype and that only a breeding experiment is really telling.

Fourth, genes affecting division have pleiotropic effects on it. Present opinions on gene action require unifunctionality. Consequently, explanations of pleiotropy consistent with this opinion must accomodate it in terms of multiple interactions rather than multiple actions. Earlier we drew attention to some of the difficulties which exist in trying to analyse causal sequences. Thus on page 62 we illustrated this problem by considering the common correlation between asynapsis and spindle anomalies. In a recent reinvestigation of asynaptic maize, MILLER (1963), like the earlier workers, found such a correlation. But, since elongate spindles were found in cells and plants showing only slight asynapsis, and at second division as well, MILLER concluded that the spindle anomaly was not a direct effect of asynapsis. In terms of immediate causation, therefore, failure of pairing and spindle elongation are separate effects of the so-called asynaptic gene. Another effect was found which could not possibly be dependent on asynapsis since it preceded it; it was the formation of syncytes owing to pre-meiotic abnormalities. These occur with low frequency in normal maize but they were found in 6 of the 14 asynaptic plants examined. However, in describing these as pleiotropic effects of the asynaptic gene, we must acknowledge the possibility that the presence of this gene may be nothing more than a marker of ancestry.

Examples of mutations which have a major effect on both the course of division and on some exophenotypic property are not common, though a careful search might well expose them. But among the classic examples is that of the recessive "claret" mutation in *Drosophila simulans*. This affects eye colour and it has also widespread effects on meiosis but only in the female. It therefore serves too as an example of the second point we made above.

Fifth, genes affecting division, like others, show different degrees of penetrance and expressivity. Both these properties are apparent, for example, in asynaptic maize (Table 97). Thus, the expression of mutations with major effect in this direction also, is affected by a background of modifying elements.

Sixth, related aspects of the phenotype can be independently controlled. Thus Rees and Thompson (1956) working with inbred rye showed that stability, as measured by the variation in respect of chiasma frequency between plants, between cells and between bivalents, is genotypically controlled. But the control shown at these three levels is not the same and that determining the mean chiasma frequency per plant is different again. Thus, the variances and the means can, to some extent, show independence in heredity.

Seventh, the preparatory events of interphase are, if anything, even more precisely timed than the division events themselves. And some, at least, of these preparations depend not on control mechanisms acting after protein synthesis but on some system which regulates the activity of the genes themselves.

Eighth, and this is something which only a study of endophenotypic properties can reveal, the events of meiosis have no adaptive significance for the individual in which they occur. Therefore, they have to be selected in retrospect.

Half of biology is tied-up in these generalisations but let us deal more specifically with the regulation of meiosis. The study of a sequence tempts one to infer a linear chain of cause and effect where an effect becomes a cause. But studies of cause and effect are likely to lead one into endless and ever increasing or decreasing circles so we must be content with an explanation at some level.

In order to distinguish the facultative from the obligatory relationships one must be able to interfere, or observe Nature's interference, with a sequence. And when the evidence of anomalous sequences (in normal individuals, mutants, hybrids or organisms subject to treatment) is taken into account it is surprising how few obligatory relationships there are. And most of these seem to run backwards. For example, where chiasma formation occurs, we can confidently predict that it was preceded by pairing. But pairing is not sufficient for chiasma formation. At a different level we seem justified in concluding that where a pool of deoxyribosidic material accumulates in the post-meiotic anther, there was prior DNA-ase activity but the presence of DNA-ase does not assure a deoxyribosidic pool.

Admittedly, some sequences, once initiated, tend to run to completion. Indeed, in general, the whole division sequence seems to proceed normally in the presence of inhibitors of energy metabolism though cells may be arrested in the antephase. Again, there are few inhibitors (short of those which eventually kill the cell) which can prevent the completion of mitotic prophase once it has started. In fact, one generally finds that the chromosomes tend to complete their morphological changes whether an effective division is achieved or not.

Be these as they may, it is clear even from the few examples given in Table 96 that the standard meiotic sequence does not owe its apparent inevitability to a chain of cause and effect. The events of the sequence are, to a large extent, independently controlled. It would seem, therefore, that sequence, synchrony, concordance—indeed all those properties which make for the proper completion of meiosis—owe their precise timing not to the dependence of the events on each other but rather to their common subjugation to some clock external to them.

It is common for cytologists to explain meiotic breakdown in terms of timing errors. Some have considered explanations in terms such as these to be facile. Perhaps they are in so far as the cytologist is usually concerned with the timing of visible events. It is quite clear, however, that the crux of morphogenesis is not the simple question of information transfer; it is in those regulatory mechanisms which ensure that there is a proper time and place for everything. And these are more elusive than pimpernels.

So far as division, even entry into division, is concerned there are many requirements and presumably there must be as many stages at which gene (or gene-product) regulation could be effective. Cells could be held up at these steps and released or "triggered" into division from them. And, as always, it is easier to analyse the components and see how they can be put apart than to understand how they are held together and co-ordinated.

Development is currently being discussed in the terms of communication engineering and computer programming though, for the most part, this is simply decanting old wine into new bottles. There is, however, little to suggest that the order to divide triggers a pre-programmed sub-programme. On the contrary, the cell seems to be in the position of Eurydice or Lot's wife and cannot refrain from looking back.

Triggers suggest themselves most forcibly in tissues showing synchronised division. Thus, in the anther we see three main zones—the wall, in which divisions soon cease; the tapetum which, in lilies, undergoes synchronous mitoses during the meiotic period and thirdly the pollen mother cells which divide meiotically and with considerable synchrony. The dividing cells show synchrony not only in respect of the division stages themselves but, as we have seen, in the preparatory events as well. We can hardly doubt, therefore, that there is a certain uniformity of material necessary for division. This, in turn, can be attributed either to easy communication between the cells or, more likely in our view, to equal but independent contact with the internal environment. Quite apart from other considerations, this preference seems to be demanded by the greater synchrony which obtains between cells at the same transverse level in different pollen sacs compared with that between cells at different levels in the same sac. The tapetum, too, represents a synchronous tissue but it and the sporogenous cells have cycles which are different in time and kind. This also suggests that direct communication between cells is not responsible for synchrony because it is difficult to conceive of a message which passes between all the tapetal cells, for example, more easily and rapidly than it does between adjacent cells of different kinds.

It is interesting to note that the very close synchrony of the pollen mother cells is not presaged by a close concordance in the pre-meiotic mitoses. In fact, the cells are brought into synchrony from asynchrony. To this extent, natural synchronisation can be compared with that which can be induced artificially in such organisms as *Tetrahymena* and *Amoeba*. But there the similarity ends for natural synchrony of division is correlated with uniformity of growth. Artificial synchrony, on the other hand, depends on holding-up some cells, so that younger or slower ones can catch-up with them, or alternatively setting-back cells in proportion to their age or speed.

The occurrence of well-circumscribed systems within the anther requires an inference of inequality as does the differentiation itself. It is quite clear from the well-known grafting experiments on *Stentor* and *Acetabularia* that messages relating to time of divsion pass between nuclei and cytoplasm. And a protein which migrates from one resting nucleus to another and from dividing nuclei to the cytoplasm and back again has been described in *Amoeba* (PRESCOTT 1963, GOLDSTEIN 1963). But if there are triggers they are not universal ones as is clear from GELFANT's (1963) work on mitosis in mouse epidermis. And, if there are pre-programmed sub-programmes in relation to division, then each cell has many of them.

Normally one must regard the cell as the unit of entry into division though in the case of plasmodia and certain cellular tissues, a larger unit can be considered as we have discussed above. But the evidence from binucleate cells shows that the cell need not be the unit. For example, JAIN (1958) described abnormal binucleate PMC's induced in *Lolium* by heat treatment in which the two nuclei, one central and one peripheral, showed signs of entry into meiosis at the same time. But the concordance was shortlived. A similar situation has been described in an oat hybrid where asynchrony became apparent by zygotene or even earlier (HOLDEN and MOTTA 1956). In both these cases, the peripherally placed nucleus did not proceed beyond very early prophase. It would appear, therefore, that the conditions and materials which determine entry into division are not the same as those which allow its continuation. It would appear too that the two classes of materials have different distributions within the cell for in both *Lilium* and *Avena* differentiation was apparent at an early, pre-meiotic stage, there being no nucleolus in the peripheral nuclei. Indeed, there may be a third class of materials necessary for completion of division for, in *Lolium*, even the central nucleus showed failure of pairing and failure of anaphase separation.

A study of other normal and anomalous situations allows one to go further. Our markers are such that the unit of entry into division cannot be something smaller than a whole nucleus. But it is quite clear that the nucleus is not a unit in respect of either preparation for, or continuation of, prophase. In the case of DNA synthesis in the sperm mother cells of grasshoppers, different chromosomes within the one nucleus show more asynchrony than different cells in the same cyst. In meiocytes of male grasshoppers asynchrony of condensation is also apparent. This aspect of discordance in particular has led to the notion that regions which differ in their condensation cycles differ in kind from each other. The distinction is pin-pointed by the coining of the nouns euchromatin and heterochromatin. Clearly these have overtones which are not present in the words used to describe the condition of pycnosity or the property of allocycly. It is difficult to avoid the implications of the nouns in many cases but in others it is quite clear that only differences in position or developmental history are involved (Table 98). Chance too and, perhaps, differentiation or paramutation may also play their part.

Table 98. *Examples of Intra-nuclear Differential Behaviour at Meiosis within Chromosomes and between Both Homologous and Non-homologous Chromosomes.*

Type	Differential Chromosomes	Differential in Respect of:
1. *Allium cepa* × *fistulosum*	II^{cepa}/II^{fistulosum}	Chiasma localisation
2. *Crepis multiflora* × *zacintha*	II^{multiflora}/II^{zacintha}	Rates of contraction and anaphase-I separation
3. Many *Heteroptera*	Autsomes/sex and m-chromosomes	Chiasma formation and condensation
	Sex chromosomes/m-chromosomes	Post versus pre-reduction
4. *Sciara* ♂	$(X_m + A_m + L)/(X_p + A_p)$	A-I disjunction
	$X_m/(A_m + L)$	A-II disjunction
5. *Miastor* ♂	$6\,S/(6\,S + 36\,E)$	Condensation, polarisation and disjunction
6. *Miastor* ♀	$12\,S/36\,E$	Pairing and condensation
7. *Phenococcus* ♂	Paternal/maternal	Condensation and division
8. *Tribolium confusum*	X$^{pairing\ seg.}$/Y$^{pairing\ seg.}$	Condensation
9. Sex ratio *Drosophila*	Y/Autosomes	Condensation and division
10. *Zea mays*	H-knob chromosomes/others	Neo-centric activity

It would, appear therefore, that we have to contend with—or, rather, the cell has to co-ordinate—effectors of various kinds. These are expected to differ in the time and place of their production which may be outside the cells, tissues or even the organs they effect. They can be expected to differ too in the amount and the means of their movement, i. e. in their communication channels. We have to consider also the competence of different cells and nuclei or different chromosomes and parts of them to react to these morphogenetic substances.

At present we are not in a position to elucidate the matter but if our existing knowledge is put down it does at least serve to define the limits of our ignorance. Indeed, the analyst finds himself in much the same position as the policeman who religiously records every piece of evidence, however small and apparently trivial, because he cannot know until the mystery is solved which items are required for its solution.

All hypotheses which aim to explain chromosome behaviour at meiosis must, of course, be based on the behaviour of the chromosomes. But their behaviour varies. Indeed, it is doubtful if there is a single aspect of chromosome behaviour, however common, which does not admit of exception. This is true of even the almost-always relationship between nuclear division and prior DNA-synthesis. Perhaps, therefore, we should refrain from citing examples of anomalous behaviour as arguments against hypotheses designed to account for phenomena which the anomalous examples do not show.

Codetta

In this review we have been concerned with the cause, the course and the control of meiosis rather than its consequences. These, of course, are known in qualitative terms even to the most elementary student of biology. But he, like some who should know better, often regards meiosis—like a sexual reproduction—to be one system and in isolation. It is, however, subject to variation as is the other major component of the genetic system, namely the breeding system. And the variation in both, being the subject of genotypic control, is also subject to selection. We must conclude that they are selected in relation to each other, their mutual selection being responsible for their mutual adaptation (DARLINGTON 1956, 1958). We cannot consider the matter here because it clearly requires an account of breeding systems. Indeed it is for this reason that we have made no attempt to deal with the phylogenetic aspects of the meiotic system.

This review will, however, have served its purpose if it helps those who come after to find their way across the breaking-barriers between the plant and animal kingdoms and across the cellular labrynths which meet at what DARLINGTON has called the microscopic middle. We only wish that our own contribution to the solution of the problems raised could have been greater. But, perhaps, defining our own limits and ignorance may serve to indicate the points from which abler hands and minds can now proceed.

References

AHLOOWAHLIA, B. S., 1962: Study of a translocation in diploid rye. Genetica 33, 128–144. 1963. Study of a translocation in tetraploid rye. Genetica 33, 207–221.

AKSTEIN, E., 1962: The chromosomes of *Aedes aegypti*, and of some other species of Mosquitoes. Bull. Res. Council Israel B, Zool. 11, 146–155.

ALEXANDER, D. E., 1956: Further evidence for crossing over between non-homologous chromosomes during megasporogenesis of haploids. Maize Genetics Coop. News Letter 30, 37–38.

ALEXANDER, M. L., 1952: The effect of two pericentric inversions upon crossing-over in *Drosophila melanogaster*. Univ. Texas Publ. 5204, 219–226.

ALFERT, M., 1955: Cytochemical detection of a difference in protein composition between micronucleus and macronucleus of *Tetrahymena pyriformis*. Anat. Rec. 122, 428–429.

— and I. GESCHWIND, 1953: A selective staining method for the basic proteins of cell nuclei. Proc. Nat. Acad. Sci. U.S. 39, 991–995.

ANDERSON, E., 1947: A case of asyndesis in *Picea albies*. Hereditas (Lund) 33, 301–347.

ANKEL, W. E., 1930: Die atypische Spermatogenese von *Janthina (Prosohranchia Stenoglossa)*. Z. Zellforsch. 11, 491–608.

ANSLEY, H. R., 1954: A cytological and cytophotometric study of alternative pathways of meiosis in the house centipede *Scutigera forceps* (RAFINESQUE). Chromosoma (Berlin) **6**, 656–695.
— 1958: Histones of mitosis and meiosis in *Loxa flavicolis* (Hemiptera). J. Biophys. Biochem. Cytology **4**, 59–62.
AR-RUSHDI, A. H., 1957: The cytogenetics of variegation in a species hybrid in *Nicotiana*. Genetics **42**, 313–325.
— 1963: The cytology of achiasmatic meiosis in the female *Tigriopus* (Copepoda). Chromosoma (Berlin) **13**, 526–539.
ASTBURY, W. T., 1934: X-ray studies of protein structure. Cold Sprg. Har. Symp. Quant. Biol. **2**, 15–27.
AVANZI, M. G., 1950: Frequenza e tipi di aberrazzione cromosomiche indotte da alcuni derivati della naftalene. Caryologia (Pisa) **3**, 165–180.

BAJER, A., 1963: Observations on dicentrics in living cells. Chromosoma (Berlin) **14**, 18–30.
— and G. ÖSTERGREN, 1961: Centromere-like behaviour of non-centromere bodies I. Neo-centric activity in chromosome arms at mitosis. Hereditas (Lund.) **47**, 563–598.
BAKER, W. K., 1958: Crossing-over in heterochromatin. Amer. Nat. **92**, 59–60.
BARBER, H. N., 1940: The suppression of meiosis and the origin of diplochromosomes. Proc. Roy. Soc. (London), Bl. **128**, 170–185.
BARR, M., 1959: Sexual dimorphism in interphase nuclei. Amer. Jour. Hum. Genet. **12**, 118–127.
BARTON, D. W., 1949: Pachytene morphology of the tomato chromosome complement. Amer. Journ. Bot. **37**, 639–643.
— 1951: Localised chiasmata in the differentiated chromosomes of the tomato. Genetics **36**, 374–381.
BATTAGLIA, E., 1949: Mutazione cromosomiche in *Scilla peruviana* L. Caryologia (Pisa) **1**, 144–174.
— 1956: A new type of segregation of the sex-chromosomes in *Dysdercus koenigii* Fabr. (Hemiptera-Pyrrhocoridae). Caryologia (Pisa) **8**, 205–213.
BAUER, H., 1931: Die Chromosomen von *Tipula paludosa* Meig. in Eibildung und Spermatogenese. Z. Zellforsch. u. Mikr. Anat. **14**, 138–193.
— 1932: Die Histologie des Ovars von *Tipula paludosa* Meig. Zeit. Zool. **143**, 53–76.
— 1946a: Karyologie und Zytogenetik der Tiere. Fiat. Rev. Biol. **2**, 61–75.
— 1946b: Gekoppelte Vererbung bei *Phyrne fenestralis* und die Beziehung zwischen Faktorenaustausch und Chiasmabildung. Biol. Zbl. **65**, 108–115.
— 1952: Die Chromosomen im Soma der Metazoen. Verh. dtsch. Zool. Ges. 1952, 252–268.
— and W. BEERMANN, 1952: Der Chromosomencyclus der Orthocladiinen (Nematocera, Diptera). Z. Naturforsch. **7b**, 557–563.
— R. DIETZ, and CH. RÖBBELEN, 1961: Die Spermatocytenteilungen der Tipuliden III. Mitteilung. Das Bewegungsverhalten der Chromosomen in Translokationsheterozygoten von *Tipula oleracae*. Chromosoma (Berlin) **12**, 116–189.
BAYREUTHER, K., 1955: Holokinetische Chromosomen bei *Haematopinus suis* (Anoplura, Haematopinidae). Chromosoma (Berlin) **7**, 260–270.
BEADLE, G. W., 1930: Genetical and cytological studies of mendelian asynapsis in *Zea Mays*. Cornell Univ. Agric. Exp. Stat. Mem. **129**, 1–23.
— 1932: A gene in *Zea Mays* for failure of cytokinesis during meiosis. Cytologia **3**, 142–155.
— 1934: Crossing over in attached-X triploids of *Drosophila melanogaster*. Genetics **29**, 277–309.
— 1935: Crossing over near the spindle attachment of the X-chromosomes in attached-X triploids of *Drosophila melanogaster*. Genetics **20**, 179–191.
BEERMANN, S., 1959: Chromatin-Diminution bei Copepoden. Chromosoma (Berlin) **10**, 504–514.
BEERMANN, W., 1954: Weibliche Heterogametie bei Copoden. Chromosoma (Berlin **6**, 381–396.
— 1956: Inversions-Heterozygotie und Fertilität der Männchen von *Chironomus* Chromosoma (Berlin) **8**, 1–11.
BĚLAŘ, K., 1923: Untersuchungen an *Actinophrys sol* EHRENBERG I. Die Morphologie des Formwechsels. Arch. Protistenk. **46**, 1–96.
— 1926: Der Formwechsel der Protistenkerne. Ergebn. Zool. **6**, 1.
— 1928: Die cytologischen Grundlagen der Vererbung. Borntraeger, Berlin.

Bělař, K., 1929: Beiträge zur Kausakanalyse der Mitose. II Untersuchungen an den Spermatocyten von *Chorthippus (Stenobothus) lineatus* Panz. Archiv. f. Entwicklungsmechanik der Organismen **18**, 359–484.

Bell, S. L., and E. D. Garber, 1961: The genus Collinsia XII. Cytogenetic studies of interspecific hybrids involving species with sessile flowers. Bot. Gaz. **122**, 210–218.

Bell, S., and S. Wolff, 1964: Studies on the mechanism of the effect of fluorodeoxyuridine on chromosomes. Proc. Nat. Acad. Sci. (Wash.) **51**, 195–202.

Belling, J., 1931: Chiasmas in flowering plants. Univ. Calif. Publ. Bot. **16**, 311–338.

— and A. F. Blakeslee, 1924: The configurations and sizes of the chromosomes in the trivalents of 25-chromosome Daturas. Proc. Nat. Acad. Sci. (Wash.) **10**, 116–120.

— — 1927: The assortment of chromosomes in haploid Daturas. La Cellule **37**, 353–365.

Bendich, A., 1952: Studies on the metabolism of the nucleic acids. Exper. Cell Res., Suppl. **2**, 181–191.

Bergmann, B., 1941: Studies on the embryo sac mother cell and its development in *Hieracium* sub genus *Archieracium*. Sv. Bot. Tidskr. **35**, 1–42.

Block, D. P., and G. C. Godman, 1955: Evidence of differences in the desoxyribonucleoprotein complex of proliferating and non-dividing cells. J. Biophys. Biochem. Cytology **1**, 531–550.

Bosemark, N. O., 1950: Accessory chromosomes in *Festuca pratensis* IV. Cytology and inheritance of small and large accessory chromosomes. Hereditas (Lund) **42**, 235–259.

— 1957: Further studies on accessory chromosomes in grasses. Hereditas (Lund) **48**, 236–297.

Bowen, R. H., 1922: Studies on insect spermatogenesis IV. The phenomenon of polymegaly in the sperm cells of the family Pentatomidae. Proc. Amer. Acad. Arts Sciences **57**, 391–423.

Brachet, J., 1957: Biochemical cytology. Academic Press Inc., N.Y.

Brink, R. A., and D. C. Cooper, 1935: A proof that crossing over involves an exchange of segments between homologous chromosomes. Genetics **20**, 22–35.

Brock, R. D., 1954: Fertility in *Lilium* hybrids. Heredity (Lond.) **8**, 409–420.

— 1955: Chromosome balance and endosperm failure in hyacinths. Heredity (Lond.) **9**, 199–222.

Brown, S. W., 1949: The structure and meiotic behaviour of the differentiated chromosomes of tomato. Genetics **34**, 437–461.

— 1954: Mitosis and meiosis in *Luzula campestris* DC. Univ. Calif. Publ. Bot. **27**, 231–278.

— 1963: The *Comstockiella* system of chromosome behaviour in the armoured scale insects (Coccoidea: Diaspididae). Chromosoma (Berlin) **14**, 360–406.

— and W. A. Nelson-Rees, 1961: Radiation analysis of a lecanoid genetic system. Genetics **46**, 983–1007.

— and W. Welshons, 1955: Maternal ageing and somatic crossing-over of attached X-chromosomes. P.N.A.S. (Wash.) **41**, 209–215.

— and D. Zohary, 1955: The relationship of chiasmata and crossing-over in *Lilium formosanum*. Genetics **40**, 850–873.

Bullough, W. S., 1952: The energy relations of mitotic activity. Biol. Rev. Cambridge philos. Soc. **27**, 133–168.

— and M. Johnson, 1951: The energy relations of mitotic activity in adult mouse epidermis. Proc. Roy. Soc. London B. **138**, 562–575.

Burnham, C. R., 1946: A gene for "long" chromosomes in barley. Genetics **31**, 212–213 (Abst.).

— 1950: Chromosome segregation in translocations involving chromosome—6 in maize. Genetics **35**, 446–481.

— 1956: Chromosomal interchange in plants. Bot. Rev. **22**, 419–552.

— 1962: Discussions in cytogentics. Burgess Pub. Co.

Bytinski-Salz, H., 1934: Verwandtschaftsverhältnisse zwischen den Arten der Gattungen *Celerio* und *Pergesa* nach Untersuchungen über die Zytologie und Fertilität ihrer Bastarde. Biol. Zbl. **54**, 300–313.

Callan, H. G., 1941: The sex-determining mechanism of the earwig, *Forficula auricularia*. J. Genetics **41**, 349–379.

— 1942: Heterochromatin in *Triton*. Proc. Roy. Soc. London B **130**, 324–335.

— 1949a: Chiasma interference in diploid, tetraploid and interchange spermatocytes of the earwig *Forficula auricularia*. J. Genetics **49**, 209–213.

CALLAN, H. G., 1949b: "Chromosomes" pp. 70–88 in New Biology 7.
— 1963: The nature of lampbrush chromosomes, Internat. Rev. Cytol. 15, 1–34.
— and P. A. JACOBS, 1957: The meiotic process in *Mantis religiosa* L. males. J. Genetics 55, 200–217.
— and L. LLOYD, 1960: Lampbrush chromosomes of crested newts *Triturus cristatus* (Laurenti). Phil. Trans. Roy. Soc. B 243, 135–219.
— and G. MONTALENTI, 1947: Chiasma interference in mosquitoes. J. Genetics 48, 119–134.
— and H. SPURWAY, 1951: A study of meiosis in interracial hybrids of the newt *Triturus cristatus*. J. Genetics 50, 235–249.
CAROTHERS, E. E., 1941: Interspecific hybridisation in the Acrididae (*Trimerotropis citrina* × *T. maritima*). Genetics 26, 144.
CARSON, H. L., 1946: The selective elimination of inversion dicentric chromatids during meiosis in the eggs of *Sciara impatiens*. Genetics 31, 95–113.
CASE, M. F., and N. H. GILES, 1958: Evidence from tetrad analysis for both normal and aberrant recombination between allelic mutants in *Neurospora crassa*. Proc. Nat. Acad. Sci. (Wash.) 44, 378–390.
CASTRO, D. DE, 1950: Notes on two cytological problems of the genus *Luzula* DC. Genetica Iberica (Lisbon) 2, 201–209.
— A. CAMARA, and N. MALHEIROS, 1949: X-rays in the centromere problem of *Luzula purpurea* Link. Genetica Iberica (Lisbon) 1, 49–54.
— M. NORONHA-WAGNER, and A. CAMARA, 1954: Two X-ray induced translocations in *Luzula purpurea*. Genetica Iberica (Lisbon) 6, 3–18.
CATCHESIDE, D. G., 1932: The chromosomes of a new haploid *Oenothera*. Cytologia (Tokyo) 4, 68–113.
— 1950: The B-chromosomes of *Parthenium argentatum*. Genetica Iberica (Lisbon) 2, 139–148.
CAVALLI-SVORZA, L. L., and J. L. JINKS, 1956: Studies on the genetic system of *Escherichia coli*. J. Genetics 54, 87–112.
CHANDLER, C., W. M. PORTERFIELD, and A. B. STOUT, 1937: Microsporogenesis in diploid and triploid types of *Lilium tigrinum* with special reference to abortions. Cytologia (Tokyo) Fujii Jub. vol, 756–784.
CHANDRA, H. S., 1962: Inverse meiosis in triploid females of the mealy bug, *Planococcus citri*. Genetics 47, 1441–1454.
CHAYEN, J., 1953: Ascorbic acid and its intracellular localisation with special reference to plants. Internat. Rev. Cytol. 2, 77–131.
CHIN, T. C., 1943: Cytology of the autotetraploid rye. Bot. Gaz. 104, 627–632.
CHRISTENSEN, B., 1960: A comparative cytological investigation of the reproductive cycle of an amphimictic diploid and a parthenogentic triploid form of *Lumbricillus lineatus* (O. F. M.) (Oligochaeta, Enchytraeidae). Chromosoma (Berlin) 11, 365–379.
— 1961: Studies on cyto-taxonomy and reproduction in the Enchytraeidae. Hereditas (Lund) 47, 385–449.
CLARK, F. J., 1940: Cytogenetic studies of divergent meiotic spindle formation in *Zea mays*. Amer. J. Bot. 27, 547–559.
CLEVELAND, L. R., 1938: Origin and development of the achromatic figure. Biol. Bull. 74, 41–55.
— 1949: The whole life cycle of chromosomes and their coiling systems. Trans. Amer. Phil. Soc. N. S. 39, 1–100.
COLEMAN, L. C., 1941: The relation of chromocenters to the differential segments in *Rhoeo discolor* Hance. Amer. J. Bot. 28, 742–748.
— 1943: Chromosome structure in the Acrididae with special reference to the X-chromosome. Genetics 28, 2–8.
— 1948: The cytology of some western species of *Trimerotropis* (Acrididae). Genetics 33, 519–528.
— and B. B. HILLIARY, 1941: The minor coil in meiotic chromosomes and associated phenomena as revealed by the Feulgen technique. Amer. J. Bot. 28, 464–469.
COOPER, K. W., 1941: Bivalent structure in the fly *Melophagus ovinus* L. (Pupipara, Hippoboscidae). Proc. Nat. Acad. Sci. (Wash.) 27, 109–114.
— 1944: Analysis of meiotic pairing in *Olfersia* and consideration of the reciprocal chiasmata hypothesis of sex chromosome conjugation in male *Drosophila*. Genetics 29, 537–568.
— 1945: Normal segregation without chiasmata in female *Drosophila melanogaster*. Genetics 30, 472–484.

COOPER, K. W., 1951: Compound sex chromosomes with anaphasic precosity in the male mecopteran *Boreus brumalis* Fitch. J. Morph. **89**, 37–57.
— S. ZIMMERING, and J. D. KRIVSCHENKO, 1955: Inter-chromosomal effects and segregation. Proc. Nat. Acad. Sci. (Wash.) **41**, 911–914.
CREIGHTON, H. B., and B. McCLINTOCK, 1931: A correlation of cytological and genetical crossing over in *Zea mays*. Proc. Nat. Acad. Sci. (Wash.) **17**, 492–497.
CREW, F. A. E., and P. C. KOLLER, 1936: Genetical and cytological studies of the intergenric hybrid of *Cairina moschata* and *Anas platyryncha platyryncha*. Proc. Roy. Soc. Edinburgh **56**, 210–241.
CROUSE, H. V., 1954: X-ray breakage of lily chromosomes at first meiotic metaphase. Science **119**, 485–487.
— 1960: The controlling element in sex chromosome behaviour in *Sciara*. Genetics **45**, 1429–1443.
— 1961: Irradiation of condensed meiotic chromosomes in *Lilium longiflorum*. Chromosoma (Berlin) **12**, 190–214.

DALTON, H. C., and J. Hall, 1950: Gene action in the axolotl. Yr. Book Carn. Instit. Wash. **49**, 188.
D'AMATO, F., 1949: Risulti di una analisi carioembriologica in una popolazione di *N. fragans*. Caryologia (Piza) **1**, 194–200.
— 1950: The chromosome breaking activity of chemicals as studied by the *Allium cepa* test. Publ. Staz. Zool. Napoli **22** (Suppl.), 158–170.
—1951: Mutazioni chromosomiche spontanee in *Pisum*. Caryologia (Piza) **3**, 285–293.
DARLINGTON, C. D., 1926: Chromosome studies in the Scilleae. J. Genetics **16**, 237–251.
— 1928: Studies in *Prunus* I and II. J. Genetics **19**, 213–256.
— 1931: Meiosis. Biol. Rev. **6**, 221–264.
— 1932: The control of the chromosomes by the genotype and its bearing on some evolutionary problems. Amer. Nat. **96**, 25–51.
— 1934: Anomalous chromosome pairing in the male *Drosophila pseudoobscura*. Genetics **19**, 95–118.
— 1935a: The internal mechanics of the chromosomes II. Prophase pairing at meiosis in *Fritillaria*. III. Relational coiling and crossing-over in *Fritillaria*. Proc. Roy. Soc. (London) B **118**, 59–73 and 74–96.
— 1935b: The time, place and action of crossing-over. J. Genetics **31**, 185–212.
— 1936a: The internal mechanics of the chromosomes V. Relational coiling of chromatids at meiosis. Cytologia (Tokyo) **7**, 248–255.
— 1936b: Crossing-over and its mechanical relationships in *Chorthippus* and *Stauroderus*. J. Genetics **33**, 465–500.
— 1937: Recent Advances in Cytology. 2nd edition. London, Churchill.
— 1939a: The Evolution of Genetic Systems. 1st edition. Cambridge, Univ. Press.
— 1939b: Misdivision and the genetics of the centromere. J. Genetics **37**, 341–364.
— 1939c: The genetical and mechanical properties of the sex chromosomes. V. *Cimex* and the Heteroptera. J. Genetics **39**, 101–137.
— 1940a: Prime variables of meiosis. Biol. Rev. **15**, 307–322.
— 1940b: The origin of iso-chromosomes. J. Genetics **39**, 306–308.
— 1941: Polyploidy, crossing-over and heterochromatin in *Paris*. Ann. Bot. N. S. **5**, 203–216.
— 1949: The working units of heredity. Hereditas (Suppl. vol.), 189–200.
— 1956a: Natural populations and the breakdown of classical genetics. Proc. Roy. Soc. (London) B **145**, 350–364.
— 1956b: Chromosome Botany. Allen and Unwin, London.
— 1957: Messages and movements in the cell. Conference on Chromosomes, Wageningen. Tjeenk Willink, Zwolle.
— 1958: The Evolution of Genetic Systems. 2nd edition. Oliver and Boyd, Edinburgh.
— 1960: Chromosomes and the theory of heredity. Nature **187**, 892–895.
— and A. A. MOFFETT, 1930: Primary and secondary chromosome balance in *Pyrus*. J. Genetics **22**, 130–151.
— and S. O. S. DARK, 1932: The origin and behaviour of chiasmata II. *Stenobothrus parallelus*. Cytologia (Tokyo) **3**, 169–185.
— and P. C. KOLLER, 1934: The genetical and mechanical properties of the sex-chromosomes I. *Rattus norvegicus* ♂. J. Genetics **29**, 159–173.
— and A. E. GAIRDNER, 1937: The variation system in *Campanula persicifolia*. J. Genetics **35**, 97–128.
— and P. T. THOMAS, 1937: The breakdown of cell-division in a *Festuca-Lolium* derivative. Ann. Bot. **1**, 747–762.

DARLINGTON, C. D., and M. B. UPCOTT, 1939: The measurement of packing and contraction in chromosomes. Chromosoma (Berlin) 1, 23–32.
— and L. F. LACOUR, 1940: Nucleic acid starvation of chromosomes in *Trillium*. J. Genetics 40, 185–213.
— and P. T. THOMAS, 1941: Morbid mitosis and the activity of inert chromosomes in *Sorghum*. Proc. Roy. Soc. (London) B 130, 127–150.
— and M. B. UPCOTT, 1941: The activity of inert chromosomes in *Zea mays*. J. Genetics 41, 275–296.
— and L. F. LACOUR, 1941: The genetics of embryo sac development. Ann. Bot., N. S. 5, 547–562.
— and P. C. KOLLER, 1947: The chemical breakage of chromosomes. Heredity (London) 1, 187–221.
— and L. F. LACOUR, 1950: Hybridity selection in *Campanula*. Heredity (London) 4, 217–248.
— and L. F. LACOUR, 1952: The effects of irradiation on meiosis. Heredity (London) 6, (Suppl.), 41–55.
— and A. P. WYLIE, 1953: A dicentric cycle in *Narcissus*. Heredity (London) 6, (Suppl.), 197–213.
— and A. HAQUE, 1955: The timing of mitosis and meiosis in *Allium ascalonicum*: a problem of differentiation. Heredity (London) 9, 117–127.
— and M. KEFALLINOU, 1957: Correlated chromosome aberrations at meiosis in *Gasteria*. Chromosoma (Berlin) 8, 364–370.
— and C. G. VOSA, 1963: Bias in the internal coiling direction of chromosomes. Chromosoma (Berlin) 13, 609–622.
DAS, K., 1955: Cytogenetic studies of partial sterility in an X-ray irradiated barley. Indian Jour. Genetics and Plt. Breeding 15, 99–111.
DAVIDSON, D., 1957: The irradiation of dividing cells I. The effects of X-rays on prophase chromosomes. Chromosoma (Berlin) 9, 39–60.
— 1964: Review of "Chromosome Marker." New Phytologist 63, 125–127.
DE, D. N., 1961: Autoradiographic studies of nucleoprotein metabolism during the division cycle. The Nucleus 4, 5–24.
DEUFEL, J., 1951: Untersuchungen über den Einfluß von Chemikalien und Röntgenstrahlen auf die Mitose von *Vicia faba*. Chromosoma (Berlin) 4, 239–272.
DHILLON, T. S., and E. D. GARBER, 1962: The genus *Collinsia* XVI. Supernumerary chromosomes. Amer. J. Bot. 49, 168–170.
DIETZ, R., 1954: Multiple Geschlechtschromosomen bei dem Ostracoden *Notodromonas monacha*. Chromosoma (Berlin) 6, 397–418.
— 1955: Zahl und Verhalten der Chromosomen einiger Ostracoden. Naturforsch. 10b, 92–95.
— 1958: Multiple Geschlechtschromosomen bei den Cypriden Ostracoden, ihre Evolution und ihr Teilungsverhalten. Chromosoma (Berlin) 9, 359–440.
— and H. A. WENT, 1966: Chromosome Movement. Protoplasmatologia (in prepn.).
DIRSH, V. M., 1958: Revision of the genus *Eyprepocnemis* Fieber, 1853 (Orthoptera: Acridoidea) Proc .Roy. Ent. Soc. (London) B 27, 33–45.
DOBZHANSKY, TH., 1934: Studies on hybrid sterility I. Spermatogenesis in pure and hybrid *Drosophila pseudoobscura*. Z. Zellforsch. 21, 169–223.
DOERMANN, A. H., 1963: Recombination in bacteriophage T 4 and the problem of high negative interference Proc. XIth Int. Congr. Genetics 2, 69–79.
DOWRICK, G. J., 1953: The chromosomes of *Chrysanthemum* III. Meiosis in *C. atratum*. Heredity (London) 7, 219–226.
DOYLE, G. G., 1963: Preferential pairing in structural heterozygotes of *Zea mays*. Genetics 48, 1011–1027.
DYER, A. F., 1963: Allocyclic segments of chromosomes and the structural heterozygosity that they reveal. Chromosoma (Berlin) 13, 545–576.

EDGAR, R. S., and G. M. STEINBERG, 1958: On the origin of high negative interference over short segments of the genetic structure of bacteriophage T 4. Virology 6, 115–128.
EINSET, J., 1943: Chromosome length in relation to transmission frequency of maize trisomes. Genetics 28, 349–364.
EKLUNDH-EHRENBERG, C., 1949: Studies on asynapsis in the elm *Ulmus glabra* Huds. Hereditas (Lund) 35, 1–26.
EMERSON, S., 1956: Notes on the identification of different causes of aberrant tetrad ratios in *Saccharomyces*. Comptes Rendus Lab. Carls. 26, 71–86.
— 1963: Meiotic recombination in fungi with special reference to tetrad analysis. Methodology in Basic Genetics, 167–208. Holden-Day Inc., San Francisco.

Emmons, L. R., and L. Husted, 1962: The sex bivalent of the golden hamster. J. Heredity **53**, 227–232.

Emsweller, S. L., and H. A. Jones, 1938: Crossing-over, fragmentation and formation of new chromosomes in an *Allium* species hybrid. Bot. Gaz. **99**, 729–772.

— — 1945: Further studies on the chiasmata of the *Allium cepa* × *A. fistulosum* hybrid and its derivatives. Amer. J. Bot. **32**, 370–379.

Endrizzi, J. E., 1957: The cytology of two hybrids of *Gossypium*. J. Heredity **48**, 221–226.

— 1958: The orientation of interchange complexes and quadrivalents in *Gossypium hirsutum* and *Eusorghum*. Cytologia (Tokyo) **23**, 362–371.

— 1962: The diploid-like cytological behaviour of tetraploid cotton. Evolution **16**, 325–329.

— and L. L. Phillips, 1960: A hybrid between Gossypium arboreum L. and G. raimondii Ulb. Canad. J. Genet. and Cytol. **2**, 311–319.

Ephrussi-Taylor, H., 1960: On the biological functions of DNA. 10th Symp. of the Soc. for Gen. Micro. ("Microbial Genetics") 132–154. C. U. P.

Erickson, R. O., 1947: Respiration of developing anthers. Nature (London) **159**, 275.

— 1948: Cytological and growth correlations in the flower bud and anther development of *Lilium longiflorum*. Amer. J. Bot. **35**, 729–739.

Erlanson, E. W., 1931: Chromosome organisation in *Rosa*. Cytologia (Tokyo) **2**, 256–282.

Evans, H. J., 1960: Supernumerary chromosomes in wild populations of the snail *Helix pomatia* L. Heredity (London) **15**, 129–138.

— 1962: Chromosome aberrations induced by ionising radiations. Internat. Rev. Cytol. **13**, 221–321.

— and J. Savage, 1963: The relation between DNA Synthesis and chromosome structure as resolved by X-ray damage. J. Cell.-Biol. **18**, 525–540.

Fahmy, O. G., 1949: The mechanisms of chromosome pairing during meiosis in male *Apolipthisa subincana* (Mycetophilidae, Diptera). Genetics **49**, 246–263.

Federley, H., 1931: Chromosomenanalyse der reziproken Bastarde zwischen *Pygaera pigra* und *P. curtula* sowie ihrer Rückkreutzungsbastarde. Z. Zellforschg. **12**, 772–816.

Fogwill, M., 1958: Differences in crossing-over and chromosome size in the sex cells of *Lilium* and *Fritillaria*. Chromosoma (Berlin) **9**, 493–504.

Forbes, C., 1960: Non-random assortment in primary nondisjunction in *Drosophila melanogaster*. Proc. Nat. Acad. Sci. (Wash.) **46**, 222–225.

Ford, C. E., T. C. Carter, and J. Hamerton, 1956: Cytogenetics of a mouse translocation. Heredity (London) **10**, 284.

— J. L. Hamerton, and G. B. Sharman, 1957: Chromosome polymorphism in the common shrew. Nature (London) **180**, 392–393.

Foster, T. S., and H. Stern, 1959: The accumulation of soluble deoxyriboside compounds in relation to nuclear division in anthers of *Lilium longiflorum*. J. Biophys. Biochem. Cytology **5**, 187–192.

Franchi, L. L., and A. M. Mandl, 1962: The ultrastructure of oogonia and oocytes in the foetal and neonatal rat. Proc. Roy. Soc. (London) B **157**, 99–114.

Frankel, O. H., 1949a and b: A self propagating structural change in *Triticum*. I. Duplication and crossing-over. Heredity (London) **3**, 163–194. II. The reproductive cycle. Heredity (London) **3**, 293–317.

Freese, E., 1957: The correlation effect for a histidine locus of *Neurospora crassa*. Genetics **42**, 671–684.

— 1958: The arrangement of DNA in the chromosome. Cold. Sprg. Har. Symp. Quant. Biol. **23**, 13–18.

Fröst, S., 1959: The cytological behaviour and mode of transmission of accessory chromosomes in *Plantago serraria*. Hereditas (Lund.) **45**, 191–210.

Frost, J. N., 1961: Preferential segregations in triploid females. Genetics **46**, 373–392.

Gajewski, W., 1954: An amphiploid hybrid of *Geum urbanum* L. and *G. molle* Vis. et Panc. Acta Soc. Botan. Polon. **23**, 259–278.

Gall, J. G., 1963: Kinetics of deoxyribonuclease action on chromosomes. Nature (Lond.) **198**, 36–38.

Garber, E. D., 1948: A reciprocal translocation in *Sorghum vesicolor* Anderss. Amer. J. Bot. **35**, 295–297.

— 1954: Cytotaxonomic studies in the genus *Sorghum* III. The polyploid species of the Subgenera *Para-sorghum* and *Stipo-sorghum*. Bot. Gaz. **115**, 336–242.

GARBER, E. D., 1956: The genus *Collinsia* I. The chromosome number and chiasma frequency of species in the two sections. Bot. Gaz. **118**, 71–73.

— and T. S. DHILLON, 1961: The genus *Collinsia* XVIII. A cytogenetic study of radiation-induced reciprocal translocations in *C. heterophylla*. Genetics **47**, 461–467.

GEITLER, L., 1937: Cytogenetische Untersuchungen an natürlichen Populationen von *Paris quadrifolia*. Z. indukt. Abst. u. Vererb. **73**, 182–197.

— 1938: Weitere cytologische Untersuchungen an natürlichen Populationen von *Paris quadrifolia*. Z. indukt. Abst. u. Vererb. **75**, 161–190.

GELFANT, S., 1963: A new theory on the mechanism of cell division. Symp. Int. Soc. Cell Biol. **2** (Cell Growth and Cell division), 229–259.

GERSHENSON, S., 1935: The mechanism of non-disjunction in the CLB stock of *Drosophila melanogaster*. J. Genetics **30**, 115–125.

GERSTEL, D. U., 1953: Chromosome translocations in interspecific hybrids of the genus *Gossypium*. Evolution **7**, 234–244.

GILES, N., 1943: The origin of iso-chromosomes at meiosis. Genetics **28**, 512–524.

GODWARD, M. B. E., 1961: Meiosis in *Spirogyra crassa*. Heredity **16**, 53–62.

GOLDSTEIN, L., 1963: RNA and protein in nucleocytoplasmic interactions. Symp. Int. Soc. Cell Biol. **2** (Cell Growth and Cell Division), 129–149.

GOTTSCHALK, W., 1951: Untersuchungen am Pachytän normaler und röntgenbestrahlter PMZ von *Solanum lycopersicum*. Chromosoma (Berlin) **4**, 298–341.

GOWEN, J. W., 1933: Meiosis as a genetic character in *Drosophila melanogaster*. J. Exp. Zool. **65**, 83–106.

GREEN, M. M., 1963: Pseudoalleles and recombination in *Drosophila*. Methodology in Basic Genetics, 279–290. Holden-Day Inc., San Francisco.

GRELL, E. H., 1963: Distributive pairing of compound chromosomes in females of *Drosophila melanogaster*. Genetics **48**, 1217–1229.

GRELL, K. G., 1940: Der Kernphasenwechsel von *Stylocephalus* (*Stylorhynchus*) *longicollis* F. stein. (Ein Beitrag zur Frage der Chromosomenreduktion der Gregarinen.) Arch. Protistenk. **94**, 161–200.

— 1958: Untersuchungen über Fortpflanzung und Sexualität der Foraminiferen II. *Rubratella intermedia*. Arch. Protistenk. **102**, 291–308.

GRELL, R. F., 1957: Non-random assortment of non-homologous chromosomes (Abst.). Genetics **42**, 374.

— 1959a: Non-random assortment of non-homologous chromosomes in *Drosophila melanogaster*. Genetics **44**, 421–435.

— 1959b: Effect of chromosome-4 on segregation of sex chromosomes in *Drosophila melanogaster* (Abst.). Genetics **44**, 514.

— 1962a: A new hypothesis on the nature and sequence of meiotic events in the female of *Drosophila melanogaster*. Proc. Nat. Acad. Sci. (Wash.) **48**, 165–172.

— 1962b: A new model for secondary nondisjunction: the role of distributive pairing. Genetics **47**, 1737–1754.

— and E. H. GRELL, 1960: The behaviour of non-homologous chromosomal elements involved in non-random assortment in *Drosophila melanogaster*. Proc. Nat. Acad. Sci. (Wash.) **46**, 51–57.

GRÖBER, K., 1961: Multivalente Assoziationen in diploiden Pollenmutterzellen der Tomate (*Lycopersicon esculentum* Mill) und ihre Bedeutung für das Problem der Chromosomengrundzahl. Die Kulturpflanze **9**, 146–162.

GUSTAFFSON, A., 1939: The interrelation of meiosis and mitosis I. The mechanism of agamospermy. Hereditas (Lund) **25**, 289–322.

— 1942: Meiosis und Mitosis. Eine Erklärung der meiotischen Erscheinungen bei *Hieracium*. Chromosoma (Berlin) **2**, 367–387.

— 1946: Apomixis in the higher plants. Part I. The mechanism of apomixis. Kgl. Fysiograf. Sallskap. Lund Handl (N. F.). Afd. 2, **43**, 1—66.

— 1947: Apomixis in the higher plants. Part II. The causal aspects of apomixis. Acta. Univ. Lund N. F. Avd. **43**, 71–178.

— 1948: Apomixis in higher plants. Part III. Biotype and species formation. Lunds Univ. Arsskr. **44**, 183—370.

— 1948: Polyploidy life-form and vegetative reproduction. Hereditas (Lund) **34**, 1–22.

HACKMAN, W., 1948: Chromosomenstudien an Araneen mit besonderer Berücksichtigung der Geschlechtschromosomen. Acta. Zool. Fenn. **54**, 1—101.

HAGA, T., 1937: Karyotypic polymorphism in *Paris hexaphylla*. Cham., with special reference to its origin and to the meiotic chromosome behaviour. Cytologia (Tokyo) Fujii Jub. vol., 681–700.

Haga, T., 1944: Meiosis in *Paris* I. Mechanism of chiasma formation. Fac. Sci. Hokkaido Imp. Univ. Ser. V, Bot. 5, 121–198.
— 1953: Meiosis in *Paris* II. Spontaneous breakage and fusion of chromosomes, Cytologia (Tokyo) 18, 50–66.
Hair, J. B., 1954: The origin of new chromosomes in *Agropyron*. Heredity (Suppl. Vol.) 6, 215–233.
Håkansson, A., 1950: Spontaneous chromosome variation in the roots of a species hybrid. Hereditas (Lund) 36, 39–59.
— 1957: Meiosis and pollen mitosis in rye plants with many accessory chromosomes. Hereditas (Lund) 43, 603–620.
— 1958: Holocentric chromosomes in *Eleocharis*. Hereditas (Lund) 44, 531–540.
Haldane, J. B. S., 1931: The cytological basis of genetic interference. Cytologia (Tokyo) 3, 54–65.
Hallka, O., 1959: Studies on the chromosomes of the Hemiptera, Homoptera, Auchenorrhynchya. Ann. Acad. Sci. Fenn. A IV 43, 1–72.
Hamerton, J. L., 1958: Mammalian sex chromosomes. Symposium on Nuclear Sex, 25–28. Interscience Pub., N. Y.
Haque, A., 1952: The irradiation of meiosis in *Tradescantia*. Heredity (London) 6, Suppl., 57–75.
Hayman, D. L., 1955: Chromosome behaviour of the univalents in two *Phalaris* hybrids. Austrae. J. Biol. Sci. 8, 241–252.
Heenen, W. K., 1963: Extensive chromosome breakage occurring spontaneously in a certain individual of *Elymus farcatus* (= *Agropyron junceum*). Hereditas 49, 1–32.
— 1963: Meiosis in the interspecific hybrid *Elymus farcatus* and *E. repens* (= *Agropyron junceum* × *A. repens*). Hereditas 49, 107–118.
Heilborn, O., 1936: The mechanism of so-called secondary association between chromosomes. Hereditas (Lund) 22, 167–188.
Helwig, E. R., 1955: Spermatogenesis in hybrids between *Circotettix verruculatus* and *Trimerotropis suffusa* (Orthoptera: Oedipodidae). Univ. Colorado Stud. Ser. Biol. 10, 49–64.
Henderson, S. A., 1961: The chromosomes of the British Tetrigidae (Orthoptera). Chromosoma (Berlin) 12, 553–572.
— 1962: Temperature and chiasma formation in *Schistocerca gregaria* II. Cytological effects at 40° C and the mechanism of heat-induced univalence: Chromosoma (Berlin) 13, 437–463.
— 1963: Chiasma distribution at diplotene in a locust. Heredity (London) 18, 173–190.
— 1965: Chromosome behaviour in diploid and tetraploid cells of *Sphodromantis gastrica* and its bearing on chromosome evolution in the mantids. Chromosoma (Berlin) 16, 192–221.
— and T. Parsons, 1963: The chromosomes of eleven species of Tipulid. Caryologia (Piza) 16, 337–346.
Hennen, S., 1963: Chromosomal and embryological analyses of nuclear changes occurring in embryos derived from transfers of nuclei between *Rana pipiens* and *Rana sylvatica*. Develop. Biol. 6, 133–183.
Hertwig, O., 1908: Über neue Probleme der Zellenlehre. Arch. Zellforsch. 1, 1–32.
Hexter, W. M., 1963: Nonreciprocal events at the garnet locus in *Drosophila melanogaster*. P.N.A.S. (Wash.) 50, 372–379.
Hirai, H., 1956: Compound sex-chromosomes with distance pairing in the male neuropteran *Plethosmyhis decoratus*. Anat. Zool. Japan 29, 155–160.
Holden, J. H. W., and M. Mota, 1956: Non-synchronised meiosis in binicleate pollen mother cells of an *Avena* hybrid. Heredity (London) 10, 109–117.
Hoover, M. E., 1938: Cytogenetic analysis of nine inversions in *Drosophila melanogaster*. Z. indukt. Abst. u. Vererb. 74, 420–434.
Hotta, Y., and H. Stern, 1961a: Deamination of deoxycytidine and 5-methyl deoxycytidine in developing anthers of *Lilium longiflorum* (var. Croft). J. Biophys. Biochem. Cytol. 9, 279–284.
— — 1961b: Transient phosphorylation of deoxyribosides in regulation of DNA synthesis. J. Biophys. Biochem. Cytol. 11, 311–319.
Hrishi, N. J., and A. Müntzing, 1960: Structural heterozygosity in *Secale Kuprijanovii*. Hereditas (Lund) 46, 745–752.
Hu, C.-H., 1960: Karyological studies in haploid rice plants IV. Chromosome morphology and intragenome pairing in haploid plants of *Oryza* glaberrima as compared with those in O. sativa. Cytologia (Tokyo) 25, 437–449.

Hughes-Schrader, S., 1943a: Polarisation, kinetochore movements and bivalent structure in the meiosis of male mantids. Biol. Bull. (Woods Hole) **85**, 265–300.
— 1943b: Meiosis without chiasmata in diploid and tetraploid spermatocytes of the mantid *Callimantis antillarum* Saugsture. J. Morph. **73**, 111–140.
— 1947: Reversion of XO to XY sex chromosome mechanism in a phasmid. Chromosoma (Berlin) **3**, 52–65.
— 1948: Cytology of coccids (Coccoidea-Homoptera). Advances in Genetics **2**, 127–203.
— and F. Schrader, 1961: The kinetochore of the Hemiptera. Chromosoma (Berlin) **12**, 327–350.
Huskins, C. L., 1937: The internal structure of chromosomes—a statement of opinion. Cytologia (Tokyo) Fuyii Jubilee Vol. **2**, 1015–1022.
— 1941: The coiling of chromonemata. Cold Spr. Harb. Symp. Quant. Biol. **9**, 13–17.
— and G. B. Wilson, 1938: Probable causes of the changes in direction of the major spiral in *Trillium erectum* L. Ann. Bot. N. S. **2**, 281–292.

Inamdar, N. B., 1949: A note on the re-orientation within the spindle of the sex-trivalent in a mantid. Biol. Bull. (Woods Hole) **97**, 300.
Itoh, H., 1933: The sex chromosomes of a stone fly *Acroneuria jezoensis* Okamoto (Plecoptera). Cytologia (Tokyo) **4**, 427–443.
Ivanov, M. A., 1938: Experimental production of haploids in *Nicotiana rustica* L. Genetica **20**, 295–397.

Jackson, R. C., and P. Newmark, 1960: Effects of supernumerary chromosomes on production of pigment in *Haplopappus gracilis*. Science **4**, 1316–1317.
Jacob, F., and J. Monod, 1961: Genetic regulatory mechanisms in the synthesis of proteins. J. Mol. Biol. **3**, 318–356.
Jain, H. K., 1957: Effect of high temperature on meiosis in *Lolium*: nucleolar inactivation. Heredity (London) **11**, 23–36.
— and S. L. Basak, 1963: Genetic interpretation of chiasmata in *Delphinium*. Genetics **48**, 329–339.
Jande, S. S., 1959: Chromosome number and sex mechanism in twenty-seven species of Indian Heteroptera. Res. Bull. (N. S.) Panj. Univ. **10**, 215–217.
— 1960: Pre-reductional sex chromosomes in the family Tingidae (Gymnocerata-Heteroptera). The Nucleus **3**, 209–214.
Janssens, F. A., 1924: La chiasmatypie dans les insectes. La Cellule **34**, 135–359.
John, B., 1957a: The chromosomes of zooparasites. I. *Acanthocephalus ranae* (Acanthocephala: Echinorhynchidae). Chromosoma (Berlin) **8**, 730–738.
— 1957b: The chromosomes of Zooparasites II. *Oswaldocruzia filiformis* (Nematoda: Trichostrongylidae). Chromosoma (Berlin) **9**, 61–68.
— 1957c: XY-segregation in the crane fly *Tipula maxima* (Diptera: Tipulidae). Heredity (London) **11**, 209–215.
— and K. R. Lewis, 1957: Studies on *Periplaneta americana* I. Experimental analysis of male meiosis. Heredity (London) **11**, 1–9.
— — 1958: Studies on *Periplaneta americana* III. Selection for heterozygosity. Heredity (London) **12**, 185–197.
— — 1959: Selection for interchange heterozygosity in an inbred culture of *Blaberus discoidalis*. Genetics **44**, 251–267.
— — 1960: Nucleolar controlled segregation of the sex chromosomes in beetles. Heredity (London) **15**, 431–439.
— — 1965: Genetic Speciation in the grasshopper *Eyprepocnemis plorans*. Chromosoma (Berlin) **16**, 308–344.
— — and S. A. Henderson, 1960: Chromosome abnormalities in a wild population of *Chorthippus brunneus*. Chromosoma (Berlin) **11**, 1–20.
— and S. A. Henderson, 1962: Asynapsis and polyploidy in *Schistocera paranensis*. Chromosoma (Berlin) **13**, 111–147.
— and G. M. Hewitt, 1963: A spontaneous interchange in *Chorthippus brunneus* with extensive chiasma formation in an interstitial segment. Chromosoma (Berlin) **14**, 638–650.
Johnsson, H., 1941: Cytological studies in the genus *Alopecurus*. Lunds. Univ. Arsskr. N. F. Avd. 2, 37, nr. 3, 1–43.
Jones, K., 1956a and b: Species differentation in *Agrostis* I. Cytological relations in *Agrostis canina* L. Journ. Genetics **54**, 370–376. II. The significance of chromosome pairing in the tetraploid hybrids of *Agrostis canina* subsp. *montana* Hartm., *A. tennus* Sibth. and *A. stolonifera* L. J. Genetics **54**, 377–393.

Jones, K., 1962: Chromosomal status, gene exchange and evolution in *Dactylis*. II. The chromosomal analysis of diploid, tetraploid and hexaploid species and hybrids. Genetica **32**, 272–295.
— 1963: Ployploidy and the interchange heterozygote. Proc. XIth Int. Congr. Genetics **1**, 119.
— 1964: Chromosomes and the nature and origin of *Anthoxanthum odoratum* L. Chromosoma (Berlin) **15**, 248–274.

Kasha, K. J., 1961: Inversion in chromosome 5. Barley Newsletter **4**, 16.
Katayama, Y., 1935: Karyological comparisons of haploid plants from octoploid Aegilotricum and diploid wheat. Jap. Jour. Bot. **7**, 349–380.
Kattermann, G., 1939: Ein neuer Karotyp bei Roggen. Chromosoma (Berlin) **1**, 284–299.
Kayano, H., 1957: Cytogenetic studies in *Lilium callosum* III. Preferential segregation of a supernumerary chromosome in EMC's. Proc. Jap. Acad. **33**, 553–558.
— 1959: Chiasma studies in structural hybrids I. Heteromorphic bivalent in *Lilium callosum*. The Nucleus **2**, 47–50.
— 1960a: Chiasma studies in structural hybrids III. Reductional and equational separation in *Disporum sessile*. Cytologia (Tokyo) **25**, 461–467.
— 1960b: Chiasma studies in structural hybrids IV. Crossing-over in *Disporum sessile*. Cytologia (Tokyo) **25**, 468–475.
— and K. Nakamura, 1960: Chiasma studies in structural hybrids V. Heterozygotes for a centric fusion and for a translocation in *Acrida lata*. Cytologia (Tokyo) **25**, 476–480.
Keeffe, M. M., 1948: A reinvestigation of chromosome coiling in *Trillium*. Amer. J. Bot. **35**, 434–440.
Kellenberger, G., M. L. Zichichi, and J. J. Weigle, 1961: Exchange of DNA in the recombination of bacteriophage λ. Proc. Nat. Acad. Sci. (Wash.) **47**, 869–878.
Kempanna, C., 1963: Investigation into the genetic regulation of meiotic chromosome behaviour in *Triticum aestivum* L. Ph.D. Thesis. Univ. of Cambridge.
— and R. Riley, 1964: Secondary association between genetically equivalent bivalents. Heredity (London) **19**, 289–299.
Keyl, H.-G., 1955: Der Formenwechsel der Chromosomen in der Spermatogenese von *Bithynia tentaculata* (L). Chromosoma (Berlin) **7**, 387–419.
— 1956: Beobachtungen über die ♂-Meiose der Muschel *Sphaerium corneum*. Chromosoma (Berlin) **8**, 12–17.
— 1957: Zur Karyologie der Hydrachnellen (Acarina). Chromosoma (Berlin) **8**, 719–729.
Kihlman, B. A., 1955: Chromosome breakage in *Allium* by 8-EOC and X-rays. Exper. Cell Res. **8**, 345–368.
— 1963: The effect of 5-halogenated deoxyuridines on the frequency of X-ray-induced chromosomal abberations in *Vicia faba*. Hereditas **49**, 353–370.
— and A. Levan, 1951: Localised breakage in *Vicia faba*. Hereditas (Lund) **37**, 382–388.
Kikudome, G. Y., 1959: Studies on the phenomenon of preferential segregation in maize. Genetics **44**, 815–831.
Kimber, G., 1961: Basis of the diploid-like meiotic behaviour of polyploid cotton. Nature **191**, 98–100.
King, G. C., 1960: The cytology of desmids: the chromosomes. New Phyt. **59**, 65–72.
Kitani, Y., 1962: Three kinds of transreplication in *Sordaria fimicola*. Jap. J. Genetics **37**, 131–146.
— L. S. Olive, and A. S. El-Ani, 1962: Genetics of *Sordaria fimicola* V. Aberrant segregation at the G-locus. Amer. J. Bot. **49**, 697–706.
Klingstedt, H., 1937: On some tetraploid spermatocytes in *Chrysochroan dispar* (Orthoptera). Mem. Soc. Fauna et Flora Fennica **12**, 194–209.
— 1939: Taxonomic and cytological studies on grasshopper hybrids. J. Genetics **37**, 389–420.
Koeperich, J., 1930: Etude comparee du noyau des chromosomes et de leur relations avec le cytoplasma. La Cellule **39**, 307–399.
Koller, P. C., 1934: The movements of chromosomes within the cell and their dynamic interpretation. Genetica **16**, 447–466.
— 1938a: The genetical and mechanical properties of the sex chromosomes IV. The golden hamster. J. Genetics **36**, 177–195.
— 1938b: Asynapsis in *Pisum sativum*. J. Genetics **36**, 275–306.
— 1941: The genetical and mechanical properties of the sex chromosomes VII. *Apodemus sylvaticus* and *A. hebridensis*. J. Genetics **41**, 375–390.
— 1953: A Dicentric chromosome in a rat tumour. Heredity **6** (Supp. Vol.), 181–196.

KOMAI, T., and T. TAKAHU, 1942: On the effect of the X-chromosome on crossing-over in *Drosophila virilis*. Cytologia (Tokyo) **12**, 357–365.

KURITA, M., 1953: A study of chromosomes in *Northoscordum fragrans*. Mem. Ehime Univ. Sect. II Science **1**, No. 4.

KUSANGI, A., 1962: The *Luzula*-type meiosis. C. I. S. **3**, 6–7.

— and N. TANAKA, 1959: Cytological studies on *Luzula* chromosomes I. The centromere problem. Jap. J. Genetics **36**, 169–173.

KUWADA, Y., and T. NAKAMURA, 1933: Behaviour of chromonemata in mitosis I. Observation of pollen mother cells in *Tradescantia reflexa*. Mem. Coll. Sci. Kyoto Imp. Univ. B **9**, 129–139.

— — 1934: Behaviour of chromonemata in mitosis IV. Double refraction of chromosomes in *Tradescantia reflexa*. Cytologia (Tokyo) **6**, 78–86.

LACHANCE, L. E., J. G. RIEMANN and D. E. HOPKINS, 1964: A reciprocal translocation in *Cochlyomyia hominivorax* (Diptera: Calliphoridae). Genetic and cytological evidence for preferential segregation in males. Genetics, **49**, 959–972.

LACOUR, L. F., 1952: The *Luzula* system analysed by X-rays. Heredity (London) **6**, 77–81.

— and A. RUTISHAUSER, 1954: X-ray breakage experiments with endosperm I. Subchromatid breakage. Chromosoma (Berlin) **6**, 696–709.

— and S. R. PELC, 1958: Effect of colchicine on the utilisation of labelled thymidine during chromosome reproduction. Nature (London) **182**, 506–508.

— — 1959: Effect of colchicine on the utilisation of thymidine labelled with tritium during chromosomal reproduction. Nature (London) **183**, 1455–1456.

LANTZ, L. A., and H. G. CALLAN, 1954: Phenotypes and spermatogenesis of interspecific hybrids between *Triturus cristatus* and *T. marmoratus*. J. Genetics **52**, 165–185.

LAWRENCE, C. W., 1961a: The effect of the irradiation of different stages in microsporogenesis on chiasma frequency. Heredity (London) **16**, 83–89.

— 1961b: The effect of radiation on chiasma formation in *Tradescantia*. Radiation Botany **1**, 92–96.

— 1963: The orientation of multiple associations resulting from interchange heterozygosity. Genetics **48**, 347–350.

LAWRENCE, W. J. C., 1931: The secondary association of chromosomes. Cytologia (Tokyo) **2**, 352–384.

LE CALVEZ, J., 1947: Morphologie et comportement des chromosomes dans la spermatogenese de quelques mycetophilides. Chromosoma (Berlin) **3**, 137–165.

— 1950: Recherches sur les Foraminiferes II. Place de la méiose et sexualité. Arch. Zool. exp. gén. **87**, 211–214.

LEDERBERG, J., 1955: Recombination mechanism in bacteria. J. Cell. Comp. Physiol. **45** (Suppl. 2), 75–107.

LESLEY, M. M., and H. B. FROST, 1927: Mendelian inheritance of chromosome shape in *Matthiola*. Genetics **12**, 449–460.

LEVAN, A., 1933: Cytological studies in *Allium* IV. *Allium fistulosum*. Sv. Bot. Tidskr. **27**, 211–232.

— 1935: Cytological studies in *Allium* VI. The chromosome morphology of some species of *Allium*. Hereditas (Lund) **20**, 289–330.

— 1936: Die Zytologie von *Allium cepa* × *fistulosum*. Hereditas (Lund) **21**, 195–214.

— 1939: The effect of colchicine on meiosis in *Allium*. Hereditas (Lund) **25**, 9–26.

— 1940: Meiosis of *Allium porrum*, a tetraploid species with chiasma localisation. Hereditas (Lund) **26**, 454–462.

— and S. L. EMSWELLER, 1938: Structural hybridity in *Northoscordum fragans*. J. Her. **29**, 291–294.

— and J. H. TJIO, 1948: Induction of chromosome fragmentation by phenol. Hereditas (Lund) **34**, 453–480.

— — 1951: Penicillin in the *Allium* test. Hereditas (Lund) **37**, 306–324.

LEVINE, R. P., 1962: Genetics. Holt, Rinehart and Winston Inc., N.Y.

— and E. E. LEVINE, 1955: Variable crossing over arising in different strains of *Drosophila pseudoobscura*. Genetics **40**, 399–405.

LEVITAN, M., 1958: Studies of linkage in populations II. Recombination between linked inversions of *D. robusta*. Genetics **43**, 620–633.

LEWIS, D., 1963: "Mark of the Master." Nature (London) **199**, 518.

LEWIS, H., 1951: The origin of supernumerary chromosomes in natural populations of *Clarkia elegans*. Evolution **5**, 142–157.

Lewis, K. R., 1961: The genetics of bryophytes. Trans. Brit. Bryolog. Soc. 4, 111–130.
— and B. John, 1957a: Studies on *Periplaneta americana* II. Interchange hetero-zygosity in isolated populations. Heredity (London) 11, 11–22.
— — 1957b: The organisation and evolution of the sex multiple in *Blaps mucronata*. Chromosoma (Berlin) 9, 69–80.
— — 1959: Breakdown and restoration of chromosome stability following in-breeding in a locust. Chromosoma (Berlin) 10, 689–618.
— — 1963a: Chromosome Marker. Churchill, London.
— — 1963b: Spontaneous interchange in *Chorthippus brunneus*. Chromosoma (Berlin) 14, 618–637.
— and G. C. Scudder, 1958: The chromosomes of *Dicranocephalus agilis* (Hemiptera: Heteroptera). Cytologia (Tokyo) 23, 92–104.
Li, H., W. Pao, and C. H. Li, 1945: Desynapsis in the common wheat. Amer. J. Bot. 32, 92–101.
Lima-de-Faria, A., 1954: Chromosome gradient and chromosome field in *Agapanthus*. Chromosoma (Berlin) 6, 330–370.
— 1956: The role of the kinetochore in chromosome organisation. Hereditas 42, 85–160.
— 1959: Differential uptake of tritiated thymidine into hetero and euchromatin in *Melanoplus* and *Secale*. J. Biophys. Biochemic. Cytology 6, 457–466.
— 1962: Metabolic DNA in *Tipula oleracea*. Chromosoma (Berlin) 13, 47–59.
— and P. Sarvella, 1958: The organisation of telomeres in species of *Solanum*, *Salvia*, *Scilla*, *Secale*, *Agapanthus* and *Ornithogalum*. Hereditas (Lund) 44, 337–346.
— and K. Borum, 1962: The period of DNA synthesis prior to meiosis in the mouse. J. Cell. Biol. 14, 381–388.
Lima-de-Faria, A., 1954, and T. Nordqvist, 1962: Disintegration of H³ labelled spermatocytes in *Melanoplus differentialis*. Chromosoma (Berlin) 13, 60–66.
— and S. Bose, 1963: The role of telomeres at anaphase. Chromosoma (Berlin) 13, 315–327.
Lin, T. P., 1954: The chromosomal cycle in *Parascaris equorum* (*Ascaris megalo-cephala*): Oogenesis and diminution. Chromosoma (Berl.) 6, 175–198.
Lindegren, C. C., 1949: The Yeast Cell. Educ. Pub., St. Louis.
— 1953: Gene conversion in *Saccharomyces*. J. Genetics 51, 625–637.
Lindsley, D. L., and E. Novitski, 1958: Localisation of factors responsible for the kinetic activity of the X-chromosome of *Drosophila melanogaster*. Genetics 43, 547–563.
Linnert, G., 1955: Die Struktur der Pachytänchromosomen in Euchromatin und Heterochromatin und ihre Auswirkung auf die Chiasmabildung bei *Salvia*-Arten. Chromosoma (Berlin) 7, 90–128.
Longley, A. E., 1945: Abnormal segregation during megasporogensis in maize. Genetics 30, 100–113.
Loveless, A., 1953: Chemical and biochemical problems arising from the study of chromosome breakage by alkylating agents and heterocyclic compounds. Heredity 6 (Suppl.), 293–298.
— and S. Revell, 1949: New evidence on the mode of action of mitotic poisons. Nature (London) 164, 938.
Lyon, M. F., 1961: Gene action in the X-chromosome of the mouse (*Mus musculus* L.). Nature (London) 190, 372–373.
— 1963: Attempts to test the inactive-X theory of dosage compensation in mammals. Genet. Res. (Camb.) 4, 93–103.

McClintock, B., 1929: A cytological and genetical study of triploid maize. Genetics 14, 180–222.
— 1931: Cytological observations of deficiences involving known genes, translocations and an inversion in *Zea mays*. Missouri Agr. Exp. Sta. Res. Bull. 163, 1–30.
— 1933: The association of non-homologous parts of chromosomes in the mid-pro-phase of meiosis in *Zea mays*. Z. Zellforsch. u. Mikr. Anat. 19, 191–237.
— 1938: The production of homozygous deficient tissues with mutant character-istics by means of the aberrant behaviour of ring-shaped chromosomes. Genetics 23, 315–376.
— 1941: The stability of broken ends of chromosomes in *Zea mays*. Genetics 26, 234–282.
— 1942: The fusion of broken ends of chromosomes following nuclear fusion. Proc. Nat. Acad. Sci. (Wash.) 28, 458–463.
— 1943: Studies with broken chromosomes. Yr. Book Carn. Inst. Wash. 42, 148–150.
— 1945: Neurospora I. Preliminary observations of the chromosomes of *Neurospora crassa*. Amer. J. Bot. 32, 671–678.

McCollum, G. D., 1958: Comparative studies of chromosome pairing in natural and induced tetraploid *Dactylis*. Chromosoma (Berlin) **9**, 571–605.

Maeda, T., 1937: Chiasma studies in *Allium fistulosum, Allium cepa* and their F_1, F_2 and backcross hybrids. Jap. J. Genetics **13**, 146–159.

Magoon, M. L. and K. H. Shambulin-Gappa, 1963: Cyto-morphological studies of some species and hybrids in the Eu-Sorghums. Chromosoma (Berl.) **14**, 572–588.

Maguire, M. P., 1960: The relation of reduplication, recombination, synapsis and chromosome coiling. Genet. Res. (Camb.) **1**, 487–488.

Maheshwari, P., 1950: An Introduction to the Embryology of Angiosperms. McGraw Hill Book Co. Inc., Toronto.

Makino, S., and K. Kanô, 1947: Association of the XY chromosomes with the nucleolus in the auxocytes of some neuropterous insects. La Kromosomo **3-4**, 135–136.

— and E. Momma, 1950: Observations on the structure of grasshopper chromosomes subjected to a new aceto-carmine treatment. J. Morph. **86**, 229–252.

Mancino, G., 1961: Anomalie meiotiche nella spermatogenesi dell'ibrido *Triturus italicus* ♀ × *Triturus vulgaris* ♂. Boll. di Zool. **28**, 691–701.

— and V. Scalli, 1961: Anomalie spermatogenetische in un esemplare di *Triturus helveticus*. Archiv. Zool. Ital. **46**, 149–166.

Manna, G. K., 1958: Cytology and inter-relationships between various groups of Heteroptera. Proc. X[th] Int. Entomol. Congr. Canada **2**, 919–934.

— 1962: A further evaluation of the cytology and inter-relationships between various groups of Heteroptera. The Nucleus **5**, 7–28.

Manton, I., 1945: Chromosome length at early meiotic prophase in *Osmunda*. Ann. Bot. **9**, 155–178.

— and J. Smiles, 1943: Observations on the spiral structure of somatic chromosomes in *Osmunda* with the aid of ultra-violet light. Ann. Bot. **7**, 195–212.

Marks, G. E., 1957a: The cytology of *Oxalis dispar* (Brown). Chromosoma (Berlin) **8**, 650–670.

— 1957b: Telocentric chromosomes. Amer. Nat. **91**, 223–232.

Marquardt, H., 1938: Die Meiosis von *Oenothera* I. Z. Zellforsch. u. Mikr. Anat. **27**, 159–209.

— 1948: Das Verhalten rontgeninduzierter Viererring mit großen interstitiellen Segmenten bei *Oenothera hookeri*. z. Ind. Abst. Vererb., **82**, 415–429.

Mather, K., 1937: The determination of position in crossing-over II. The chromosome length—chiasma frequency relation. Cytologia (Tokyo) Fujii Jub. Vol. 514–526.

— 1938: Crossing-over. Biol. Rev. (Camb.) **13**, 252–292.

— 1940: The determination of position in crossing over III. The evidence of metaphase chiasmata. J. Genetics **39**, 205–223.

— and L. H. A. Stone, 1933: The effect of X-radiation upon somatic chromosomes. J. Genetics **28**, 1–24.

Matsuura, H., 1935: Chromosome studies on *Trillium kamtschaticum* Pall. II. The direction of coiling of the chromonema within the first meiotic metaphase of the PMC. Jour. Fac. Sci. Hokkaido Imp. Univ. **3**, 233–250.

— 1937a: Chromosome studies on *Trillium kamtschaticum* Pall. III. The mode of chromatid disjunction at the first meiotic metaphase of the PMC. Cytologia (Tokyo) **8**, 142–177.

— 1937b: Chromosomes studies on *Trillium kamtschaticum* Pall. IV. Further studies on the direction of coiling of the chromonema within the first meiotic chromosomes. Cytologia (Tokyo) **8**, 178–184.

— and T. Haga, 1942: Chromosome studies on *Trillium kamtschaticum* Pall. X. On the origin of the chiasma. Cytologia (Tokyo) **12**, 397–417.

— and M. Kurabayashi, 1951: Chromosome studies on *Trillium kamtschaticum* Pall. and its allies XXIV. The association of kinetochores of non-homologous chromosomes at meiosis. Chromosoma (Berlin) **4**, 273–283.

— T. Saho, S. Tanifuji, and M. Iwabuchi, 1962: Chromosome studies on *Trillium kamtschatikum* Pall. and its allies XXIX. Effects of ATP and DNA on the rejoining of chromosome breaks induced by X-raying. J. Fac. Sci. Hokkaido Univ. series V, vol. VIII, 173–200.

— and S. Tanifuji, 1962: Chromosome studies on *Trillium kamtschaticum* Pall. and its allies XXVIII. Modifying effects of chloramphenicol on X-ray induced chromosome aberrations. J. Fac. Sci. Hokkaido Univ. Series V, vol. VIII, 157–172.

Matthey, R., 1936: La formule chromosomiale et les hétérochromosomes chez les *Apodemus* européens. Z. Zellforsch. u. Mikr. Anat. **25**, 501–515.

Matthey, R., 1952: Chromosomes de Muridae (Microtinae et Cricetinae). Chromosoma (Berlin) **5**, 113–138.
— 1961: Cytologie comparée des Cricetinae palearctiques et américains. Rev. Suisse Zool. **68**, 41–61.
— 1962: Etudes sur les chromosomes d'*Ellobius lutescens* Th. (*Mammalia-Muridae-Microtinae*) Cytogenetics **1**: 180–195.
— 1963: Polymorphisme chromosomique intraspecifique et intra individuel chez *Acomys minous* Bate (Mammalia—Rodentia—Muridae). Etude cytologiques des hybrides *Acomys minous* ♂ × *Acomys cahirinus* ♀. Le mechanisme des fusions centriques. Chromosoma (Berlin) **14**, 468–497.
— and J. Aubert, 1947: Les chromosomes des Plécoptères. Bull. Biol. **81**, 202–246.
Mazia, D., 1961: Mitosis and the physiology of cell division. The Cell. vol. III, 77–412.
Melander, Y., 1950: Accessory chromosomes in animals, especially in *Polycelis tenuis*. Hereditas (Lund) **36**, 261–296.
— 1963a: Chromatid tension and fragmentation during the development of *Calliphora erthyrocephala* Meig. (Diptera). Hereditas (Lund) **49**, 91–106.
— 1963b: Cytogenetic aspects of embryogenesis in *Paludicola*, Tricladida. Hereditas (Lund) **49**, 119–166.
— 1963c: The role of a secondary constriction in a tumour chromosome. Hereditas (Lund) **49**, 241–273.
Mesa, A., 1961: Morfologia falica y cariologia de *Neuquenina fictor* (Rehn) (Orthoptera-Acridordea). Com. Zool. del Mus. de His. Nat. des Montevideo **5**, 1–11.
Meselson, M., and F. W. Stahl, 1958: The replication of DNA in *E. coli*. Proc. Nat. Acad. Sci. (Wash.) **44**, 671–682.
— and J. J. Weigle, 1961: Chromosome breakage accompanying genetic recombination in bacteriophage. Proc. Nat. Acad. Sci. (Wash.) **47**, 857–868.
Metz, C., 1933: Monocentric mitosis with segregation of chromosomes in *Sciara* and its bearing on the mechanism of mitosis. Biol. Bull (Woods Hole) **54**, 333–347.
— and J. F. Nonidez, 1921: Spermatogenesis in the fly *Asilus sericeus* Say. J. Exp. Zool. **32**, 165–185.
— — 1923: Spermatogenesis in *Asilus notatus* Wied (Diptera). Arch. f. Zellforsch. u. Mikr. Anat. **17**, 438–449.
Meurman, O., 1929: Association and types of chromosomes in *Aucuba japonica*. Hereditas (Lund) **12**, 179–209.
Meyer, G. F., 1960: The fine structure of spermatocyte nuclei of *Drosophila melanogaster*. Proc. Europ. Reg. Congr. Electron Microscopy.
— 1963: Die Funktionsstrukturen des Y-Chromosoms in den Spermatocytenkernen von *Drosophila hydei*, *D. neohydei*, *D. repleta* und einigen anderen *Drosophila*-Arten. Chromosoma (Berlin) **14**, 207–255.
Michaelis, A., 1959: Über das Verhalten eines Ringchromosoms in der Mitose und Meiose von *Antirrhinum majus* L. Chromosoma (Berlin) **10**, 144–162.
Michie, D., 1953: Affinity: a new genetic phenomenon in the house mouse. Nature (London) **171**, 26–27.
Miller, O. L., 1963: Cytological studies in asynaptic maize. Genetics **48**, 1445–1466.
Mitchell, M. B., 1955a: Aberrant recombination of pyridoxine mutants of *Neurospora*. Proc. Nat. Acad. Sci. (Wash.) **41**, 215–226.
— 1955b: Further evidence of aberrant recombination in *Neurospora*. Proc. Nat. Acad. Sci. (Wash.) **41**, 935–937.
Mitra, S., 1958: Effects of X-rays on chromosomes of *Lilium longiflorum* during meiosis. Genetics **43**, 771–789.
Moens, P. B., 1964: A new interpretation of meiotic prophase in *Lycopersicon esculentum* (Tomato). Chromosoma (Berl.) **15**, 231–242.
Moffat, A. A., 1932: Chromosome studies in *Anemone* I. A new type of chiasma behaviour. Cytologia (Tokyo) **4**, 26–37.
Moh, C. C., and R. A. Nilan, 1954: "Short" chromosome—a mutant barley induced by atomic bomb irradiations. Cytologia (Tokyo) **19**, 48–53.
Monesi, V., 1962: Autoradiographic study of DNA synthesis and the cell cycle in spermatogonia and spermatocytes of mouse testis using tritiated thymidine. J. Cell Biol. **14**, 1-18.
Morgan, L. V., 1933: A closed X-chromosome in *Drosophila melanogaster*. Genetics **18**, 250–283.
Morgan, T. H., C. B. Bridges, and J. Schultz, 1933: Constitution of the germinal material in relation to heredity. Carnegie Inst. Wash. Yr. Book **32**, 298–302.
— — — and A. H. Sturtevant, 1925: The genetics of *Drosophila*. Bibliogr. Genetica **2**, 1–262.

MORRISON, J. W., 1953a: Effect of X-rays in *Triticum*. Heredity (London) **6** (Suppl. Vol.), 83–91.
— 1953b: Chromosome behaviour in wheat monosomics. Heredity **7**, 203–217.
— 1953c: Pollen formation in pentaploid and near pentaploid wheat hybrids. Heredity **7**, 419–428.
— 1954a: A dicentric wheat chromosome in division. Can. J. Bot. **32**, 491–502.
— 1956: Chromosome behaviour and fertility of Tetra Petkus rye. Canad. J. agr. Sci. **36**, 157–165.
— and T. RAJATHY, 1960: Chromosome behaviour in autotetraploid cereals and grasses. Chromosoma (Berlin) **11**, 297–309.
MOSES, M. J., 1958: The relation between the axial complex of meiotic prophase chromosomes and chromosome pairing in a salamander (*Plethodon cinereus*). J. Biophysic. Biochem. Cytol. **4**, 633–638.
MULLER, H. J., 1916: The mechanism of crossing over I–IV. Amer. Nat. **50**, 193–221, 284–305, 350–366, 421–434.
MÜNTZING, A., 1946: Cytological studies of extra fragment chromosomes in rye III. The mechanism of non-disjunction at the pollen mitosis. Hereditas (Lund) **32**, 97–119.
— 1951: Cytogenetic properties and practical value of tetraploid rye. Hereditas (Lund) **37**, 17–84.
— 1958: A new category of chromosomes. Proc. X[th] Int. Congr. Genetics **1**, 453–476.
— and A. LIMA-DE-FARIA, 1952: Pachytene analysis of a deficient accessory chromosome in rye. Hereditas (Lund) **38**, 1–10.

NABOURS, R. K., 1929: The genetics of the Tettigidae. Biblio. Genetica V. **5**, 27–104.
— 1947: The grouse locusts. J. Kans. Ent. Soc. **20**, 127–141.
NAGAO, S., 1935: Distribution of pollen grains in certain triploid and hypertriploid *Narcissus* plants. Jap. J. Genetics **11**, 1–5.
NARBEL-HOFSTETTER, M., 1964: Meiosis in Parthenogenetic forms. Protoplasmatologia (in preparation).
NASATIR, M., and H. STERN, 1959: Changes in the activities of aldolase and of D-glyceraldehyde-3-phosphate dehydrogenese during the mitotic cycle in microspores of *Lilium longiflorum*. J. Biophys. Biochem. Cytol. **6**, 189–192.
NAWASCHIN, M., 1933: Altern der Samen als Ursache der Chromosomenmutationen. Planta **20**, 233–243.
NAYLOR, B., and H. REES, 1958: Chromosome size in *Lolium temulentum* and *L. perenne*. Nature (London) **181**, 854–855.
NEBEL, B. R., 1937: Chromosome structure XII. Further radiation experiments with *Tradescantia*. Amer. J. Bot. **24**, 365–372.
— 1941: Structure of *Tradescantia* and *Trillium* chromosomes with particular emphasis on number of chromonemata. Cold. Sprg. Harb. Symb. Quant. Biol. **9**, 7–12.
— 1959: On the structure of mammalian chromosomes during spermatogenesis and after radiation with special reference to cores. 4th Int. Conf. Electron Microscopy. **2**, 227–230. Springer-Verlag, Berlin.
— and M. L. RUTTLE, 1935: Chromosome structure IX. *Tradescantia reflexa* and *Trillium erectum*. Amer. J. Bot. **23**, 652–663.
— and E. M. COULON, 1962: The fine structure of chromosomes in pigeon spermatocytes. Chromosoma (Berlin) **13**, 272–291.
NELSON, O. E., 1962: The waxy locus in maize I. Intralocus recombination frequency estimates by pollen and by conventional analysis. Genetics **47**, 737–742.
NICHOLS, C., 1941: Spontaneous chromosome aberrations in *Allium*. Genetics **26**, 89–100.
NICKLAS, R. B., 1961a: Recurrent pole to pole movements of the sex chromosome during prometaphase—I in *Melanoplus differentialis* spermatocytes. Chromosoma (Berlin) **12**, 97–115.
— 1961b: The relationship between DNA content and alternative meiotic patterns in certain discocephalinids (Pentatomidae; Heteroptera). J. Biophys. Biochem. Cytol. **9**, 486–490.
NIGON, V., 1949: Modalités de la reproduction et déterminisme du sexe chez quelques nématodes libres. Ann. Sci. Natur. (Paris) **11**, 1–132.
NODA, S., 1960: Chiasma studies in structural hybrids II. Reciprocal translocation in *Lilium maximowiczii*. Cytologia (Tokyo) **25**, 456–460.
— 1961: Chiasma studies in structural hybrids VII. Reciprocal translocation in *Scilla scilloides*. Cytologia (Tokyo) **26**, 74–77.
NORDENSKIÖLD, H., 1961: Tetrad analysis and the course of meiosis in three hybrids of *Luzula campestris*. Hereditas (Lund) **47**, 203–238.
— 1962: Studies of meiosis in *Luzula purpurea*. Hereditas (Lund) **48**, 503–519.

Novitski, E., 1951: Non-random disjunction in *Drosophila*. Genetics **36**, 267–280.
— 1952: The genetic consequences of anaphase bridge formation in *Drosophila*. Genetics **37**, 270–287.
— 1955: Genic measures of centromere activity in *Drosophila melanogaster*. J. Cell comp. Physiol. **45** (Suppl. 2), 151–169.
— and I. Sandler, 1957: Are all products of spermatogenesis regularly functional. Proc. Nat. Acad. Sci. (Wash.) **43**, 318–324.
Nur, U., 1962: Meiotic behaviour of an unequal bivalent in the grasshopper *Calliptamus palaestinensis* Bdhr. Chromosoma (Berlin) **12**, 272–279.
— 1963: A mitotically unstable supernumerary chromosome with an accumulation mechanism in a grasshopper. Chromosome (Berlin) **14**, 407–422.

Ogawa, K., 1962: Unusual features of chromosomes in *Thereuonema hilgendorfi*. C. I. S. **3**, 5–6.
Ohno, S., and B. M. Cattanach, 1962: Cytological study of an X-autosome translocation in *Mus musculus*. Cytogenetics **1**, 129–140.
— W. D. Kaplan, and R. Kinosita, 1956: Concentration óf RNA on the heteropycnotic XY bivalent of the rat. Exper. Cell Res. **11**, 520–526.
— — — 1957a: Conjugation of the heteropycnotic X and Y chromosomes of the rat spermatocyte. Exper. Cell Res. **12**, 395–426.
— — — 1957b: Heterochromatic regions and nucleolus organisers in chromosomes of the mouse *Mus musculus*. Exper. Cell Res. **13**, 358–364.
— — — 1959: On the end to end association of the X and Y chromosomes of *Mus musculus*. Exper. Cell Res. **18**, 282–290.
— and C. Weiler, 1961: Sex chromosome behaviour pattern in germ and somatic cells of *Mesocricetus auratus*. Chromosoma (Berlin) **12**, 362–373.
— — 1962: Relationship between large Y-chromosome and side-by-side pairing of the XY-bivalent observed in the chinese hamster, *Cricetulus griseus*. Chromosoma (Berlin) **13**, 106–110.
— J. Jainchill, and C. Stenius, 1963: The creeping role (*Microtus oregoni*) as a gonosomic mosaic I. The OY/XY constitution of the male. Cytogenetics **2**, 232–239.
Oksala, T., 1943: Zytologishe studien an Odonaten I. Ann. Acad. Sci. Fenn. Ser. A IV, **4**, 1–63.
— Chiasma formation and chiasma interference in the Odonata. Hereditas (Lund.) **38**, 449–480.
— 1958: Chromosome pairing, crossing-over and segregation in meiosis in *Drosophila melanogaster* females. Cold. Sprg. Har. Symp. Quant. Biol. **23**, 197–210.
Olive, L. S., 1959: Aberrant tetrads in *Sordaria fimicola*. Proc. Nat. Acad. Sci. (Wash.) **45**, 727–732.
Östergren, G., 1943: Elastic chromosome repulsions. Hereditas (Lund) **29**, 444–450.
— 1947: Heterochromatic B-chromosomes in *Anthoxanthum*. Hereditas (Lund) **33**, 261–296.
— 1948: Chromosome bridges and breaks by coumarin. Bot. Notis. (Lund) **4**, 376–380.
— 1951: The mechanism of co-orientation in bivalents and multivalents. The theory of orientation by pulling. Hereditas (Lund) **37**, 85–156.
— 1957: The course of meiosis in a case of abnormally prolonged chromatid pairing. Arkiv. f. Zool. **11**, 130–131.
— and R. Prakken, 1946: Behaviour on the spindle of the actively mobile chromosome ends of rye. Hereditas (Lund) **32**, 473–494.
— and E. Vigfusson, 1953: On position correlations of univalents and quasibivalents formed by sticky univalents. Hereditas (Lund) **39**, 33–50.
— and T. Wakonig, 1954: True or apparent sub-chromatid breakage and the induction of labile states in cytological chromosome loci. Bot. Notis. (Lund) **4**, 357–375.
Owen, A. R. G., 1949: A possible interpretation of the apparent interference across the centromere found by Callan and Montalenti in *Culex pipiens*. Heredity (London) **3**, 357–367.

Parsons, P. A., 1959: Possible affinity between linkage groups V and XIII of the house mouse. Genetica **29**, 304–311.
Pastor, J. B., and H. G. Callan, 1952: Chiasma formation in spermatocytes and oocytes of the turbellarian *Dendrocoelum lacteum*. J. Genetics **50**, 449–454.
Pätau, K., 1941: Cytologischer Nachweis einer positiven Interferenz über das Centromer (Der Paarungskoeffizient I). Chromosoma (Berlin) **2**, 36–63.
— 1948: X-segregation and heterochromasy in the spider *Aranea reaumuri*. Heredity (London) **11**, 77–100.

PEACOCK, W. J., 1961: Sub-chromatid structure and chromosome duplication in *Vicia faba*. Nature (London) **191**, 832–833.

— 1963: Chromosome duplication and structure as determined by autoradiography. Proc. Nat. Acad. Sci. (Wash.) **49**, 793–801.

PERKINS, D. D., 1962: The frequency in *Neurospora* tetrads of multiple exchanges within short intervals. Genet. Res. (Camb.) **3**, 315–327.

PERSON, C., 1955: An analytical study of chromosome behaviour in a wheat haploid. Can. J. Bot. **33**, 11–30.

PHILLIPS, L. L., 1964: Cytogenetical evidence on the question of affinity in cotton. Heredity **19**, 21–26.

PHILP, J., and C. L. HUSKINS, 1931: The cytology of *Matthiola incana* R. Br. especially in relation to the inheritance of double flowers. J. Genetics **25**, 359–404.

PIZA, S. de T., 1943: Meiosis in the male of the brazilian scorpian, *Tityus bahiensis*. Rev. Agric. Sao Paulo **18**, 249–276.

— 1946: Uma nova modalidade de sexodeterminação no grilo sub-americano *Eneoptera surinamensis*. Ann. Esc. Agric. Queiroz **3**, 69–88.

— 1958: Normally dicentric insect chromosomes. Proc. X[th]. Congr. Entomol. **2**, 945–951.

POLL, H., 1920: Pflaumischling. Arch. mikroskop. Anat. Festschr. Hertwrg, 365–458.

POLLISTER, A. W., 1952: Photomultiplier apparatus for microspectrophotometry of cells. Lab. Invest. **1**, 106–114.

— and P. F. POLLISTER, 1943: The relation between centriole and centromere in atypical spermatogenesis of viviparid snails. Ann. N. Y. Acad. Sci. **45**, 1–48.

PONTECORVO, G., 1943: Meiosis in the striped hamster (*Cricetulus griseus* Milne-Edw.) and the problem of heterochromatin in mammalian sex chromosomes. Proc. Roy. Soc. Edin. B. **62**, 32–42.

— 1959: Trends in Genetic Analysis. Oxford Univ. Press.

— and H. J. MULLER, 1941: The lethality of dicentric chromosomes in *Drosophila*. Genetics **26**, 165.

PRAKKEN, R., 1943a: A new trisomic *Matthiola*-type. Hereditas (Lund) **28**, 297–305.

— 1943b: Studies of asynapsis in rye. Hereditas (Lund) **29**, 475–495.

— and A. MÜNTZING, 1942: A meiotic peculiarity in rye, simulating a terminal centromere. Hereditas (Lund) **28**, 441–482.

PRENSKY, W., 1962: Uptake of thymine-methyl-H³ by pachytene chromosomes of *Lilium longiflorum*. Genetics **47**, 977.

PRESCOTT, D. M., 1963: RNA and protein replacement in the nucleus during growth and division and the conservation of components in the chromosome. Symp. Inter. Cell Biol. **2**, 111–128. Academic Press, N. Y.

PRICE, S., 1958: Chiasma terminalisation in a structural heterozygote in *Secale*. Genetics **44**, 705–712.

PRITCHARD, R. H., 1960a: Localised negative interference and its bearing on models of gene recombination. Genet. Res. (Camb.) **1**, 1–24.

— 1960b: The bearing of recombination analysis at high resolution on genetic fine structure in *Aspergillus nidulans* and the mechanism of recombination in higher organisms. 10th Symp. Soc. Gen. Microbiol. ("Microbiol Genetics") **1**, 155–180. C. U. P.

— 1963a: Mitotic recombination in fungi. Methodology in Basic Genetics, 228–243.

— 1963b: The relationship between conjugation, recombination and DNA synthesis in *Escherichia coli*. Proc. XI. Int. Congress Genetics **2**, 55–68.

RAMAER, H., 1935: Cytology of *Hevea*. Genetica **17**, 193–194.

RAMEL, C., 1962: Interchromosomal effects of inversions in *Drosophila melanogaster*. II. Non-homologous pairing and segregation. Hereditas (Lund) **48**, 59–82.

RASCH, E., H. SWIFT, and R. M. KLEIN, 1959: Nucleoprotein changes in plant tumour growth. J. Biophys. Biochem. Cytol. **6**, 11–34.

RAVIN, A. W., 1961: The genetics of transformation. Adv. in Genetics **10**, 61–163.

RAY-CHAUDHURI, S. P., 1961: Induction of chromosome aberrations in the spermatocytes of grasshoppers. The Nucleus **4**, 47–66.

REDFIELD, H., 1955: Recombination increase due to heterologous inversions and the relation to cytological length. Proc. Nat. Acad. Sci. (Wash.) **41**, 1084–1091.

— 1957: Egg mortality and interchromosomal effects on recombination. Genetics **42**, 712–728.

REES, H., 1952: Asynapsis and spontaneous chromosome breakage in *Scilla*. Heredity (London) **6**, 89–97.

— 1953: Centromere control of chromosome splitting and breakage. Heredity (London) **6** (Suppl. Vol.), 236–245.

Rees, H., 1955: Genotypic control of chromosome behaviour in rye I. Inbred lines. Heredity (London) **9**, 93–116.
— 1957: Genotypic control of chromosome behaviour in rye. IV. The origin of new variation. Heredity (London) **11**, 185–193.
— 1962: Developmental variation in the expressivity of genes causing chromosome breakage in rye. Heredity (London) **17**, 427–437.
— 1961: Genotypic control of chromosome behaviour. Bot. Rev. **27**, 288–318.
— and A. Jamieson, 1954: A supernumerary chromosome in *Locusta*. Nature (London) **173**, 43.
— and J. B. Thompson, 1955: Localisation of chromosome breakage at meiosis. Heredity (London) **9**, 399–407.
— — 1956: Genotypic control of chromosome behaviour in rye III. Chiasma frequency in homozygotes and heterozygotes. Heredity (London) **10**, 409–424.
— and B. Naylor, 1960: Developmental variation in chromosome behaviour. Heredity (London) **15**, 17–28.
— and S. Sun, 1965: Chiasma frequency and the disjunction of interchange associations in rye. Chromosoma (Berlin) **16**, 500–510.
Revell, S., 1947: Controlled X-segregation at meiosis in *Tegenaria*. Heredity (London) **1**, 337–347.
— 1954: A new hypothesis for "chromatid" changes. Proc. Radiobiology Symposium (Liege 1954), 243–253. Butterworth, London.
— 1959: The accurate estimation of chromatid breakage, and its relevance to a new interpretation of chromatid aberrations induced by ionising radiations. Proc. Roy. Soc. (London) **B 150**, 563–589.
Rhoades, M. M., 1942: Preferential segregation in maize. Genetics **27**, 395–407.
— 1952: Preferential segregation in maize. Heterosis, chapt. **4**, 66–80. Towa State College Press.
— 1961: Meiosis. The Cell, vol. III, 1–75. Academic Press, N. Y.
— and E. Dempsey, 1953: Cytogenetic studies of deficient-duplicate chromosomes derived from inversion heterozygotes in maize. Amer. J. Bot. **40**, 405–424.
Rhoades, M. M., and W. E. Kerr, 1959: A note on centromere organisation. Proc. Nat. Acad. Sci. (Wash.) **35**, 129–132.
— and H. Vilkommerson, 1942: On the anaphase movement of chromosomes. Proc. Nat. Acad. Sci. (Wash.) **28**, 433–436.
Ribbands, C. R., 1937: The consequences of structural hybridity at meiosis in *Lilium testaceum*. J. Genetics **35**, 1–24.
— 1941: Meiosis in Diptera I. Prophase associations of non-homologous chromosomes and their relation to mutual attraction between centromeres, centrosomes and chromosome ends. J. Genetics **41**, 411–442.
Richardson, M. M., 1935: Meiosis in *Crepis* II. Failure of pairing in *Crepis capillaris* (L) Wallr. J. Genetics **31**, 119–143.
Rickards, G. K., 1964: Some theoretical aspects of selective segregation in interchange complexes. Chromosoma (Berlin) **15**, 140-155.
Rieger, R., 1957: Inhomologenpaarung und Meioseablauf bei haploiden Formen von *Antirrhinum majus* L. Chromosoma (Berlin) **9**, 1–38.
Riley, R., 1960a: The secondary pairing of bivalents with genetically similar chromosomes. Nature (London) **185**, 751–752.
— 1960b: The diploidisation of wheat. Heredity (London) **15**, 407–429.
— and V. Chapman, 1937: Haploids and polyhaploids in *Aegilops* and *Triticum*. Heredity **11**, 195–207.
— — 1958: Genetic control of the cytologically diploid behaviour of hexaploid wheat. Nature (London) **182**, 713–715.
— and Kimber, 1964: Haploid Angiosperms. Bot. Rev. **29**, 480–531.
— and V. Chapman, 1961: Origin of genetic control of diploid-like behaviour of polyploid wheat. J. Heredity **52**, 22–25.
Ris, H., 1957: Chromosome structure. Pp. 23–62 in "The Chemical Basis of Heredity." Ed. Wn. D. McElroy and B. Glass. Baltimore.
Rizet, G., P. Lissouba, and J. Mosseau, 1960: Les mutations d'ascospore chez l'ascomycète *Ascobulus immersus* et l'analyse de la structure fine des gènes. Bull. Soc. Franc. Physiol. Veg. **6**, 175–193.
Roman, H., 1963: Genic conversion in fungi. Methodology in Basic Genetics, 209–227.
Roseweir, J., and H. Rees, 1962: Fertility and chromosome pairing in autotetraploid rye. Nature (London) **195**, 203–204.
Rothfels, K. H., 1950: Chromosome complement, polyploidy and supernumeraries in *Neopodismopsis abdominalis* (*Acrididae*). J. Morph. **87**, 287–316.

ROTHFELS, K. H., and I. TESHIMA, 1964: Homologies among heterologous chromosomes of a grasshopper *Chloealtis conspersa*. Canad. J. Genet. Cytol., **6**, 243.

RUDKIN, D., and H. A. GRIECH, 1962: On the persistence of oocyte nuclei from fetus to maturity in the laboratory mouse. J. Cell Biol. **12**, 169–175.

RUSSEL, L. B., 1961: Genetics of mammalian sex chromosomes. Science **33**, 1795–1803.

— J. W. BANGHAM, and C. L. SAYLORS, 1962: Delimitation of chromosomal regions involved in V-type position effects from X-autosome translocations in the mouse. Genetics **47**, 981–982.

RUTISHAUSER, A., 1960: Telocentric fragment chromosomes in *Trillium grandiflorum*. Heredity (London) **15**, 241–246.

— 1960: Zur Genetik überzähliger Chromosomen. Archiv der Julius Klaus-Stiftung für Vererbungsforschung, Sozialanthropologie und Rassenhygiene **35**, 440–458.

— 1963: Genetics of B-chromosomes. Proc. XIth Int. Congr. Genetics **1**, 118–119.

SACHS, L., 1953: The giant sex chromosomes in the mammal *Microtus agrestis*. Heredity (London) **7**, 227–238.

— 1954: Sex linkage and the sex chromosomes in man. Ann. Eugen. (London) **18**, 255–261.

SAEZ, F. A., 1957: An extreme karyotype in an orthopteran insect. Amer. Nat. **91**, 259–264.

— and A. DÍAZ, 1960: Neo-X neo-Y system of sex determination in *Xyleus laevipes* (Orthoptera: Romaleinae). Tex. Rep. Biol. Med. **18**, 116–128.

SANDLER, L., and E. NOVITSKI, 1956: Evidence for genetic homology between chromosomes I and IV in *Drosophila melanogaster*, with a proposed explanation for the crowding effect in triploids. Genetics **41**, 189–193.

— Y. HIRAIZUMI, and I. SANDLER, 1959: Meiotic drive in natural populations of *Drosophila melanogaster* I. The cytogenetic basis of segregation-distortion. Genetics **44**, 233–250.

— and Y. HIRAIZUMI, 1960: Meiotic drive in natural populations of *Drosphila melanogaster* V. On the nature of the SD region. Genetics **45**, 1671–1689.

SATINA, S., and A. F. BLAKESLEE, 1937a: Chromosome behaviour in triploids of *Datura stramonium* I. The male gametophyte. Amer. J. Bot. **24**, 518–527.

— — 1937b: Chromosome behaviour in triploid *Datura* II. The female gametophyte. Amer. J. Bot. **24**, 621–627.

— — and A. G. AVERY, 1938: Chromosome behaviour in triploid *Datura* III. The seed. Amer. J. Bot. **25**, 595–602.

SATO, D., 1942: Karyotype alteration and phylogeny V. New types of SAT-chromosomes in *Northoscordum* and *Nerine*. Cytologia (Tokyo) **12**, 170–178.

— and A. ASANO, 1951: Basikaryotype analysis in *Northoscordum fragans* (2n = 19). Bot. Mag. Tokyo **64**, 209–214.

SAUERLAND, H., 1956: Quantitative Untersuchungen von Röntgeneffekten nach Bestrahlung verschiedener Meiosistadien bei *Lilium candidum* L. Chromosoma (Berlin) **7**, 627–654.

SAX, K., 1938: Chromosome aberrations induced by X-rays. Genetics **23**, 494–516.

— 1941: The behaviour of X-ray induced chromosomal aberrations in *Allium* root tip cells. Genetics **26**, 418–425.

— and L. M. HUMPHREY, 1934: Structure of meiotic chromosomes in microsporogenesis of *Tradescantia*. Bot. Gaz. **96**, 353–362.

— and E. D. KING, 1955: An X-ray analysis of chromosome duplication. Proc. Nat. Acad. Sci. (Wash.) **41**, 150–155.

SCHNEIDER, W. C., and J. ROTHERHAM, 1958: Phosphorus compounds in animal tissues. J. Biol. Chem. **233**, 948–953.

SCHOLL, H., 1953: Ein Beitrag zur Kenntnis der Spermatogenese der Mallophagen. Chromosoma (Berlin) **7**, 271–274.

SCHRADER, F., 1931: The chromosome cycle of *Protortonia primitiva* (Coccidae) and a consideration of the meiotic division apparatus in the male. Z. Wiss. Zool. **138**, 386–408.

— 1935: Notes on the mitotic behaviour of long chromosomes. Cytologia (Tokyo) **6**, 422–430.

— 1960: Cytological and evolutionary implications of aberrant chromosome behaviour in the harlequin lobe of some Pentatomidae (Heteroptera). Chromosoma (Berlin) **11**, 103–128.

— and C. LEUCHTENBERGER, 1950: A cytochemical analysis of the functional interrelations of various cell structures in *Arvelius albomaculatus* (de Geer). Exper. Cell Res. **1**, 421–452.

SCHULTZ, J., and H. REDFIELD, 1951: Interchromosal effects on crossing-over in *Drosophila*. Cold Sprg. Har. Symp. Quant. Biol. **16**, 175–197.

Schwartz, D., 1953a: The behaviour of an X-ray induced ring chromosome in maize. Amer. Nat. 87, 19–28.
— 1953b: Evidence for sister-strand crossing over in maize. Genetics 38, 251–260.
— 1954: Studies on the mechanism of crossing-over. Genetics 39, 692–700.
Sears, E. R., 1954: The aneuploids of common wheat. Miss. Agr. Exp. Sta. Res. Bull. 572.
— and C. A. Camara, 1952: A transmissible dicentric chromosome. Genetics 37, 125–135.
— and M. Okomata, 1958: Intergenomic chromosome relationships in hexaploid wheat. Proc. Xth Int. Congr. Genetics 2, 258–259.
Seiler, J., 1921: Geschlechtschromosomen-Untersuchungen an Psychiden I. Experimentelle Beeinflussung der geschlechtsbestimmenden Reifeteilung bei Talaeoporia tubulosa Retz. Arch. Zellforsch. 15, 249–268.
Serra, J. A., 1955: Chemistry of the nucleus. Encyclopaedia of Plant Physiology 1, 413–444. Springer-Verlag, Berlin.
Shah, S. S., 1963: Studies on supernumerary chromosomes in the genus Dactylis. Chromosoma (Berlin) 14, 162–185.
Sharma, G. P., S. S. Jande, and K. K. Tandon, 1959: Cytological studies on the Indian spiders IV. Chromosome complement and meiosis in Senelops radiatus (Selenopidae) and Leucauge decorata (Tetiagnathidae) with special ref. to the $X_1 X_2 X_3$ O-type of male sex determining mechanism. Res. Bull. Panjab Univ. Sci. 10, 73–80.
— and S. Singh, 1957: Cytological studies on the Indian spiders I. Chromosome complement and male meiosis in Stegodyphus specificus. Res. Bull. Panjab Univ. Zool. 120, 389–393.
Sharman, G. B., and H. N. Barber, 1952: Multiple sex-chromosomes in the marsupial, Potorus. Heredity (London) 6, 345–355.
Shaver, D. L., 1963: The effect of structural heterozygosity on the degree of preferential pairing in allotetraploids of Zea. Genetics 48, 515–524.
Shaw, G. W., 1958: Adhesion loci in the differentiated heterochromatin of Trillium species. Chromosoma (Berlin) 9, 292–304.
Sherman, F., and H. Roman, 1963: Evidence for two types of allelic recombination in yeast. Genetics 48, 255–261.
Shimakura, K., 1957: The chromosome movements during the spindle formation as observed in the living spermatocytes of some grasshoppers. Proc. Internat. Genet. Symp. (Tokyo-Kyoto), 126–128.
Skovsted, A., 1937: Cytological studies in cotton IV. Chromosome conjugation in inter-specific hybrids. J. Genetics 34, 97–134.
Slizynski, B. M., 1960: Sexual dimorphism in mouse gametogenesis. Gent. Res. (Camb.) 1, 477–486.
Smith, B., 1955: Sex chromosomes and natural polyploidy in Rumex hastalulus. J. Heredity 46, 226–232.
Smith, S. G., 1952a: The evolution of heterochromatin in the genus Tribolium (Tenebrionidae: Coleoptera). Chromosoma (Berlin) 4, 585–610.
— 1952: The cytology of some tenebrionoid beetles (Coleoptera). J. Morph. 91, 325–363.
— 1953: A pseudo-multiple sex chromosome mechanism in an Indian gryllid. Chromosoma (Berlin) 5, 555–573.
— and D. E. Maxwell, 1953: Post-reduction of the X-chromosome and complete chiasma interference in the Lampyridae (Coleoptera). Can. J. Zool. 31, 179–192.
Smith-White, S., and A. McCusker, 1960: Spontaneous chromosome breakage in Astroloma pinifolium. Proc. Lin. Soc. N. S. W. 85, 142–153.
Snell, G. D., 1946: An analysis of translocations in the mouse. Genetics 31, 157–180.
Snoad, B., 1963: Tomato chromosomes and their relationship to the linkage maps. Proc. XI. Int. Congress Genetics 1, 117.
Solari, A. J., 1964: The morphology and ultrastructure of sex vesicle in the mouse. Exper. Cell Res. 36, 160–168.
Soost, R. K., 1951: Comparative cytology and genetics of asynaptic mutants in Lycopersicum esculentum Mill. Genetics 36, 410–434.
Soriano, J. D., 1957: The genus Collinsia IV. The cytogenetics of colchicine-induced reciprocal translocations in C. heterophylla. Bot. Gaz. 118, 139–145.
Sparrow, A. H., 1942: The structure and development of the chromosome spirals in microspores of Trillium. Can. J. Res. D. 20, 257–266.
— C. L. Huskins, and G. B. Wilson, 1941: Studies on the chromosome spiralisation cycle in Trillium. Can. J. Res. 19, 323–350.
— M. J. Moses, and R. Steele, 1952: A cytological and cytochemical approach to an understanding of radiation damage in dividing cells. Brit. J. Radiobiol. 26, 182–188.

SPIELER, R. A., 1963: Genic control of chromosome loss and non-disjunction in *Drosophila melanogaster*. Genetics **48**, 73–90.

SPURWAY, H., and H. G. CALLAN, 1960: The vigour and male sterility of hybrids between the species *Triturus vulgaris* and *T. helveticus*. J. Genetics **57**, 84–118.

STAIGER, H., 1954: Der Chromosomendimorphismus beim Prosobranchier *Purpurea lapillus* in Beziehung zur Ökologie der Art. Chromosoma (Berlin) **6**, 419–478.

— and H. GLOOR, 1952: Mitosehemmung und Polyploidie durch einen Letalfaktor (Lpl = lethal polyploid) bei *Drosophila hydei*. Chromosoma (Berlin) **5**, 221–245.

STEBBINS, G. L., 1950: Variation and Evolution in Plants. Columbia Univ. Press N.Y.

— 1958: The inviability, weakness and sterility of interspecific hybrids. Adv. in Genetics **9**, 147–215.

— J. I. VALENCIA, and R. M. VALENCIA, 1946: Artificial and natural hybrids in the Gramineae, tribe Hordeae II. *Agropyron, Elymus* and *Hordeum*. Amer. J. Bot. **33**, 579–586.

STEINBERG, A. G., and F. C. FRASER, 1944: Studies on the effects of X-chromosome inversions on crossing-over in the third chromosome of *Drosophila melanogaster*. Genetics **29**, 83–103.

STEINBERG, C. M., and R. S. EDGAR, 1961: On the absence of high negative interference in triparental crosses. Virology **15**, 511–512.

STERN, C., 1931: Zytologisch-genetische Untersuchungen als Beweise für die Morgansche Theorie des Faktorenaustausches. Biol. Zentral **51**, 547–587.

— 1936: Somatic crossing over and segregation in *Drosophila melanogaster*. Genetics **21**, 625–730.

STERN, H., 1946: The permeability of cells undergoing nuclear divisions. Trans. Roy. Soc. Canada **40**, 141–148.

— 1956: The physiology of cell division. Ann. Rev. Plant Physiol. **7**, 91–114.

— 1958: Variations in sulphydryl concentration during microsporocyte meiosis in anthers of *Lilium* and *Trillium*. J. Biophys. Biochem. Cytol. **4**, 157–161.

— 1960a: Aspects of deoxyriboside metabolism in relation to the mitotic cycle. Ann. N.Y. Acad. Sci. **90**, 440–454.

— 1960b: Developing cell systems and their control. 18th Growth Symposium, 135–165. Ronald Press, N.Y.

— 1961: Periodic induction of deoxyribonuclease activity in relation to the mitotic cycle. J. Biophys. Biochem. Cytol. **9**, 271–277.

— 1962: Function and reproduction of chromosomes. Physiol. Rev. **42**, 271–296.

— and Y. HOTTA, 1963: Facets of intracellular regulation of meiosis and mitosis. Cell growth and cell division. Symp. Internat. Soc. Cell Biol. **2**, 57–76.

— and P. L. KIRK, 1948: The oxygen consumption of the microspores in *Trillium* in relation to the mitotic cycle. J. gen. Physiol. **31**, 243–248.

— and S. TIMONEN, 1954: The position of the cell nucleus in pathways of hydrogen transfer: cytochrome C, flavoproteins, glutathione and ascorbic acid. J. gen. Physiol. **38**, 41–52.

STICH, H. F., 1962: Variations of the DNA content in embryonal cells of *Cyclops strenuus*. Exper. Cell Res. **26**, 136–143.

STONE, W., and I. THOMAS, 1935: Crossover and disjunctional properties of X-chromosome inversions in *Drosophila melanogaster*. Genetica **17**, 170–184.

STURTEVANT, A. H., 1919: Contributions to the genetics of *Drosophila melanogaster* III. Inherited linkage variations in the second chromosome. Carnegie Inst. Wash. Pub. **278**, 305–341.

— and G. W. BEADLE, 1936: The relations of inversions in the X-chromosome of *Drosophila melanogaster* to crossing-over and disjunction. Genetics **21**, 554–604.

— and TH. DOBZHANSKY, 1936: Geographical distribution and cytology of "sex ratio" in *Drosophila pseudoobscura* and related species. Genetics **21**, 554–604.

SUGINO, Y., 1957: Deoxycytidinediphosphatecholine, a new deoxyriboside compound. J. Amer. Chem. Soc. **79**, 5074–5075.

SUN, S., 1963: Cytogenetic studies in *Secale*. Ph. D. Thesis. University of Wales.

SUOMALAINEN, E., 1953: The kinetochore and bivalent structure in the Lepidoptera. Hereditas (Lund) **39**, 88–96.

— and O. HALKKA, 1963: The mode of meiosis in the *Psyllina*. Chromosoma (Berlin) **14**, 498–510.

SUOMALAINEN, H. O. T., 1952: Localisation of chiasmata in the light of observations on the spermatogenesis of certain Neuroptera. Ann. Zool. Soc. Vanamo **15**, 1–104.

SUZUKI, D., 1963: Interchromosomal effects on crossing-over in *Drosophila melanogaster* II. A re-examination of X-chromosome inversion effects. Genetics **48**, 1605–1617.

Suzuki, S., 1951: Cytological studies in Spiders I. A comparative study of the chromosomes in the family Argiopidae. J. Sci. Hiroshima Univ. B I (Zool.) **12**, 67–98.

Swanson, C. P., 1942: Some considerations on the phenomenon of chiasma terminalisation. Amer. Nat. **76**, 593–610.

— 1943a: Differential sensitivity of prophase pollen tube chromosomes to X-rays and ultra-violet radiation. J. gen. Physiol. **26**, 485–494.

— 1943b: The behaviour of meiotic prophase chromosomes as revealed through the use of high temperature. Amer. J. Bot. **30**, 422–428.

— 1947: X-ray and ultra-violet studies on pollen tube chromosomes II. The quadripartite structure of the prophase chromosomes of *Tradescantia*. Proc. Nat. Acad. Sci. (Wash.) **33**, 229–232.

— 1957: Cytology and Cytogenetics. Prentice Hall, New Jersey.

Swift, H., 1950: The constancy of deoxyribosenucleic acid in plant nuclei. Proc. Nat. Acad. Sci. (Wash.) **36**, 643–654.

Takats, S. T., 1959: Chromatin extrusion and DNA transfer during microsporogenesis. Chromosoma (Berlin) **10**, 430–453.

— 1962: An attempt to detect utilisation of DNA breakdown products from the tapetum for DNA synthesis in the microspores of *Lilium longiflorum*. Amer. J. Bot. **49**, 748–758.

Tanaka, N., 1941: Chromosome studies in Cyperaceae VIII. Meiosis in diploid and tetraploid forms of *Carex siderosticta* Hance. Cytologia (Tokyo) **11**, 282–310.

— and M. Kohno, 1961: Breakage in tritium labelled *Tradescantia* chromosomes. Proc. Symp. on Genetic Effects of Radiation. Jap. J. Genetics **36** (Suppl. Vol.) 88–93.

Tandon, S. L., and B. M. Kapoor, 1963: Contribution to the cytology of endosperm in some angiosperms II. *Northoscordum fragrans* Kunth. Caryologia (Piza) **16**, 377–395.

Taylor, E. W., 1959: Dynamics of spindle formation and its inhibition by chemicals. J. Biophys. Biochem. Cytol. **6**, 193–196.

Taylor, J. H., 1957: The time and mode of duplucation of chromosomes. Amer. Nat. **91**, 209–221.

— 1958a: Sister chromatid exchanges in tritium-labelled chromosomes. Genetics **43**, 515–529.

— 1958b: The organisation and duplication of genetic material. Proc. X[th] Int. Congr. Genetics **1**, 63–78.

— 1959: Autoradiographic studies of the organisation and mode of duplication of chromosomes. Symp. on Mol. Biol. Univ. Chicago, 304–320.

— 1962: Chromosome reproduction. Internat. Rev. Cytol. **13**, 39–73.

— 1963: Molecular genetics, part 1, chapt. 2. The replication and organisation of DNA in chromosomes, pp. 65–111. Academic Press, N.Y.

— and R. D. MacMaster, 1954: Autoradiographic and microphotometric studies of deoxyribose nucleic acid during microgametogenesis in *Lilium longiflorum*. Chromosoma (Berlin) **6**, 489–521.

— W. F. Haut, and J. Tung, 1962: Effects of fluorodeoxyuridine on DNA replication, chromosome breakage and reunion. Proc. Nat. Acad. Sci. (Wash.) **48**, 190–198.

Thomas, P. T., and S. H. Revell, 1946: Secondary association and heterochromatic attraction I. *Cicer arietinum*. Ann. Bot. **9**, 159–164.

Thompson, J. B., 1956: Genotypic control of chromosome behaviour in rye II. Disjunction at meiosis in interchange heterozygotes. Heredity (London) **10**, 99–108.

Togby, H. A., 1943: A cytological study of *Crepis fuliginosa, C. neglecta* and their F_1 hybrid and its bearing on the mechanism of phylogenetic reduction in chromosome number. J. Genetics **45**, 67–111.

Trujillo, J. M., C. Stenius, L. C. Christian, and S. Ohno, 1962: Chromosomes of the horse, the donkey and the mule. Chromosoma (Berlin) **13**, 243–248.

Tsuchiya, T., 1962: Haploid plants in barley. C. I. S. (Tokyo) **3**, 14–15.

Ueshima, N., 1963: Chromosome behaviour in the *Cimex pilosellus* complex (Cimicidae: Hemiptera). Chromosoma (Berlin) **14**, 511–521.

Ullerich, F.-H., 1961: Achiasmate Spermatogenese bei der Skorpionsfliege *Panorpa* (Mecoptera). Chromosoma (Berlin) **12**, 215–232.

Upcott, M. B., 1935: The cytology of triploid and tetraploid *Lycopersicum esculentum*. J. Genetics **31**, 1–19.

— 1937: The genetic structure of *Tulipa* II. Structural hybridity. J. Genetics **34**, 339–398.

— 1938: The internal mechanics of the chromosomes VI. Relic and relational coiling in pollen grains. Cytologia (Tokyo) **8**, 398–407.

VAARAMA, A., 1950: Cases of asyndesis in *Matricaria indora* and *Hyoscyamus niger*. Hereditas (Lund) **36**, 342–362.
— 1954: Cytological observations on *Pleurozium schreberi* with special reference to centromere evolution. Ann. Bot. Soc. Vaanamo **28**, 1–59.
VAN BRINK, J. M. and B. KIAUTA, 1964: Notes on chromosome behaviour in the spermatogenesis of the damselfly *Enallagma cyathigerum* (Charp.) (Odonata: Coenagrionidae), Genetics **35**, 171–174.
VASEK, F. C., 1962: "Multiple spindle" — a meiotic irregularity in *Clarkia exilis*. Amer. J. Bot., **49**, 536–539.
VILKOMERSON, H., 1950: The unusual meiotic behaviour of *Elymus wiegandii*. Exper. Cell Res. **1**, 534–542.
VIRKKI, N., 1959: Neo-XY-mechanism in two scarabaeoid beetles, *Phanaeus vindex* MacL. (Scarabaeidae) and *Dorcus parallelopipedus* L. (Lucanidae). Hereditas (Lund.) **45**, 481–494.
— 1961: Non-conjugation and late-conjugation of the sex-chromosomes in the beetles of the genus *Alagoasa* (Chrysomelidae: Alticinae). Ann. Acad. Sci. Fenn. A IV. Biol. **54**, 1–22.
— 1963: On the cytology of some neotropical Cantharoids (Coleoptera). Ann. Acad. Sci. Fenn. A IV Biol. **65**, 1–16.
VITAGLIANO, G., 1947: La spermatogenesi e la distribuzione dei chiasma in *Asellus aquaticus* L. Publ. della Staz. Zool. Napoli **21**, 164–182.

WALD, H., 1936: Cytologic studies on the abnormal development of the eggs of the claret mutant type of *Drosophila simulans*. Genetics **21**, 264–281.
WAHRMAN, J., and U. RITTE, 1963: Crossing-over in the sex bivalent of male mammals. Proc. XIth Int. Congr. Genetics **1**, 125.
WALLACE, M. E., 1953: Affinity a new genetic phenomenon in the house mouse. Nature (London) **171**, 27–28.
— 1958: Experimental evidence for a new genetic phenomenon. Phil. Trans. Roy. Soc. (London) **B 241**, 211–254.
— 1960a: Possible cases of affinity in cotton. Heredity (London) **14**, 263–274.
— 1960b: A possible case of affinity in tomatoes. Heredity (London) **14**, 275–283.
WALTERS, J. L., 1956: Spontaneous meiotic chromosome breakage in natural populations of *Paeonia californica*. Amer. J. Bot. **43**, 342–354.
WALTERS, M. S., 1951: Spontaneous chromosome breakage and atypical chromosome movement in meiosis of the hybrid *Bromus marginatus* × *B. pseduolaevipes*. Genetics **37**, 8–25.
— 1957: Studies of spontaneous chromosome breakage in interspecific hybrids of *Bromus*. Univ. Calif. Pub. Bot. **28**, 335–447.
WATSON, J. D., and H. G. CALLAN, 1963: The form of bivalent chromosomes in newt oocytes at first metaphase of meiosis. Quart. Jour. Micro. Sci. **104**, 281–295.
WENT, H. A., 1959a: Some immunochemical studies on the mitotic apparatus of the sea urchin. J. Biophysic. Biochem. Cytology **5**, 353–356.
— 1959b: Studies on the mitotic apparatus of the sea urchin by means of antigen-antibody reactions in agar. J. Biophysic. Biochem. Cytology **6**, 447–455.
WESTERGAARD, M., 1964: Studies on the mechanism of crossing-over I. Theoretical considerations. Compt. Rendus des Trav. Lab. Carlsberg **34**, 359–405.
WHITE, M. J. D., 1936: Chiasma-localisation in *Mecostethus grossus* L. and *Metrioptera brachyptera* L. (Orthoptera). Z. für Zellforsch. Mikr. Anat. **24**, 128–135.
— 1940a: The heteropycnosis of sex chromosomes and its interpretation in terms of spiral structure. J. Genetics **40**, 67–82.
— 1940b: A translocation in a wild population of grasshoppers. J. Heredity **31**, 137–140.
— 1941: The evolution of the sex chromosomes I. The XO and $X_1 X_2$ Y mechanisms in praying mantids. J. Genetics **42**, 143–172.
— 1946: The spermatogenesis of hybrids between *Triturus cristatus* and *Triturus marmoratus* (Urodela). J. Exp. Zool. **102**, 179–204.
— 1951a: A cytological survey of wild populations of *Trimerotropis* and *Circotettix* (Orthoptera, Acrididae) II. Racial differentation in *T. sparsa*. Genetics **36**, 31–53.
— 1951b: Cytogenetics of orthopteroid insects. Adv. Genetics **4**, 267–330.
— 1954a: Animal Cytology and Evolution. C. U. P.
— 1954b: An extreme form of chiasma localisation in a species of *Bryodema* (Orthoptera, Acrididae). Evolution **8**, 350–358.
— 1957: Cytogenetics of the grasshopper *Moraba scurra* I. Meiosis of interracial and interpopulation hybrids. Austral. J. Zool. **5**, 285–304.

White, M. J. D., 1959: Telomeres and terminal chiasmata—a reinterpretation. Biol. Contribs., Univ. of Texas Pub. No. 5914, 107–111.
— 1961: The role of chromosomal translocations in urodele evolution and speciation in the light of work on grasshoppers. Amer. Nat. 95, 315–321.
— 1962: A unique type of sex-chromosome mechanism in an Australian mantid. Evolution 16, 75–85.
— 1965: Chiasmatic and achiasmatic meiosis in african Eumastacid grasshoppers. Chromosoma (Berlin) 16, 271–307.
— and F. H. W. Morley, 1955: Effects of pericentric rearrangements on recombination in grasshopper chromosomes. Genetics 40, 604–619.
Whitehouse, H. L. K., 1963: A theory of crossing-over by means of hybrid deoxyribonucleic acid. Nature (London) 199, 1034–1040.
Wilson, E. B., 1910: Studies on chromosomes VI. A new type of chromosome combination in Metapodius. J. Exper. Zool. 9, 53–78.
— 1925: The Cell in Development and Heredity. MacMillan, New York.
Wilson, G. B., and E. R. Boothroyd, 1944: Temperature-induced differential contraction in the somatic chromosomes of Trillium erectum. Can. J. Res. C 22, 105–119.
— and C. L. Huskins, 1939: Chromosome and chromonema length during meiotic coiling in Trillium erectum L. Ann. Bot. N. S. 3, 257–270.
— and I. Hutcheson, 1941: Further studies on changes in the major coil of the chromosomes of Trillium erectum L. Can. J. Res. C 19, 383–390.
— A. Sparrow, and V. Pond, 1959: Sub-chromatid rearrangements in Trillium erectum L. I. Origin and nature of configurations induced by ionising radiation. Amer. J. Bot. 46, 309–316.
Wilson, J. Y., 1959: Cytogenetics of triploid bluebells, Endymion nonscriptus (L.) Garcke and E. hispanicus (Mill.) Chouard. Cytologia (Tokyo) 23, 435–446.
Winge, O., and C. Roberts, 1954: Causes of deviations from 2:2 segregations in the tetrads of monohybrid yeasts. Comptes Rendus des travaux du Laboratoire Carlsberg Serie Physiol. 25, 285–329.
Wimber, D. E., 1959: Chromosome breakage produced by tritium-labelled thymidine in Tradescantia paludosa. Proc. Nat. Acad. Sci. (Wash.) 45, 839–846.
— and W. Prensky, 1963: Autoradiography with meiotic chromosomes of the male newt (Triturus viridescens) using H^3-thymidine. Genetics 48, 1731–1738.
Winkler, H., 1930: Die Konversion der Gene. Gustav Fischer, Jena.
Wolf, B. E., 1941: Die Chromosomen in der Spermatogenese einiger Nematoceren. Chromosomes (Berlin) 2, 192–246.
— 1950: Die Chromosomen in der Spermatogenese der Dipteren Phryne und Mycetobia. Chromosoma (Berlin) 4, 148–204.
— 1960: Zur Karyologie der Eireifung und Furchung bei Cloeon diperum L. (Bengtsson) (Ephemerida, Baetididae). Biol. Zentral. 79, 153–198.
Wolff, S., 1960: Radiation studies on the nature of chromosome breakage. Amer. Nat. 94, 85–93.
— and H. E. Luippold, 1964: Chromosome splitting as revealed by combined X-ray and labelling experiments. Exper. Cell Res. 34, 548–556.
Woods, P. S., and M. U. Schairer, 1959: Distribution of newly synthesised deoxyribonucleic acid in dividing chromosomes. Nature (London) 183, 303–305.

Yamamoto, Y., 1934: Reifungsteilungen bei einer asynaptischen Pflanze von Rumex acetosa L. Bot. Zool. 2, 1160–1168.
Yerganian, G., 1959: Chromosomes of the chinese hamster, Cricetulus griseus I. The normal complement and identification of sex chromosomes. Cytologia (Tokyo) 24, 66–75.

Zen, S., 1961: Chiasma studies in structural hybrids VI. Heteromorphic bivalent and reciprocal translocation in Allium fistulosum. Cytologia (Tokyo) 26, 67–73.
Zimmering, S., 1955: A genetic study of segregation in a translocation heterozygote in Drosophila. Genetics 40, 809–825.
Zohary, D., 1955: Secondary centric activity in Lilium formosanum. Amer. Nat. 89, 50–52.

Appendix

The literature survey on which this monograph was based was completed early in 1964. A number of important papers have subsequently appeared which either enlarge upon or, in some cases, even contradict statements which appear in the text. We have included brief notes on these so that the reader may be aware of the current status of the issues in question.

Note 1 (see pg. 45) — BALDWIN and SHOOTER (1963) have evidence that the DNA of the bacterium *Escherichia coli* certainly replicates semi-conservatorily. Using hybrid DNA labelled with 5-BU in one sub-unit only, they have shown by measuring optical and buoyant density at ambient pH that this DNA melts in the manner expected if the sub-units are single polynucleotide chains.

> BALDWIN, R. L., and E. M. SHOOTER, 1963: The alkaline transition of BU-containing DNA and its bearing on the replication of DNA. J. Mol. Biol. **7**, 511—526.

Note 2 (see pg. 142) — In the olive scale, *Parlatoria oleae* (2n = 8), NUR (1965) finds that the D^H chromosomes replicate but do not pair while the D^E chromosomes do not replicate but do pair. Meiosis is consequently achiasmate (see pg. 20).

> NUR, U., 1965: A modified Constockiella system in the olive scale insect, *Parlatoria oleae* (Homoptera: Coccida). Chromosoma (Berl.) (in press).

Note 3 (see pg. 189) — FREDGA and SANTESSON (1964) report that a visible chiasma is frequently formed between the short arms of the X and Y chromosomes in both Chinese (*Cricetulus griseus*) and European (*Cricetus cricetus*) hamsters. Sex-bivalents with the chromosomes associated end-to-end were also found in both species. This, together with the fact that in the Syrian hamster (*Mesocricetus auratus*) the sex-chromosomes are invariably associated end-to end favours the hypothesis that the sex-bivalent found in most mammals results from a terminalised chiasma between minute homologous segments.

> FREDGA, K., and B. SANTESSON, 1964: Male meiosis in the Syrian, Chinese and European hamsters. Hereditas (Lund) **52**, 36—48.

Note 4 (see pg. 205) — In the past much has been made of the existence of chiasma-like configurations not only in the neurocyte mitoses of male and female larvae of *Drosophila* (KAUFMANN 1934) but also in the meiosis of *Drosophila* males (COOPER 1949). Indeed on the basis of these claims many have argued that chiasmata can arise without antecedent crossing-over. We had deliberately omitted both claims since in our opinion they are based on observations necessitating a strong element of subjective interpretation. Two recent studies have utilised much improved techniques. First, SLIZYNSKI (1964) has concluded that what earlier had been claimed as chiasmata are in reality but surface associations of homologous chromosomes. Second, PEACOCK and ERICKSON (1965) state that none of the associations which they observed between homologous chromosomes at 'diplotene' of male meiosis "seemed to be equivalent to the chiasmata reported by COOPER but divergence of opinion may rest on semantic rather than observational differences".

21*

Cooper, K. W., 1949: The cytogenetics of meiosis in *Drosophila melanogaster*. J. Morph. **84**, 81—122.

Kaufmann, B. P., 1934: Somatic mitoses in *Drosophila melanogaster*. J. Morph. **56**, 125—155.

Peacock, W. J., and J. Erickson, 1965: Segregation-distortion and regularly non-functional products of spermatogenesis in *Drosophila melanogaster*. Genetics **51**, 313—328.

Slizynski, B. M., 1964: Chiasmata in spermatocytes of *Drosophila melanogaster*. Genet. Res. (Camb.) **5**, 80—84.

Note 5 (see pg. 220) — Lawrence (1965) has examined the influence of non-lethal doses of radiation on recombination in the arg 1–paba 2 region of linkage group I in *Chlamydomonas reinhardi*. It was shown that recombination was influenced by irradiation at two short stages during the germination of the zygote. The pattern of response is thus strikingly similar to that of chiasma frequency in the *Lilium/Tradescantia* experimental series.

Lawrence, C. W., 1965: Influence of non-lethal doses of radiation on recombination in *Chlamydomonas reinhardi*. Nature (Lond.) **206**, 789—791.

Note 6 (see pg. 223) — Dewey and Humphrey (1964) claim that, in cells irradiated in the S-phase, breaks remain open and no restitution occurs for at least one hour following irradiation. In cells irradiated in G_1, however, restitution was complete within 5 mins. after irradiation.

Dewey, W. C., and R. M. Humphrey, 1964: Restitution of radiation-induced chromosomal damage in Chinese hamster cells related to the cell's life cycle. Exper. Cell Res. **35**, 262—276.

Note 7 (see pg. 244) — A reinvestigation of the cytology of both laboratory and wild 'sex-ratio' lines of *Drosophila pseudoobscura* and *D. athabasca* (Novitski, Peacock and Engel 1965) has shown that, contrary to earlier reports, no extra-replication of the X-chromosome occurs in primary spermatocytes. The X and Y chromosomes pair and separate normally at first division but the Y degenerates either early in metaphase-II or else after the autosomes have initiated their anaphase movement. In no case was there any indication that both second division cells contained an X-chromosome. Likewise sex-ratio males had a full complement of 128 sperms in each bundle with no indications of degeneration or abnormal development. These observations can be accomodated in terms of the regular non-functioning of two of the products of meiosis (see Note 8). In sex-ratio males, therefore, it is the preferential movement of the X-chromosome to the functional first anaphase pole that ensures an all female progeny.

Novitski, E., W. J. Peacock and J. Engel, 1965: Cytological basis of 'sex-ratio' in *Drosophila pseudoobscura*. Science **148**, 516—517.

Note 8 (see pg. 245) — Peacock and Erickson (1965) have, in fact, shown that Sandler's model is most certainly not valid. Meiosis and sperm development were both found to be normal in SD males of *Drosophila melanogaster* and the chromosome carrying SD+ was included in 50% of the motile sperm. But counts of stored sperm and of progeny recovered from comparable females showed that only half of the sperms are able to fertilise eggs and that the non-functional sperms all carried SD+. It was concluded that *D. melanogaster* regularly forms two functional and two non-functional sperms from each primary spermatocyte. The determination of these classes of sperm occurs at the first meiotic division where it is proposed that an inequality of the two spindle poles is present (see Drosophila Information Service **39**, 107—108, 1964).

The segregation of SD thus depends on a specific orientation at first metaphase relative to this polarity, such that the SD-bearing homologue of bivalent II moves to the pole that produces the two functional gametes. Thus SD, like the sex-ratio X, (see Note 7) passes preferentially to the functional pole at anaphase-I.

PEACOCK, W. J., and J. ERICKSON, 1965: Segregation-distortion and regularly non-functional products of spermatogenesis in *Drosophila melanogaster*. Genetics **51**, 313—328.

Note 9 (see pg. 248) — NOVITSKI (1964), too, feels that GRELL's hypothesis of distributive pairing is unnecessary and has accomodated her observations in terms of a pairing sequence which is initiated by an association of all the chromosomes in a non-specific chromocentre followed by the specific pairing of homologues.

NOVITSKI, E., 1964: An alternative to the distributive pairing hypothesis in *Drosophila*. Genetics **50**, 1449—1451.

Note 10 (see pg. 240) — It is true that NOVITSKI and SANDLER proposed as early as 1957 that not all the products of spermatogenesis are in fact functional in *Drosophila*. This concept of a regular class of non-functional sperm has recently been verified in a convincing manner by the work of PEACOCK and ERICKSON (see Note 8).

Note 11 (see pg. 286) — By a comparison of Cattanach's and Searle's X-autosome translocations, OHNO and LYON (1965) have demonstrated that, when a centromere-bearing portion of the X becomes positively heteropycnotic, it is able to induce an attached autosomal piece to assume the same state. But a non-centromere-bearing portion of the X lacks this ability.

OHNO, S., and M. F. LYON, 1965: Cytological study of Searle's X-autosome translocation in *Mus musculus*. Chromosoma (Berl.) **16**, 90—100.

Species Index

Author Index

SPRINGER-VERLAG / WIEN · NEW YORK

Fortsetzung von der 4. Umschlagseite

Chemistry of Viruses. By C. A. Knight, Berkeley (California). With 27 figures. IV, 177 pages. 8vo. 1963. Band IV. Virus. 2. S 303.—, DM 48.—, $ 12.—

The Multiplication of Viruses. By S. E. Luria, Urbana, Illinois. IV, 63 pages. — **Virus Inclusions in Plant Cells.** By Kenneth M. Smith, Cambridge. With 5 plates. 16 pages. — **Virus Inclusions in Insect Cells.** By Kenneth M. Smith, Cambridge. With 16 figures. 25 pages. — **Antibiotika erzeugende virus-ähnliche Faktoren in Bakterien.** Von Pierre Fredericq, Lüttich. 14 Seiten. Gr.-8°. 1958. Band IV. Virus. 3, 4a, 4b, 5.
S 268.—, DM 42.50, $ 10.65

Strukturtypen der Ruhekerne von Pflanzen und Tieren. Von Elisabeth Tschermak-Woess, Wien. Mit 91 Textabbildungen (427 Einzelbildern). IV, 158 Seiten. Gr.-8°. 1963. Band V. Karyoplasma (Nucleus). 1. S 353.—, DM 56.—, $ 14.—

The Nuclear Membrane and Nucleocytoplasmic Interchange. By C. M. Feldherr, Edmonton, J. G. Gall, Minneapolis. L. Goldstein, Philadelphia, C. V. Harding, New York, W. R. Loewenstein, New York, A. E. Mirsky, New York. With 32 figures. IV, 72 pages. 8vo. 1964. Band V. Karyoplasma (Nucleus). 2. S 138.60, DM 22.—, $ 5.50

Riesenchromosomen. Von Wolfgang Beermann, Tübingen. Mit 113 Textabbildungen. IV, 161 Seiten. Gr.-8°. 1962. Band VI. Kern- und Zellteilung. D. S 312.—, DM 49.50, $ 12.40

Die Amitose der tierischen und menschlichen Zelle. Von Otto Bucher, Lausanne. Mit 56 Textabbildungen. IV, 159 Seiten. Gr.-8°. 1959. Band VI. Kern- und Zellteilung. E. Amitose. 1. S 426.—, DM 67.50, $ 16.90

Les altérations de la méiose chez les animaux parthénogénétiques. Par Marguerite Narbel-Hofstetter, Lausanne (Schweiz). Avec 112 figures (686 Einzelbilder). IV, 163 pages. in-8°. 1964. Band VI. Kern- und Zellteilung. F. Die Chromosomen in der Meiose. 2. S 397.—, DM 63.—, $ 15.75

Différenciation des cellules sexuelles et Fécondation chez les Phanérogames. Par Bernard Vazart, Bondy (Seine). Avec 54 figures. IV, 158 pages. in-8°. 1958. Band VII. Befruchtung und Kernverschmelzung. 3a. S 378.—, DM 60.—, $ 15.—

Différenciation des cellules sexuelles et Fécondation chez les Cryptogames. Par Bernard Vazart, Bondy (Seine). Avec 122 figures. IV, 363 pages. in-8°. 1963. Band VII. Befruchtung und Kernverschmelzung. 3b. S 706.—, DM 112.—, $ 28.—

Protoplasmic Streaming. By Noburô Kamiya, Osaka, Japan. With 82 figures. IV, 199 pages. 8vo. 1959. Band VIII. Physiologie des Protoplasmas. 3. Motilität. a. S 472.—, DM 75.—, $ 18.75

Frost, Drought, and Heat Resistance. By J. Levitt, Columbia, Missouri. With 29 figures. IV, 87 pages. 8vo. 1958. Band VIII. Physiologie des Protoplasmas. 6. S 220.—, DM 35.—, $ 8.75

Polarität und inäquale Teilung des pflanzlichen Protoplasten. Von Erwin Bünning, Tübingen. Mit 72 Textabbildungen. IV, 86 Seiten. Gr.-8°. 1958. Band VIII. Physiologie des Protoplasmas. 9. Polarität. a. S 220.—, DM 35.—, $ 8.75

Morphology and Physiology of Plant Tumors. By Armin C. Braun and Tom Stonier, New York, N. Y. With 7 figures. IV, 93 pages. 8vo. 1958. Band X. Pathologie des Protoplasmas. 5a. S 206.—, DM 32.50, $ 8.15

Protoplasmatische Ökologie der Pflanzen. Wasser und Temperatur. Von Richard Biebl, Wien. Mit 92 Textabbildungen. IV, 344 Seiten. Gr.-8°. 1962. Band XII. Protoplasmatische Ökologie der Pflanzen. 1. S 618.—, DM 98.—, $ 24.50

Zu beziehen durch Ihre Buchhandlung